Vibration Analysis

SECOND EDITION

Vibration Analysis

Robert K. Vierck

The Pennsylvania State University

THOMAS Y. CROWELL COMPANY

HARPER & ROW, PUBLISHERS

New York   Hagerstown   San Francisco   London

Sponsoring Editor: Charlie Dresser
Project Editor: Penelope Schmukler
Designer: Helen Iranyi
Production Manager: Marion A. Palen
Compositor: Santype International Limited
Printer and Binder: The Maple Press Company
Art Studio: Danmark & Michaels, Inc.

VIBRATION ANALYSIS, Second Edition

Library of Congress Cataloging in Publication Data
Vierck, Robert K ·
  Vibration analysis.
  Includes bibliographical references.
  1. Vibration. I. Title.
QA935.V5 1978   531′.32   77-12685
ISBN 0-7002-2525-0

To my wife, Myrtle

Contents

Preface

The purpose of this revision is to bring the treatment of the subject of vibration analysis into line with present practice. At the same time, it is hoped that the clearness of presentation and understandability which was characteristic of the previous edition have been retained. Basic principles and thoroughness of explanation are again emphasized, keeping the student's viewpoint firmly in mind.

The changes occur mainly in the latter half of the book. This makes the text suitable for a second, more advanced, course and does not disturb its usefulness for a first, basic, course.

An explanation of SI units has been introduced in Chapter 1, and both engineering units and SI units are used throughout the text. This double system is followed because the United States will be in a period of transition to the metric system which will extend for several years. In this connection, it is recognized that the system of units followed, while not unimportant, is not a major concern in the subject of vibrations and, accordingly, should not be overemphasized.

For the most part, the changes in the first six chapters are not major in nature and are mainly for the purpose of clarity and understandability. In Chapter 5 the section on response to irregular forcing conditions has been replaced and represents a numerical procedure currently in use.

In Chapter 7, on two-degree-of-freedom systems, matrix notation has been introduced separately as an alternate representation of equation systems, and thus may be employed or not, depending on course arrangement and length.

The applicability of the first seven chapters to a beginning course in vibrations has been maintained. The adaptability to a first course of varying length, depending on the number of course credits available, has also been retained. Accordingly, a most abbreviated course might cover Chapters 1 through 4, treating single-degree-of-freedom systems. If more time is available, Chapter 5 on response to general forcing conditions and/or Chapter 6 on instrumentation may be included. Chapter 7 covering two-degrees-of-freedom systems, which is the simplest form of multidegree systems, can also be included to improve the adequacy of coverage of a beginning course. For a course of greater extent, Chapter 8 on multidegree-of-freedom systems would be included, in which case, Chapter 7 could be omitted if desired.

A second, more advanced, course would cover the latter portion of the text, starting with Chapter 8.

Revisions in Chapter 8 are extensive, involving the use of matrices and the resulting matrix concepts and procedures, and thus follows current treatment for such material. Numerical methods for multidegree systems have been separated from Chapter 8 and are placed in a new Chapter 9 on numerical analysis for lumped systems. The numerical analyses have also been revised and rearranged. Transfer matrices have been introduced and followed reasonably extensively, forming the basis for certain of the numerical procedures, including the Myklestad method.

The main change in Chapter 10 (formerly Chapter 9) is the simplification of the treatment of boundary conditions for continuous systems.

In Chapter 11 (formerly Chapter 10), on computer techniques, a section on electromechanical analogies has been included. The section on analog computers and computation has been rearranged and simplified. The numerical analyses related to digital computation and programs have been coordinated with corresponding developments in other parts of the text. A section on digital computation for the Continuous System Modeling Program has also been included.

An appendix has been added. This includes parts on matrices and matrix algebra, the development of Lagrange's equation, and units and dimensional values and conversions between systems of units.

Throughout the text, new illustrative examples have been introduced, problems have been revised, and many new problems have been added.

I am grateful to those who aided in the revision of this book, particularly to Dr. Vernon Neubert, The Pennsylvania State University, who is mainly responsible for the revision of Chapter 11 on computer methods; to Dr. John G. Bollinger, University of Wisconsin, and Dr. O. E. Adams, Ohio University, for their considerate reviews and suggestions, and to many students for their acceptance and response; to Mrs. Marilyn Day for the capable typing of manuscript; and especially to my wife, Myrtle, for her help and understanding.

ROBERT K. VIERCK

Preface to the First Edition

The aim of this book is to present the analysis of vibrations in a basic manner so that the reader may gain a fundamental understanding of this field and the methods which are peculiar to its investigation. The book is intended for a first course in vibrations for college seniors or beginning graduate students. The text was composed from class notes, written and revised over a period of several years and used in teaching courses in vibrations. It presumes no previous knowledge of the subject, but does expect the student to have a good foundation in calculus, differential equations, and dynamics. Although a prior course in strength of materials would be helpful, it is not necessary.

The book, in essentially its present context, was classroom tested by a rather large number of students, with good response and feedback later from graduates in industry. Subsequently, some revisions and additions were made and incorporated into the manuscript.

In writing this book, I have tried to retain the student's viewpoint. For the student's sake, each topic is presented in greater detail and length than is customary. In derivations and developments, steps which appear to be essential for continuity of understanding have been included. Reasons and meaning which seem necessary in developing an idea have been readily stated. The temptation to leave out intermediate steps and omit meaningful statements has been resisted. Naturally, these features increase the length of treatment for each topic. However, brevity of presentation is a question-able virtue in an introductory text. It is hoped that the student will find the text to be useful in his or her study, and that, indirectly, this will aid the instructor.

Some subject matter is included which is not found in the usual text on vibrations. Also, certain material is given greater coverage than is common. For example, the basic theory of instrumentation is presented in detail and explained more fully than is customary. The criteria for dynamic stability are developed for relatively simple conditions and are then extended to the general case for systems of several degrees of freedom. The principle of orthogonality is derived for the general case involving both dynamic and static coupling, and then is modified to apply to the case of static coupling only.

In the arrangement of exercises for assignment, in general, several problems in which similar situations occur are grouped together. This enables problems to be assigned in succeeding semesters, or terms, without duplication.

No claim is made of having adequately covered the entire field of vibrations. Indeed, this is impossible in a single book of reasonable length. The difficult choice faced in writing a general text on vibrations is deciding what important topics one can afford to omit. A major hope is that the text will enable readers to gain a basic understanding of the subject so that they can progress to more advanced material, and that their interest will have been stimulated to do so.

I am grateful to those who contributed toward the preparation of this book, especially to the late Dr. Joseph Marin, who encouraged the undertaking; to Dr. John G. Bollinger, University of Wisconsin, for the preparation of the material on computer methods; to a host of former students for their response and stimulating interest—particularly to Mr. M. C. Patel and Mr. J. S. Patel for their willing assistance in checking manuscript and proof; to Mrs. R. F. Trufant for the competent typing of the manuscript; and to my wife, Myrtle, for her encouragement and assistance.

ROBERT K. VIERCK

List of Symbols

Symbol	Meaning	U.S.-British Engineering Units	SI Units
a_{ij}	flexibility coefficient	in./lb	m/n
a, b, c	coefficients, constants, lengths, dimensions		
b	hysteresis damping constant		
b	natural frequency ratio		
A	area	in.^{2}	m^2
A	amplifier gain		
$B = EI$	section stiffness in bending	lb in.^{2}	N $\cdot$ m^2
A, B, C, D	constants, amplitudes		
c	wave velocity	in./sec	N/s
c	viscous damping constant	lb sec/in.	N $\cdot$ s/m
c_c	critical viscous damping constant	lb sec/in.	N $\cdot$ s/m
c_e	equivalent viscous damping constant	lb sec/in.	N $\cdot$ s/m
C	electric capacitance	farad	F
$\bar{C}$	complex amplitude		
d	dimension, diameter	in.	m
d_{ij}	tangential deviation	in.	m
D	determinant		
e	eccentricity	in.	m
e, E	electric voltage (electromotive force)	volt	V
E	tension-compression modulus of elasticity	lb/in.^{2}	Pa $=$ N/m^2
E	error or residue of Holzer calculations		
f	frequency	Hz	Hz
F	force, transmitted force	lb	N
F_d	damping force	lb	N

Symbol	Meaning	U.S.-British Engineering Units	SI Units
F_T	maximum transmitted force	lb	N
$[F]$	field transfer matrix		
g	acceleration of gravity	in./sec^2	m/s^2
G	modulus of rigidity (shearing modulus of elasticity)	lb/in.^{2}	Pa = N/m^2
h	dimension, height	in.	m
$h(t)$	response to unit impulse		
H	determinant		
i	$\sqrt{-1}$		
i, I	electric current	amp	A
i, j, k	indices		
I	moment of inertia of area	in.^{4}	m^4
I, J	mass moment of inertia	lb in./sec^2	kg·m^2
j	interval number		
J	polar moment of inertia	in.^{4}	m^4
k	spring constant or modulus, elastic constant	lb/in.	N/m
k_{ij}	stiffness coefficient	lb/in.	N/m
k_T, K_T	torsional spring constant	lb in./radian	N·m/rad
l	length	in.	m
L	electric inductance	henry	H
m	mass	lb sec^2/in.	kg
m_o	eccentric mass	lb sec^2/in.	kg
M	mass	lb sec^2/in.	kg
M	bending moment	lb in.	N·m
M_t	twisting moment	lb in.	N·m
MF	magnification factor		
M, N	coefficients, constants		
n	general number, modal number		
n	gear ratio		
N	normal force	lb	N
p, q	real numbers		
$P, P(t)$	force	lb	N
P_o	constant force, force amplitude	lb	N
$[P]$	point transfer matrix		
q	time-interval width or length	sec	s
q	normal coordinate		
q	electric charge	coulomb	C
q_k	generalized coordinate		
Q_k	generalized force		
$[Q]$	transfer matrix		
$r = \omega_f/\omega$	frequency ratio		
r, R	radius	in.	m
R	electric resistance	ohm	Ω
$\mathscr{R}$	real part of		
s	exponential coefficient, root of equation		
S	scale factor		
t	time	sec	s
t', t''	time offset	sec	s
T	kinetic energy, kinetic energy function	in. lb	J = N·m

Symbol	Meaning	U.S.-British Engineering Units	SI Units
T	torque	lb in.	$N \cdot m$
$T = T(t)$	function of t only		
$\bar{T}$	function of amplitudes		
TR	transmissibility		
u	displacement	in.	m
U	work or energy	in. lb	$J = N \cdot m$
$U = U(x)$	displacement function		
ΔU	energy change	in. lb	$J = N \cdot m$
V	potential energy, potential-energy function	in. lb	$J = N \cdot m$
V	shear force	lb	N
$\bar{V}$	function of amplitudes		
w	intensity of loading	lb/in.	N/m
w, W	weight	lb	N
x	displacement	in.	m
x_o	initial displacement	in.	m
x_a	complementary function		
x_b	particular solution		
x_λ	wavelength	in.	m
$\dot{x}$	velocity	in./sec	m/s
$\dot{x}_o$	initial velocity	in./sec	m/s
$\ddot{x}$	acceleration	in./sec^2	m/s^2
$\ddot{x}_o$	initial acceleration	in./sec^2	m/s^2
(sgn $\dot{x}$)	sign of $\dot{x}$		
X	displacement amplitude	in.	m
$X_o = P_o/k$	displacement reference	in.	m
$X = X(x)$	function of x only		
$\bar{X}$	complex amplitude		
$\dot{X}$	velocity amplitude	in./sec	m/s
$\ddot{X}$	acceleration amplitude	in./sec^2	m/s^2
y	displacement	in.	m
Y	displacement amplitude	in.	m
z	relative displacement	in.	m
$z = \omega_f t$	time related variable	radians	rad
Z	relative displacement amplitude	in.	m
Z	electric impedance	volt/amp	V/A
α_{ij}	slope influence coefficient in bending	radian	rad
α, β, γ	coefficients, constants, angles, exponents		
β	hysteresis damping coefficient		
β, γ	phase angle	radians	rad
γ	specific mass		
γ	hysteresis exponent		
δ	deflection, displacement, deformation	in.	m
δ	logarithmic decrement		
Δ	static displacement	in.	m
ε	constant		
ε	unit strain	in./in.	m/m
$\zeta = \dfrac{c}{c_c}$	damping factor		

Symbol	Meaning	U.S.-British Engineering Units	SI Units
η	constant, coefficient		
θ	slope	radians	rad
θ	angular displacement	radians	rad
$\dot{\theta}$	angular velocity	radians/sec	rad/s
$\ddot{\theta}$	angular acceleration	radians/sec^2	rad/s^2
θ_o	angular-displacement amplitude	radians	rad
λ	inverse frequency factor		
μ	coefficient of friction		
μ	mass ratio, constant		
v	number of intervals		
$[v]$	state vector (matrix)		
ξ	direct frequency ratio		
ρ	weight or mass density		
σ	exponential coefficient		
σ	unit stress	lb/in.^{2}	Pa $=$ N/m^2
τ	period	sec	s
τ	time at which impulse segment is imposed	sec	s
ϕ	phase, phase angle	radians	rad
ϕ	scale factor		
ϕ	magnetic flux	weber	Wb
χ	frequency factor		
χ	phase angle	radians	rad
ψ	phase angle	radians	rad
ω	natural circular frequency	radians/sec	rad/s
ω_d	damped natural circular frequency	radians/sec	rad/s
ω_f	forced circular frequency	radians/sec	rad/s
ω_R	Rayleigh circular frequency	radians/sec	rad/s

chapter

1

Introduction

1-1. PRELIMINARY REMARKS

The primary aim of this chapter is to introduce, in a relatively simple manner, the subject of vibrations. Some essential definitions are set down and general concepts are established. No attempt is made to be complete, as many ideas and the related terms and definitions become meaningful only when they are developed as significant features of specific topics and conditions, and consequently are left until later.

The present chapter also contains a partial review of kinetics. Again, no effort is made to be complete, but some of the important principles and relations of motion are recorded. A discussion of units and dimensional values is also included.

1-2. THE NATURE OF VIBRATIONS

The study of vibrations treats the oscillatory motion of mechanical systems and the dynamic conditions related thereto. This motion may be of regular form and repeated continuously, or it may be irregular or of a random nature. Vibrations are accompanied by, or are produced by, forces that vary in an oscillatory manner.

Although the term "vibration" usually implies a mechanical oscillation, similar conditions prevail in other areas, such as for alternating electric circuits, electromagnetic waves, and acoustics. This condition may be related, in some manner, in different fields; for example, a mechanical vibration may cause an acoustical vibration or sound. A mechanical vibration may produce an electric oscillation, or vice versa. The basic principles, analyses, mathematical formulations, and terminology for oscillatory phenomena are similar in the various fields. The vibration of mechanical systems *only* will be considered in this text.

1-3. VIBRATORY MOTION AND SYSTEMS

In order for a mechanical vibration to occur, a minimum of two energy-storage elements is required—a mass which stores kinetic energy, and an elastic member which stores potential energy. These can be represented as in Fig. 1-1, where *m* and *k* are the mass and the elastic elements, respectively. Assuming that horizontal movement is prevented, if *m* is displaced vertically from its equilibrium position and released, it will exhibit an oscillatory vertical motion. Such motion is repeated in equal time intervals and hence is said to be cyclic or periodic. If the elastic element is linear (that is, if the spring force is proportional to its deformation), then the motion curve of the mass displacement against time will be sinusoidal in form. This is called *harmonic motion* and is shown in Fig. 1-2, where *x* is displacement and *t* is time. The difference between the motions of parts (a), (b), and (c) is entirely due to the initial conditions of displacement and velocity. The maximum displacement *X* is generally referred to as the displacement *amplitude*; the term ϕ represents the *phase* or *phase angle*, and ω is a constant called the *circular frequency*.

Certain conditions produce cyclic or periodic motion that is not harmonic. A motion of this type is represented in Fig. 1-3. One complete movement of any repeated motion is called a *cycle*. The time for one cycle is termed the *period*. It is designated by τ and is generally measured in seconds. The *frequency* is the number of cycles of motion occurring in unit time. The symbol for frequency is *f*, and the most common unit is cycles per second, which are called hertz (Hz). Note that τ is the reciprocal of *f*. Thus

$$f = \frac{1}{\tau} \quad \text{and} \quad \tau = \frac{1}{f} \tag{1-1}$$

A vibration can also be of an irregular nature, such as that shown in Fig. 1-4. Here there is no repeated part to the movement, although many of the peak displacement values may occur again and again. This type of motion, for which there is no apparent pattern in the vibration record, is called a *random vibration*. A random vibration is produced by input forces of an irregular nature acting on the vibratory system. Such random forces occur in missiles and space vehicles, due to aerodynamic buffeting during launching. Packaged assemblies of structural and mechanical equipment are subjected to random forces while they are being shipped or transported.

A type of motion related to vibrations is the short-time response of a

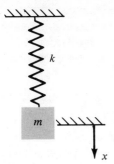

Figure 1-1

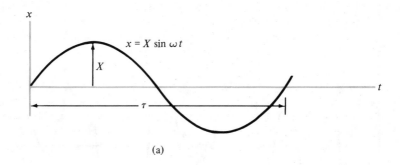

(a)

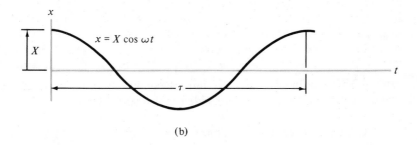

(b)

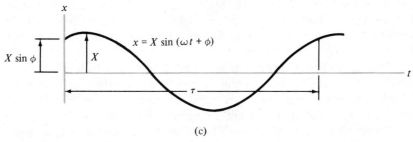

(c)

Figure 1-2

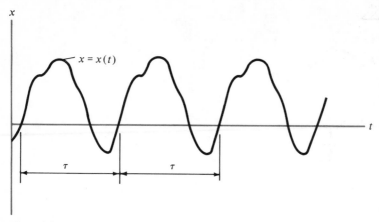

Figure 1-3

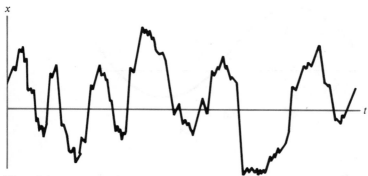

Figure 1-4

system. This motion is caused by impact conditions being imposed on a vibratory system, resulting in a sudden movement such as that shown in Fig. 1-5.

A vibratory system may be subject to resistance due to air friction, shock absorbers, and other dissipative elements. Such resistance is called *damping*, and an arrangement containing this is referred to as a damped system. Typical damped-motion curves are exhibited in Fig. 1-6. Since there is no period for that shown in part (b), this form of motion is said to be nonperiodic, or aperiodic.

The system may be acted on by an external force which is often of a repeated type that tends to maintain the oscillation. It is then designated as a forced system, and the motion is called a *forced vibration*. If no forcing condition exists, the motion is said to be a *free vibration*.

Specifying the configuration of a system may require several independent coordinates. The number of coordinates needed indicates the number

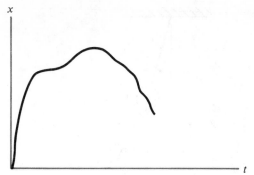

Figure 1-5

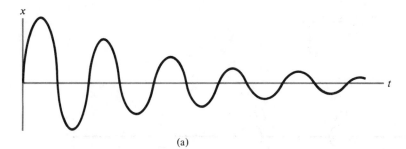

(a)

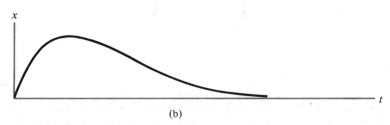

(b)

Figure 1-6

of degrees of freedom of the system. The system of Fig. 1-1 requires only one coordinate to define the motion and hence is a *single-degree-of-freedom* system. Figures 1-7(a) and (b) represent two-degree-of-freedom systems, and Fig. 1-7(c) exhibits a three-degree-of-freedom system. Systems of more than three degrees of freedom can be readily visualized. The term multidegree-of-freedom system is used as a general designation for systems having several degrees of freedom. Note that a continuously distributed mass-and-elastic system, such as a beam, would have an infinite number of degrees of freedom since, in effect, it is composed of an infinite number of mass-elastic elements.

For a system of several degrees of freedom under certain conditions the motion in each coordinate may be harmonic. This pattern of movement is

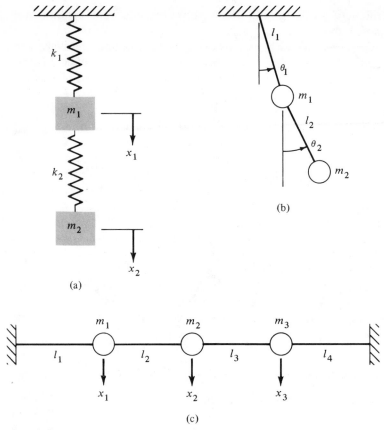

(a)

(b)

(c)

Figure 1-7

called a *principal mode* of vibration. It will be shown later that the number of such principal modes is equal to the number of degrees of freedom for the system.

In general, vibrations produce, or may be accompanied by, conditions that are undesirable. Stresses of such magnitude may result so that a machine or structure may be damaged or destroyed. Fatigue failure may occur if stress amplitudes and the number of stress cycles are sufficiently large. Large forces may be transmitted to supports or adjacent parts. Vibratory motion may develop enough amplitude to disturb the functioning of the mechanism involved. In other cases, although no critical condition develops, vibration may simply be objectionable because of the noise produced or the shaking condition transmitted. On the other hand, oscillatory motion may be desirable or necessary in order to perform some useful function, as in the shaking or stamping motions employed in various industrial operations, or in the reciprocating motion which is typical of many engines. However, these useful oscillations are likely to be accompanied by undesirable conditions of the type noted above.

For a machine or structure that is subject to vibration, the shape and arrangement of the parts may be complicated. However, it is often possible to replace the actual arrangement by a simple system composed of essential elements consisting of masses, springs, and dampers. Such a schematic representation is called an *equivalent system*. It is also referred to as a *vibration model* or mechanical oscillator. The elastic and mass elements are designated as the equivalent spring and the effective mass, respectively. The determination of the values of the equivalent spring and of the effective mass is done in a rational manner, for example, by means of equating energy for the actual and the equivalent systems. The analysis and solution procedure is then carried out for the equivalent system. The elements of the system are considered as *lumped*—the spring is treated as a purely elastic member (unless otherwise stated) having negligible mass, and the mass element is regarded as rigid. Throughout this text when various systems, consisting of spring, mass, and damping elements, are analyzed, it should be kept in mind that these are vibration models and that they merely represent equivalent systems having all the essential elements of an actual arrangement (structure or machine) which, in reality, may appear quite different from the simple figure shown.

1-4. STATEMENT OF THE VIBRATION PROBLEM

It is desirable to have some concept of the problem the vibration analyst faces. In the first place, he must recognize the possibility of a vibration occurring or developing. This often involves a machine, structure, space vehicle, or other equipment which is not in existence but which is being developed or designed. The problem must then be analyzed to determine what aspects of the vibration may be significant. This may require a complete solution—that is, determining the value of all measurements of the vibration, such as frequency, amplitudes of displacement, velocity, acceleration, transmitted force, and damping and decay rate. In other instances it may be necessary only to obtain the fundamental frequency for the system. The refinement required in the determinations also varies considerably. In some cases, great precision may not be necessary for heavy stationary machines or equipment. On the other hand, extreme refinement in calculations would be needed in connection with guidance-control devices for a missile or a space vehicle.

The vibration problem may be summarized by the following steps: (a) determination of the existence and nature of the problem, its features and aspects, and the physical components involved; (b) analysis of the problem and formulation of the governing equations; (c) solution of the governing equations; (d) calculation of the important results from the solution; (e) interpretation of the solution and results; and (f) recommendations regarding the significance of results and any changes required or desirable in the structure or equipment involved.

1-5. KINETICS; NEWTON'S LAWS OF MOTION

In the study of dynamics, including the area of vibrations, Newton's laws of motion are important. These laws represent basic relations, the acceptance of which requires fundamental experiments. Other governing relations of motion and equilibrium may then be developed from these laws. Newton's laws may be stated as follows:

First Law.   A particle remains at rest or continues to move with an unchanging velocity (constant speed along a straight line) unless acted on by an unbalanced (resultant) force.

Second Law.   The magnitude of the acceleration of a particle is proportional to the resultant force acting on it and has the same direction and sense as this force.

Third Law.   The force (action) exerted by particle A on another particle B is equal in magnitude, *opposite* in sense, and collinear to the force (reaction) exerted by B on A.

A common approach to the analysis of many vibration problems employs Newton's second law of motion and principles based on this law. For such a purpose it becomes necessary to extend and adapt the relation to the kinds of motion and conditions pertaining to large rigid bodies composed of particles. A résumé of some of the relations for rigid-body motion is given in the sections that follow. In most cases no proof is made here for these relationships. It is suggested that a recognized text on dynamics be consulted for this purpose and for a thorough review of the principles of motion.

1-6. PRINCIPLE OF MOTION FOR A PARTICLE

Consider a particle having mass m to be acted on by a three-dimensional system of concurrent forces composed of $\mathbf{F}_1, \mathbf{F}_2, \mathbf{F}_3, \ldots$, having the resultant $\mathbf{R} = \sum \mathbf{F}$, as shown in Fig. 1-8. (Boldface type here indicates that the term represents a vector.) Thus

$$\mathbf{R} = \mathbf{F}_1 + \mathbf{F}_2 + \mathbf{F}_3 + \cdots$$
$$= \sum \mathbf{F} \tag{1-2}$$

The acceleration $\mathbf{a}$ of the particle and the resultant $\mathbf{R}$ are vectors having the same direction and sense, and for which Newton's second law may be expressed as the vector relation

$$\mathbf{R} = m\mathbf{a} \qquad \text{or} \qquad \sum \mathbf{F} = m\mathbf{a} \tag{1-3}$$

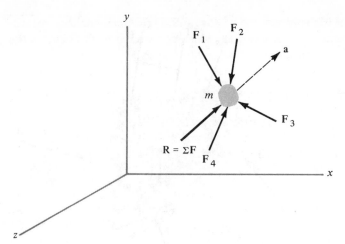

Figure 1-8

Using a Cartesian reference frame, this may be written as

$$\sum F_x = ma_x, \quad \sum F_y = ma_y, \quad \sum F_z = ma_z \qquad (1\text{-}4)$$

representing the scalar form of the equations of motion for a particle.

1-7. PRINCIPLE OF MOTION OF THE MASS CENTER

Consider a large rigid body (and thus one composed of many particles) as having a general motion. Assume that the body has a mass center G with the linear acceleration of $\mathbf{a}_G$, and that the body is acted on by a system of nonconcurrent forces $\mathbf{F}_1, \mathbf{F}_2, \mathbf{F}_3, \ldots$. The resultant for this force system, for the general case, is composed of force $\mathbf{R}$ $(= \sum \mathbf{F})$ and a couple (consisting of equal and opposite parallel forces $\mathbf{P}$) lying in a plane which does not contain $\mathbf{R}$, as represented in Fig. 1-9. Then the principle of motion of the mass center may be stated as the vector relation

$$\mathbf{R} = m\mathbf{a}_G \qquad \text{or} \qquad \sum \mathbf{F} = m\mathbf{a}_G \qquad (1\text{-}5)$$

where m is the mass of the body and the summation includes all forces acting on the body. The direction and sense of $\mathbf{R}$ and $\mathbf{a}_G$ are the same, but their line of action generally differs. Using a Cartesian coordinate reference system, this principle may be written in scalar form as

$$\sum F_x = ma_{G_x}, \quad \sum F_y = ma_{G_y}, \quad \sum F_z = ma_{G_z} \qquad (1\text{-}6)$$

1-8. PRINCIPLE OF MOTION FOR A SYSTEM OF BODIES

Consider a system of large rigid bodies, such as that shown in Fig. 1-10, to have general motion under the action of a set of forces $\mathbf{P}_1, \mathbf{N}_1, \ldots$. Some of these bodies may be in contact and some may not; there is no requirement

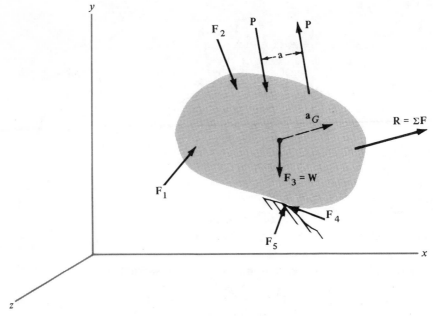

Figure 1-9

regarding this. The mass centers of the bodies are $G_1, G_2, \ldots$, having linear accelerations $\mathbf{a}_G^{(1)}, \mathbf{a}_G^{(2)}, \ldots$, respectively. (The numbered subscript or superscript here refers to the body involved.) Then for each separate body the principle of motion of its mass center can be written. Thus

$$\sum F_x^{(1)} = m_1 a_{G_x}^{(1)}$$
$$\sum F_x^{(2)} = m_2 a_{G_x}^{(2)}$$
$$\ldots \qquad \ldots$$
$$\ldots \qquad \ldots \tag{1-7}$$
$$\ldots \qquad \ldots$$
$$\sum F_x^{(n)} = m_n a_{G_x}^{(n)}$$

Adding these results in

$$\sum F_x = \sum_{i=1}^{n} m_i a_{G_x}^{(i)} \tag{1-8}$$

where $\sum F_x$ involves only forces that are external to the system, including gravity forces, since internal forces between bodies would cancel because they are equal but opposite.

The foregoing argument may be applied independently to the y and z

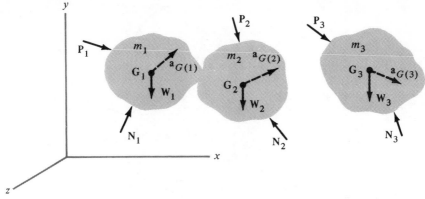

Figure 1-10

coordinate directions, resulting in the relations

$$\sum F_y = \sum_{i=1}^{n} m_i a_{G_y}^{(i)} \qquad \sum F_z = \sum_{i=1}^{n} m_i a_{G_z}^{(i)} \qquad (1\text{-}9)$$

Together, Eqs. 1-8 and 1-9 represent a principle of motion for a *system* of rigid bodies.

1-9. PRINCIPLE OF ANGULAR MOTION FOR A BODY

For a large rigid body exhibiting plane motion, a useful expression exists that involves the moments of the forces for the system, angular motion, and related quantities. Consider the body of Fig. 1-11 to have plane motion due to the action of forces **P, Q,** .... The reference point 0 serves as the origin of the coordinate axes shown and has linear acceleration a_0. The coordinates of the mass center G are x_G and y_G, as indicated. The body has the angular acceleration represented by α. Using reference center 0, the principle that governs the angular motion of the body is

$$\sum T_0 = I_0 \alpha + m a_{0y} x_G - m a_{0x} y_G \qquad (1\text{-}10)$$

where m is the mass of the body, I_0 is the moment of inertia of its mass with respect to an axis at 0 normal to the view, and $\sum T_0$ includes the moment about the axis at 0 of all forces for the system.

The relation can be simplified in form (and hence for use) by suitable selection of the reference point 0. For example, if the mass center G is chosen, Eq. 1-10 becomes

$$\sum T_G = I_G \alpha \qquad (1\text{-}11)$$

since $x_G = 0 = y_G$. In many cases, using G as the reference center is a wise selection.

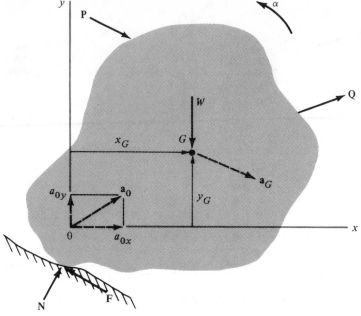

Figure 1-11

1-10. RELATION FOR MOTION OF ROTATION

The case of rotation of a rigid body represents a special case of plane motion, so that the relations of Sections 1-7 and 1-9 apply here. In Fig. 1-12 assume that the body is rotating about a fixed axis at 0 normal to the view, and has the angular acceleration α. Then, using 0 for the reference center, Eq. 1-10 becomes

$$\sum T_0 = I_0 \alpha \qquad (1\text{-}12)$$

governing the motion of rotation for the body.

1-11. UNITS AND DIMENSIONAL VALUES

The dimensional values of the various quantities treated in vibrations are important. The International System of Units (abbreviated SI) has been universally recognized and adopted by almost all major countries of the world. The United States is presently in a state of transition from the U.S.–British engineering system to the SI system. Accordingly, both systems of units will be followed in this text, in examples and in problems. This should not result in difficulty or confusion if one considers the basic units for each system. The main difference between the two systems occurs in the basic units and derived units established by Newton's second law. These are listed in the table that follows.

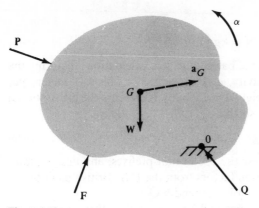

Figure 1-12

Quantity (dimensional symbol)	U.S.–British engineering unit (symbol)	SI unit (symbol)
Length, L	inch[a] (in.)	meter (m)
Time, T	second (sec)	second (s)
Mass, M	pound second squared per inch (lb sec²/in.)	kilogram (kg)
Force, F	pound (lb)	newton (N)

[a] The inch rather than the foot is used in vibration determinations.

For the U.S.–British engineering system, the quantities length, time, and force are considered as fundamental and mass is a derived quantity, being determined from $F = ma$. A unit mass is defined as the mass of a body that would be given a unit acceleration (1 inch per second squared) by a unit force (1 pound). From Newton's second law, this unit mass would be 1 pound second squared per inch (1 lb sec²/in.). That is, from $m = F/a$,

$$(\text{dimension of } m) = \frac{1 \text{ lb}}{1 \text{ in./sec}^2}$$

$$= \text{lb sec}^2/\text{in.} \qquad (1\text{-}13)$$

In the engineering system, mass is determined from body weight W divided by the acceleration of gravity g, which is taken to be 386 in./sec². Thus

$$m = \frac{W}{g} = \frac{W}{386 \text{ in./sec}^2} \qquad (1\text{-}14)$$

In the SI system, the quantities length, time, and mass are basic and force is a derived quantity which is defined by Newton's second law. Thus a unit force (1 newton) is that force which will give a unit mass (1 kilogram) a unit acceleration (1 meter per second squared). That is, from $F = ma$,

$$(1 \text{ N}) = (1 \text{ kg}) \times (1 \text{ m/s}^2)$$

and

$$N = kg \cdot m/s^2 \tag{1-15}$$

which defines a newton in terms of basic units. For calculations requiring the use of the acceleration of gravity, g is considered to be 9.81 meters per second squared (m/s^2). The weight **W** of a body represents the gravitational attraction (force) on the body so that

$$\mathbf{W} = m\mathbf{g} \quad \text{and} \quad W = mg \tag{1-16}$$

For the SI system, guidelines for the use of unit prefixes, unit designations, number grouping, and conversion factors from the U.S.–British engineering system to the SI system are given in Appendix C.

The units for various quantities will be presented as they are taken up throughout the text. These will be given in both the engineering system and the SI system. Also, the units for all quantities are contained in the list of symbols in the front of the text.

The system of using a space rather than a comma for separation in writing large numbers will be followed. Thus the number 21,347.658,94 will be written as 21 347.658 94. Four digit numbers are an exception to this, as for example 3257.4698.

chapter

2

Undamped Free Vibrations for Single-Degree-of-Freedom Systems

2-1. INTRODUCTION

The simplest vibratory system consists of an elastic member and a mass element, as represented in Fig. 2-1(a). This is a single-degree-of-freedom system since it can move in but one coordinate; that is, it requires only coordinate x to define its configuration. Since there is no external force to drive the system, the motion is designated as a free vibration. It is also undamped, as no condition is present which would inhibit the motion.

Devices of such simplicity often occur in structures and machines. In some instances the mass may be distributed but the behavior may be approximated by a concentrated or point mass. The elastic element may appear complicated but it also may be approximated by a single spring. Thus a complex arrangement may be reduced to a lumped system or may be assumed to be composed of such discrete parts, corresponding to a single-degree-of-freedom oscillator. The analysis and solution for this type of system will be set forth in the sections that follow.

2-2. ANALYSIS BY NEWTON'S PRINCIPLES

The procedure employing a free-body diagram of the mass element and Newton's second law may be used to obtain the differential equation that governs the motion of the system. Consider that a mass m is suspended by a spring having a linear constant k, representing the uniform force required for unit deformation of the spring. The units are pounds per inch ($N/m = kg/s^2$

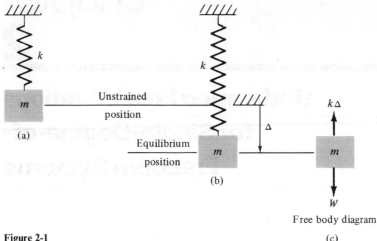

Figure 2-1

(c)

for the SI system). Also assume that the arrangement has horizontal con-straints so that m can be displaced only vertically. For the position of rest shown in Fig. 2-1(b), the static displacement is Δ. The corresponding spring force is then $k\Delta$, as shown in the free-body diagram, part (c). Writing the force equation of equilibrium results in

$$W - k\Delta = 0$$

so that

$$W = k\Delta \tag{2-1}$$

When displaced from its equilibrium position and released, the mass m will move vertically. As indicated in Fig. 2-2(a), the coordinate x defines the displacement of m, referred to its static equilibrium position. For conven-ience, the positive direction of x has been taken downward. Then, for the plus x position assumed for m, the spring force is $k(\Delta + x)$ and should be shown upward in the complete free-body diagram represented by part (b). Since the downward direction is chosen as positive for x, then the time derivatives must also have the same sign convention in the coordinate system. Thus dx/dt and d^2x/dt^2 are written as positive downward. If at any instant one of the terms x, dx/dt, or d^2x/dt^2 is negative, it must be accom-panied by a minus sign. Newton's second law of motion would then be written as

$$m\frac{d^2x}{dt^2} = -k(\Delta + x) + W \tag{2-2}$$

(Although it is not particularly significant, the relation is usually written in this order in vibration work.) Substituting Eq. 2-1 into this gives

$$m\frac{d^2x}{dt^2} = -kx \tag{2-3}$$

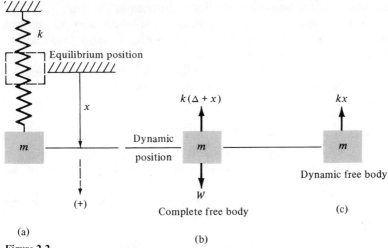

Figure 2-2

This step could have been accomplished by first canceling the equal and opposite forces $k\Delta$ and W in the complete free-body diagram, resulting in the dynamic free-body diagram shown in part (c). From this diagram Eq. 2-3 results directly. The procedure of using the dynamic free-body diagram, which omits all balancing static forces, is convenient and will generally be followed hereafter. The only difficulty that can arise is for cases where the manner in which the static forces cancel is not apparent, due to involvements of the arrangement. Significantly, Eq. 2-3 shows that the vibration takes place about the equilibrium position.

2-3. SOLUTION OF THE DIFFERENTIAL EQUATION

Equation 2-3 can be rearranged as

$$\frac{d^2x}{dt^2} + \frac{k}{m}x = 0 \qquad (2\text{-}4)$$

which can be written as

$$\frac{d^2x}{dt^2} + \omega^2 x = 0 \qquad \text{or} \qquad \ddot{x} + \omega^2 x = 0 \qquad (2\text{-}5)$$

where

$$\omega^2 = \frac{k}{m} \qquad (2\text{-}6)$$

The dot system is used, for convenience, to represent time derivatives. Thus $\dot{x} = dx/dt$ and $\ddot{x} = d^2x/dt^2$. This notation will be largely employed hereafter.

Equation 2-5 is a homogeneous linear differential equation with constant coefficients. Since it is of second order, the solution must contain two arbitrary constants. In order to satisfy the differential equation, the solution $x = x(t)$ must return to the same form in its second derivative. Either $\sin \omega t$ or $\cos \omega t$ would appear to satisfy this condition. It is then reasoned that the sum of these functions, together with appropriate constants, would represent the general solution. Thus the following is set down:

$$x = A \sin \omega t + B \cos \omega t \qquad (2\text{-}7)$$

where A and B are the required constants. Substitution will reveal that Eq. 2-7 satisfies Eq. 2-5. The arbitrary constants A and B can be determined from the initial conditions of the motion.

For an expression such as Eq. 2-7, it is always possible to replace the arbitrary constants by a new set. For example, let

$$A = C \cos \phi \qquad \text{and} \qquad B = C \sin \phi \qquad (2\text{-}8)$$

Substituting into Eq. 2-7 results in

$$x = C(\sin \omega t \cos \phi + \cos \omega t \sin \phi)$$
$$= C \sin (\omega t + \phi) \qquad (2\text{-}9)$$

where C and ϕ are the new arbitrary constants defined by Eq. 2-8, or as follows:

$$(C \cos \phi)^2 + (C \sin \phi)^2 = A^2 + B^2$$

and

$$C = \sqrt{A^2 + B^2} \qquad (2\text{-}10a)$$

$$\tan \phi = \frac{C \sin \phi}{C \cos \phi}$$

$$= \frac{B}{A} \qquad (2\text{-}10b)$$

The term ϕ is called the *phase angle* or *phase*. The solution form of Eq. 2-9 also satisfies the differential equation. Note that this form might logically have been assumed initially, in place of Eq. 2-7.

The trigonometric relations of Eq. 2-8 were chosen so that the equation form of Eq. 2-9 would result. This can be done in such a manner that the following solution forms also will be obtained:

$$x = C_1 \sin (\omega t - \alpha)$$
$$= C_2 \cos (\omega t + \beta) \qquad (2\text{-}11)$$
$$= C_3 \cos (\omega t - \gamma)$$

Each of these will satisfy the differential equation. The only difference in these expressions is in the arbitrary constant represented by the phase term. For any specified set of initial conditions, the four forms will become identical.

2-4. INTERPRETATION OF THE SOLUTION

To gain a general understanding of the solution, it is desirable to consider that m has both initial velocity and displacement. These can be expressed as

$$\left. \begin{array}{l} x = x_0 \\ \dot{x} = \dot{x}_0 \end{array} \right\} \quad \text{at } t = 0 \tag{2-12}$$

Substituting these into Eq. 2-7 and its time derivative will evaluate the arbitrary constants as

$$A = \frac{\dot{x}_0}{\omega} \quad \text{and} \quad B = x_0 \tag{2-13}$$

so that the solution becomes

$$x = \frac{\dot{x}_0}{\omega} \sin \omega t + x_0 \cos \omega t \tag{2-14}$$

Similarly, substitution of Eq. 2-13 into Eq. 2-10 and then into Eq. 2-9 gives

$$x = X \sin (\omega t + \phi) \tag{2-15}$$

where X represents the *amplitude* of the displacement, and ϕ is the phase angle as defined by

$$X = \sqrt{(\dot{x}_0/\omega)^2 + x_0^2}$$

$$\tan \phi = \frac{x_0}{\dot{x}_0/\omega} \tag{2-16}$$

The motion represented by Eqs. 2-14 and 2-15 is said to be harmonic, because of its sinusoidal form. The motion is repeated, with the time for one cycle being defined by the value of ωt equal to 2π. Thus the *period* τ, or the time for one cycle, is given by

$$\tau = \frac{2\pi}{\omega} \tag{2-17a}$$

The reciprocal of τ expresses the *frequency* f in cycles per unit time. Thus

$$f = \frac{\omega}{2\pi} \tag{2-17b}$$

Because the solution is a circular function, the term ω is designated as the *circular frequency*. It is measured in radians per second. Since $\omega = \sqrt{k/m}$, this is taken care of by expressing k as pounds per inch and m as pound seconds squared per inch. (In the SI system, the units for k are newtons per meter, which is the same as $kg \cdot m \cdot s^{-2}/m = kg/s^2$, and for m the units are kilograms.) The dimension of τ will then be seconds, and that for f will be cycles per second, or *hertz* (Hz). The frequency can also be determined from the static deflection Δ alone, based on

$$\omega = \sqrt{\frac{k}{m}} = \sqrt{\frac{g}{mg/k}} = \sqrt{\frac{g}{\Delta}} \qquad (2\text{-}18)$$

The velocity $\dot{x}$ and the acceleration $\ddot{x}$ are expressed by time derivatives of Eqs. 2-14 and 2-15. These can be written in various forms, as noted below. Thus

$$\dot{x} = \dot{x}_0 \cos \omega t - x_0 \omega \sin \omega t$$

$$= X\omega \cos (\omega t + \phi) = X\omega \sin \left(\omega t + \phi + \frac{\pi}{2}\right) \qquad (2\text{-}19)$$

$$= \dot{X} \cos (\omega t + \phi)$$

where $\dot{X} = X\omega$ represents the amplitude of the velocity. Similarly,

$$\ddot{x} = -\dot{x}_0 \omega \sin \omega t - x_0 \omega^2 \cos \omega t$$

$$= -X\omega^2 \sin (\omega t + \phi) = X\omega^2 \sin (\omega t + \phi + \pi) \qquad (2\text{-}20)$$

$$= -\omega^2 x$$

$$= -\ddot{X} \sin (\omega t + \phi)$$

where $\ddot{X} = X\omega^2$ represents the amplitude of the acceleration. Note that the velocity is ω times the displacement and leads it by 90 degrees, and that the acceleration is ω^2 times the displacement and leads it by 180 degrees.

In examining the solution to gain an understanding of the motion, it is helpful to plot the displacement, velocity, and acceleration against time. Since ω is constant, it is convenient—and hence more common—to use ωt instead of the time t axis in such diagrams. This has been carried out in Fig. 2-3. It will be observed that in the figure the positive direction for the displacement x and its derivatives has been taken upward, rather than in the downward direction used for the free-body diagram and for setting up the differential equation of motion. The curves should be studied thoroughly, noting the relation between the diagrams at maximum, minimum, and zero points, the direction of motion, the corresponding acceleration, and so on.

The phase angle ϕ simply indicates the amount by which each curve is shifted along the horizontal axis. That is, the displacement curve is shifted ahead, with respect to an ordinary sine curve, by the amount of ϕ on the ωt

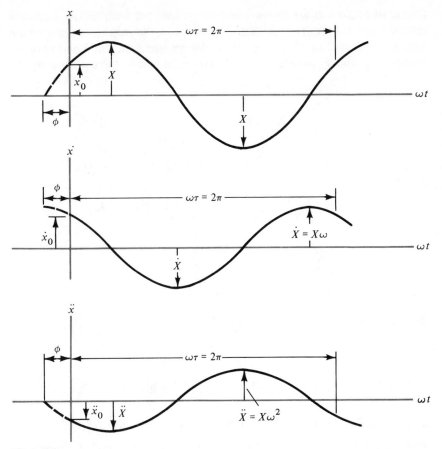

Figure 2-3

axis, or along the time axis by the amount t_0 defined by

$$\omega t_0 + \phi = 0$$

$$t_0 = -\frac{\phi}{\omega} \tag{2-21}$$

The other curves are similarly shifted by the same amount.

Somewhat simplified cases result when either the initial displacement or the velocity is zero. Thus for $x_0 = 0$ and $\dot{x}_0 \neq 0$,

$$x = X \sin \omega t \tag{2-22}$$

where $X = \dot{x}_0/\omega$. In case $x_0 \neq 0$ and $\dot{x}_0 = 0$, then

$$x = X \cos \omega t \tag{2-23}$$

with $X = x_0$.

EXAMPLE 2-1 A weight of 1.93 lb is suspended by a spring having a modulus of 30 lb/ft. The oscillatory motion of the mass has a maximum velocity measured as 15 in./sec. Determine the frequency and period of the motion, and the amplitude of the displacement and of the acceleration.

SOLUTION

$$k = \frac{30}{12} = 2.5 \text{ lb/in.}$$

$$m = \frac{W}{g} = \frac{1.93 \text{ lb}}{386 \text{ in./sec}^2} = 0.005 \text{ lb sec}^2/\text{in.}$$

$$\omega = \sqrt{\frac{k}{m}} = \sqrt{\frac{2.5}{0.005}} = \sqrt{500} = 22.36 \text{ rad/sec}$$

$$f = \frac{\omega}{2\pi} = \frac{22.36}{2\pi} = 3.56 \text{ Hz}$$

$$\tau = \frac{1}{f} = \frac{2\pi}{\omega} = 0.281 \text{ sec}$$

Since $\dot{X} = X\omega$ and $\ddot{X} = X\omega^2 = \dot{X}\omega$, then

$$X = \frac{\dot{X}}{\omega} = \frac{15}{22.36} = 0.671 \text{ in.}$$

$$\ddot{X} = \dot{X}\omega = 15 \times 22.36 = 335 \text{ in./sec}^2$$

2-5. GRAPHICAL REPRESENTATION OF MOTION

The motion defined by Eqs. 2-14 and 2-15 can be represented graphically by rotating vectors, as shown in Fig. 2-4(a). Note that x_0, $\dot{x}_0/\omega$, and X have angular positions which are fixed relative to each other, or, specifically, $\dot{x}_0/\omega$ and x_0 are at right angles and X leads $\dot{x}_0/\omega$ by the phase angle ϕ. It can be observed that the magnitude and the relative positions of these vectors are in agreement with Eq. 2-16. The system of three vectors then rotates with an angular velocity ω so that at any instant t, the angular position is defined by ωt. Each curve shown in Fig. 2-4(b) is simply the vertical projection of the corresponding vector represented in Fig. 2-4(a). The equivalence of Eqs. 2-14 and 2-15 is also verified by the geometry of Fig. 2-4(a).

The velocity and acceleration curves, as well as the displacement, can be represented in this manner. For convenience, take the case

$$x = X \sin \omega t = \frac{\dot{x}_0}{\omega} \sin \omega t$$

Then

$$\dot{x} = X\omega \cos \omega t = \dot{X} \cos \omega t$$

$$\ddot{x} = X\omega^2(-\sin \omega t) = \ddot{X}(-\sin \omega t)$$

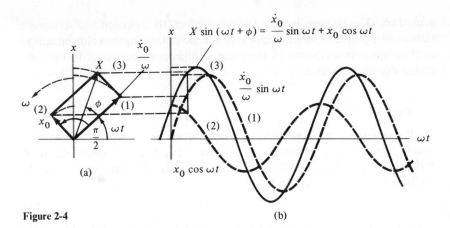

Figure 2-4

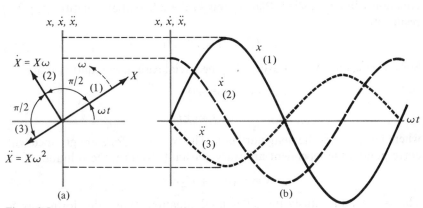

Figure 2-5

These can be graphically represented by the system of rotating vectors shown in Fig. 2-5. The velocity and acceleration vectors lead the displacement by 90 and 180 degrees, respectively, as shown. This system of three vectors, whose relative positions are fixed, rotates with a circular velocity ω. Their angular position at any time t is defined by ωt. The curves shown in Fig. 2-5(b) are then merely the vertical projections of the corresponding vectors indicated in Fig. 2-5(a).

2-6. METHODS OF SOLVING THE DIFFERENTIAL EQUATION

First Method

In Section 2-3 the solution of the differential equation was obtained essentially by intuitive reasoning. This is a common approach and is due to the fact that most differential equations are not solvable by a direct

procedure. The process is indirect, or in effect, the solution is discovered†
rather than derived. Different approaches for obtaining the solution can be
used in many cases. Some of the ideas and procedures are set forth in the
remainder of this section.

Second Method

Reasoning in the same manner as in Section 2-3 would also lead to the
conclusion that an exponential function of t returns to the same form in all
derivatives and therefore should satisfy the differential equation. Thus the
solution may be assumed to be

$$x = Ce^{st} \qquad (2\text{-}24)$$

where C is an arbitrary constant but s is to be defined so that the differential
equation will be satisfied. Placing this in the differential equation (Eq. 2-5)
results in

$$(s^2 + \omega^2)Ce^{st} = 0 \qquad (2\text{-}25)$$

Since Ce^{st} cannot vanish except for the trivial case, it is required that

$$s^2 + \omega^2 = 0$$
$$s = \pm i\omega \qquad (2\text{-}26)$$

where $i = \sqrt{-1}$ is the imaginary unit. The two roots for s are grouped with
corresponding independent arbitrary constants to form the solution

$$x = C_1 e^{i\omega t} + C_2 e^{-i\omega t} \qquad (2\text{-}27)$$

This will readily satisfy the differential equation. While this form may be
useful, it appears difficult to interpret. Although it seems to be independent
of the other solution, this is not the case. By using the Euler relation

$$e^{i\theta} = \cos \theta + i \sin \theta \qquad (2\text{-}28)$$

which can be proved by the series for the separate parts, Eq. 2-27 can be
written as

$$x = C_1(\cos \omega t + i \sin \omega t) + C_2(\cos \omega t - i \sin \omega t)$$
$$= i(C_1 - C_2) \sin \omega t + (C_1 + C_2) \cos \omega t$$
$$= A \sin \omega t + B \cos \omega t$$

This is identical to the previously obtained solution, Eq. 2-7. [The substitutions of A and B here can be justified by recognizing that C_1 and C_2 may be
conjugate complex numbers.‡ Thus taking $C_1 = a + ib$ and $C_2 = a - ib$
leads to $i(C_1 - C_2) = i(2ib) = -2b = A$, and $C_1 + C_2 = 2a = B$.]

† See R. E. Gaskell, *Engineering Mathematics* (New York: Dryden Press, 1958), p. 8.
‡ See Gaskell, *Engineering Mathematics*, p. 17.

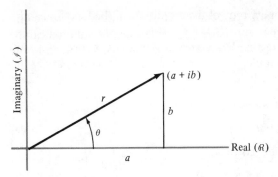

Figure 2-6

Third Method

A harmonic solution can be expressed in complex form. Consequently, the complex form of a function may be a means of determining the solution. Before proceeding with this method, pertinent definitions and concepts will be established.

The complex number $(a + ib)$, for which a and b are considered to have real values, can be represented graphically by a vector in the *complex plane*. This is also called an *Argand diagram* and is shown in Fig. 2-6. The vector may be specified either by giving the magnitudes of a and b or by giving the values of r and θ. The relation between the Cartesian pair (a, b) and the polar set (r, θ) is given by

$$a + ib = r(\cos \theta + i \sin \theta) \qquad (2\text{-}29)$$

$$= re^{i\theta} \qquad (2\text{-}30)$$

The length r of the vector is called the *absolute value*, modulus, or magnitude, and the angle θ is designated as the *argument*, angle, or amplitude. (The first term in each case will be used hereafter in order to avoid confusion. In particular, the word "amplitude" will not be employed for θ, because of its ordinary connotation in the field of vibrations.)

Complex notation may be used to express a trigonometric function such as $x = C \cos (\omega t - \gamma)$. Using the symbol $\mathscr{R}$ to mean "the real part of," then

$$x = C \cos (\omega t - \gamma) = \mathscr{R}[C \cos (\omega t - \gamma) + iC \sin (\omega t - \gamma)]$$

$$= \mathscr{R}[Ce^{i(\omega t - \gamma)}] = \mathscr{R}(Ce^{-i\gamma}e^{i\omega t})$$

$$= \mathscr{R}(\bar{C}e^{i\omega t}) \qquad (2\text{-}31a)$$

where

$$\bar{C} = Ce^{-i\gamma} = C \cos \gamma - iC \sin \gamma \qquad (2\text{-}31b)$$

The quantity $\bar{C}$ is appropriately called the *complex amplitude* of the trigonometric form of the function $x(t)$. The complex amplitude $\bar{C}$ has an absolute value C, representing the magnitude of the amplitude of the function, and γ indicating its argument.

As shown below, taking the derivatives of the complex form of a harmonic function is quite simple. In order to interpret the derivative, the trigonometric form may then be expressed. Thus

$$x = \mathscr{R}(\bar{C}e^{i\omega t}) = \mathscr{R}[C \cos(\omega t - \gamma) + iC \sin(\omega t - \gamma)]$$

$$= C \cos(\omega t - \gamma) \qquad (2\text{-}32a)$$

$$\dot{x} = \mathscr{R}(i\omega\bar{C}e^{i\omega t}) = \mathscr{R}[i\omega C \cos(\omega t - \gamma) - \omega C \sin(\omega t - \gamma)]$$

$$= -\omega C \sin(\omega t - \gamma) \qquad (2\text{-}32b)$$

$$\ddot{x} = \mathscr{R}(i^2\omega^2\bar{C}e^{i\omega t})$$

$$= \mathscr{R}(-\omega^2\bar{C}e^{i\omega t}) = \mathscr{R}[-\omega^2 C \cos(\omega t - \gamma) - i\omega^2 C \sin(\omega t - \gamma)]$$

$$= -\omega^2 C \cos(\omega t - \gamma) \qquad (2\text{-}32c)$$

The location of the real part of the trigonometric form should be noted. Also observe that $\dot{x} = \mathscr{R}(i\omega x)$ and $\ddot{x} = \mathscr{R}(i^2\omega^2 x)$, and that $\dot{x}$ leads x by 90 degrees and $\ddot{x}$ leads x by 180 degrees. Thus, in effect, taking a derivative simply multiplies by $i\omega$ and advances the vector by 90 degrees.

For Eq. 2-31 to represent a solution, the complete function—and thus the imaginary as well as the real part—must satisfy the differential equation. Accordingly, the function $x = \bar{C}e^{i\omega t}$ must satisfy Eq. 2-5. Performing this operation gives

$$(-\omega^2 + \omega^2)\bar{C}e^{i\omega t} = 0$$

so that the condition is fulfilled. The solution, expressed by the real part of the function, is then (refer to Eq. 2-31)

$$x = \mathscr{R}\bar{C}e^{i\omega t}$$

$$= C \cos(\omega t - \gamma) \qquad [2\text{-}32a]$$

The foregoing may appear to be unduly complicated and a cumbersome way of arriving at the solution. In addition, the concept may seem vague, whereas the first approach, in which a trigonometric function was assumed, appears clear-cut. For the case in hand this is true, but for more complicated problems, such as those involving more than one degree of freedom together with damping and forcing conditions, the use of complex notation greatly simplifies the solution determinations.

Other Methods

A basic way to obtain the solution to linear differential equations employs the operator notation $D = d/dt$, $D^2 = d^2/dt^2$, ... The resulting operator has a polynomial form which can be treated in an algebraic manner.† Although in some instances there are certain advantages to this method, in general it is similar to the procedure of using the factor Ce^{st}, so it will not be further discussed here.

A powerful means for solving differential equations of the type involved in vibration analyses is the Laplace transformation.‡ This is an operational method by which the differential equation is transformed into a subsidiary equation that can be treated algebraically. Inverse transformation of the solution for the subsidiary equation produces the solution to the differential equation. To gain understanding and proficiency in the use of Laplace transforms requires considerable study and practice; accordingly, the method is not included in this text.

No single procedure is best for solving all the different types of differential equations that will be encountered in vibration analysis. Hereafter the method which is most instructive will be selected whenever possible. In some instances the solution method that is the clearest and simplest in concept may become lengthy and complicated in application; in that case, for practical considerations the most efficient method will be chosen.

2-7. TORSIONAL VIBRATION

Certain structural components and a large number of machine parts exhibit torsional-vibration characteristics. Although the machine members may be complicated in form and arrangement, these can often be reduced to equivalent systems composed of shafts and disks having torsional freedom. Such a torsional model is shown in Fig. 2-7(a). It is made up of an elastic shaft and an inertial disk. The shaft is assumed to be massless. The torsional spring constant for the shaft is k_T, measured by the twisting moment per unit angle of twist, and has the dimension of pound inch per radian. The character I represents the mass moment of inertia, with units of pound inch second squared, for the disk relative to its rotation axis. [For the SI system, the units for k_T are newton meters per radian, which is the same as $\text{kg} \cdot \text{m} \cdot \text{s}^{-2} \times \text{m/rad} = \text{kg} \cdot \text{m}^2/\text{s}^2$, and for I the units are kilogram meters squared.] The angular displacement is positive, as shown, and is expressed in radians.

For an assumed positive angular displacement of the disk, the restoring torque acting on the disk would be $T = k_T \theta$ in the direction shown in the free-body diagram, Fig. 2-7(b). Writing the Newtonian relation (Eq. 1-12) for rotation about a fixed axis gives

$$I\ddot{\theta} = -k_T\theta$$

† See Gaskell, *Engineering Mathematics*, pp. 7–15.

‡ See W. T. Thomson, *Laplace Transformation*, 2d ed. (Englewood Cliffs, N.J.: Prentice-Hall, 1960). R. V. Churchill, *Operational Mathematics*, 3d ed. (New York: McGraw-Hill, 1972).

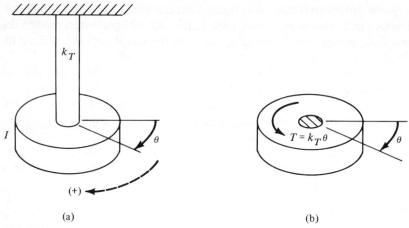

Figure 2-7

whence

$$\ddot{\theta} + \frac{k_T}{I}\theta = 0 \tag{2-33}$$

This is of the same form as Eq. 2-4 for the rectilinear case. Accordingly the solution will be

$$\theta = A \sin \omega t + B \cos \omega t$$
$$= C \sin (\omega t + \phi) \tag{2-34}$$

where

$$\omega = \sqrt{\frac{k_T}{I}}$$

depends on the physical constants of the system. Note that A and B or C and ϕ are arbitrary constants which are determined by the initial conditions of motion.

By replacing x by θ, the entire discussion of Sections 2-3 through 2-5 may be applied to the present case.

EXAMPLE 2-2 A brass gear is fastened to the lower end of a steel shaft that has its upper end fixed, thus forming a torsional arrangement similar to that of Fig. 2-7(a). The shaft is 1 in. in diameter and 40 in. long, has a shearing modulus of elasticity of 12×10^6 psi, and is to be considered weightless. When vibrating torsionally, the gear is found to have an angular-displacement amplitude of 0.05 rad and a maximum angular acceleration of 2000 rad/sec². Determine the mass moment of inertia of the gear.

SOLUTION For the shaft,

$$J = \frac{\pi}{32}d^4 = \frac{\pi}{32}1^4 = 0.0982 \text{ in.}^4$$

The torsional spring constant is defined for the shaft as the torque per unit angle of twist. Based on elastic theory, the angle of twist is expressed by $\theta = Tl/GJ$, whence, for $\theta = 1$ rad,

$$k_T = T = \frac{GJ\theta}{l} = \frac{12 \times 10^6 \times 0.0982 \times 1}{40} = 2.95 \times 10^4 \text{ lb in./rad}$$

Since $\ddot{\theta} = \omega^2\theta$,

$$\omega^2 = \frac{\ddot{\theta}}{\theta} = \frac{2000}{0.05} = 4 \times 10^4$$

From $\omega = \sqrt{k_T/I}$,

$$I = \frac{k_T}{\omega^2} = \frac{2.95 \times 10^4}{4 \times 10^4} = 0.738 \text{ lb in. sec}^2$$

EXAMPLE 2-3   The front-end torsion-rod suspension system for a small automobile is shown in Fig. 2-8. The sprung mass of the unloaded car is 1200 kg and 60% of this is supported by the front wheels (equally). When the front end is raised 11.7 cm, the front wheels leave the ground. Each front-wheel assembly, consisting of wheel, axle, brake, pivot arms, and tire has a mass of 38 kg. When fully loaded the mass of the vehicle is increased by 400 kg and the total mass is then equally supported on all four wheel suspensions. The damping effect due to shock absorbers is to be disregarded.

a. Determine the frequency of vibration of the wheel assembly. Assume the mass of the wheel assembly to be concentrated 26 cm from the torsion rod axis.

b. For the condition in which the fully loaded vehicle pivots about the rear suspension, determine the period of vibration of the front-end suspension of the car.

SOLUTION   a. The torsional spring constant can be calculated from the data for the support.

$$\Delta\theta = \frac{11.7 \text{ cm}}{26 \text{ cm}} = 0.45 \text{ rad}$$

$$\text{load } P = \frac{1200 \text{ kg} \times 0.60}{2} \times 9.81 \text{ m/s}^2 = 3531.6 \text{ N}$$

$$k_T = \frac{T}{\theta} = \frac{3531.6 \text{ N} \times 0.26 \text{ m}}{0.45 \text{ rad}} = 2040.48 \text{ N} \cdot \text{m/rad}$$

$$\omega = \sqrt{\frac{k_T}{I}} = \sqrt{\frac{2048.48 \text{ N} \cdot \text{m/rad}}{38 \text{ kg} \times (0.26 \text{ m})^2}} = 28.184 \text{ rad/s}$$

$$f = \frac{\omega}{2\pi} = \frac{28.184 \text{ rad/s}}{2\pi \text{ rad}} = 4.486 \text{ Hz}$$

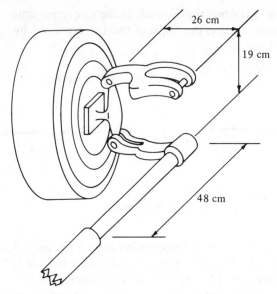

Figure 2-8

b. For the motion of the front end of the car,

$$k = \frac{1200 \text{ kg} \times 0.60 \times 9.81 \text{ m/s}^2}{0.117 \text{ m}} = 60\,369 \text{ N/m}$$

$$m = (1200 + 400) \times 0.50 = 800 \text{ kg}$$

$$\omega = \sqrt{\frac{k}{m}} = \sqrt{\frac{60\,369 \text{ N/m}}{800 \text{ kg}}} = 8.6869 \text{ rad/s}$$

$$\tau = \frac{2\pi}{\omega} = \frac{2\pi \text{ rad}}{8.6869 \text{ rad/s}} = 0.7233 \text{ s}$$

2-8. ANGULAR OSCILLATION

Of interest here is the system shown in Fig. 2-9(a), representing a simple pendulum for which the restoring force or condition is gravitational rather than elastic. The angular oscillatory motion of the system can be readily observed. Here, $W = mg$ is the weight of the small mass m fastened to the massless arm of length l, which pivots freely about the point O. When displaced from the vertical position of equilibrium, the component of W normal to the arm represents the restoring force. The mass m is considered to be small so that $I_O = ml^2$. For the positive position shown, the differential equation (refer to Eq. 1-12) for angular motion about the point O will be

$$ml^2\ddot{\theta} = -W \sin\theta \cdot l \tag{2-35}$$

$$\ddot{\theta} + \frac{g}{l}\sin\theta = 0 \tag{2-36}$$

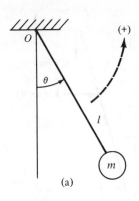
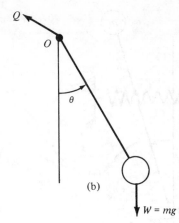

(a)

(b)

Figure 2-9

$W = mg$

The term $\sin \theta$ represents a transcendental function; hence the differential equation is nonlinear and is not amenable to a closed analytical solution. Replacing the $\sin \theta$ term by the series form results in

$$\ddot{\theta} + \frac{g}{l}\left(\theta - \frac{\theta^3}{3!} + \frac{\theta^5}{5!} - \frac{\theta^7}{7!} + \cdots\right) = 0 \tag{2-37}$$

which is also nonlinear. However, if the motion is limited to small amplitudes, then terms in the sine series of third and higher degree may be neglected, and the equation reduces to

$$\ddot{\theta} + \frac{g}{l}\theta = 0 \tag{2-38}$$

which is linear and readily solvable.

By comparison to the preceding section, it can be observed that Eq. 2-34 represents the solution for the present case, with the frequency defined by

$$\omega = \sqrt{\frac{g}{l}} \tag{2-39}$$

2-9. RESTORING CONDITION AND STABILITY

Consider the combined system represented by the inverted pendulum and symmetrical spring arrangement of Fig. 2-10(a). Assume that the springs are unstrained when the pendulum is vertical, and that the equivalent spring constant for the arrangement is k. For the positive angular position shown, the spring force is then $k(a \sin \theta)$, as shown. From the free-body diagram of Fig. 2-10(b), and assuming m to be small, then for rotation about the point O, the equation of motion will be

$$ml^2\ddot{\theta} = -(ka \sin \theta) \cdot a \cos \theta + W \sin \theta \cdot l \tag{2-40}$$

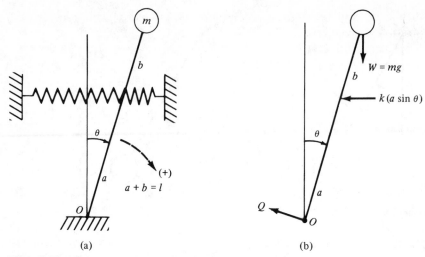

Figure 2-10

and for small oscillations, $\sin\theta \to \theta$ and $\cos\theta \to 1$, so that

$$ml^2\ddot{\theta} = -ka^2 \cdot \theta + Wl \cdot \theta \qquad (2\text{-}41)$$

This can be arranged as

$$\ddot{\theta} + \left(\frac{ka^2 - Wl}{ml^2}\right)\theta = 0 \qquad (2\text{-}42)$$

The solution to this differential equation will be

$$\theta = C\sin(\omega t + \phi) \qquad \text{for } ka^2 > Wl \qquad (2\text{-}43)$$

where

$$\omega = \sqrt{\frac{ka^2 - Wl}{ml^2}} \qquad (2\text{-}44)$$

This represents the case of a stable oscillation. However, if the limitation noted does not hold, then either equilibrium or instability will prevail.

Consider the case for which $ka^2 = Wl$. Equation 2-42 then becomes

$$\ddot{\theta} = 0 \qquad (2\text{-}45)$$

This can be solved directly by integrating twice, giving

$$\theta = C_1 t + C_2 \qquad (2\text{-}46)$$

From this, evaluating the arbitrary constants results in

$$\theta = \dot{\theta}_0 t + \theta_0 \qquad (2\text{-}47)$$

This represents the case of constant-velocity equilibrium, for which θ_0 can either have a value or be equal to zero. Since θ was assumed to be small, Eq. 2-47 specifies only the beginning of the constant-velocity motion. If $\dot{\theta}_0 = 0$, then static equilibrium exists, and the pendulum will remain in its original position, defined by θ_0.

The condition $ka^2 < Wl$ will be investigated next. For this, let

$$\frac{ka^2 - Wl}{ml^2} = -\gamma^2$$

where γ is real, so that $-\gamma^2$ is negative. Equation 2-42 then becomes

$$\ddot{\theta} - \gamma^2\theta = 0 \tag{2-48}$$

Assuming that $\theta = Ce^{st}$ and substituting gives

$$(s^2 - \gamma^2)Ce^{st} = 0$$

which requires that

$$s^2 - \gamma^2 = 0$$

$$s = \pm\gamma \tag{2-49}$$

and the solution is

$$\theta = C_3 e^{\gamma t} + C_4 e^{-\gamma t} \tag{2-50}$$

whence

$$\theta = \tfrac{1}{2}(\theta_0 + \dot{\theta}_0/\gamma)e^{\gamma t} + \tfrac{1}{2}(\theta_0 - \dot{\theta}_0/\gamma)e^{-\gamma t} \tag{2-51}$$

where γ is a positive real quantity defined by

$$\gamma = \sqrt{\frac{Wl - ka^2}{ml^2}} \tag{2-52}$$

This represents a nonoscillatory motion in which θ increases exponentially with time, providing that θ_0 and $\dot{\theta}_0$ are not zero. This defines only the beginning of the motion, as θ was assumed to be small. The arrangement is unstable. This is so, essentially, because the restoring moment of $ka^2\theta$ is less than the nonrestoring moment $-Wl\theta$. A restoring condition tends to return the system to equilibrium; a nonrestoring condition tends to move it away from the equilibrium configuration. A restoring condition (force or moment) will be recognized by its negative sign on the right-hand side of the differential equation, as in Eq. 2-41, or by its positive sign on the left-hand side, as in Eq. 2-42. Signs opposite to this will indicate a nonrestoring condition. In analyzing various systems these factors regarding restoring and nonrestoring conditions and stability should be kept in mind.

2-10. EQUIVALENT SPRINGS

In analyzing vibratory systems it is sometimes convenient to replace an elastic member by an equivalent spring. The problem is then reduced to that of a simple mass-spring model. The procedure for this involves defining the equivalent spring constant or modulus for the prototype member. This is

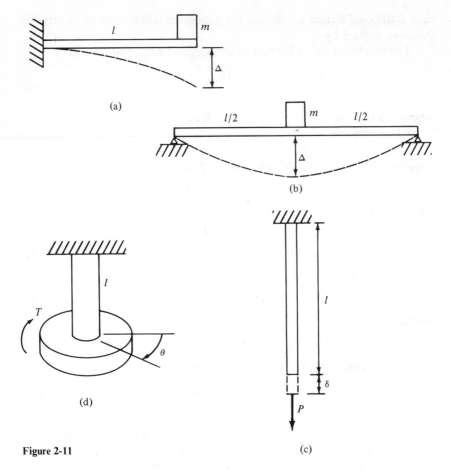

(a)

(b)

(d)

(c)

Figure 2-11

done by expressing the force or the moment for unit displacement of the member at the location of the mass or inertial element. This can be accomplished by rearranging the deflection or displacement expression for the type of member involved.

As an example of the procedure, consider the cantilever-beam arrangement shown in Fig. 2-11(a). For an elastic condition the relation for deflection Δ at the free end of a cantilever, subjected to a concentrated load P at that location, is $\Delta = Pl^3/3EI$, where E is the modulus of elasticity of the beam material, I is the moment of inertia of the section, and l is the beam length. From this, the equivalent spring constant k is

$$k = \frac{P}{\Delta} = \frac{3EI}{l^3} \tag{2-53}$$

Similarly, for the simple beam setup shown in Fig. 2-11(b), the deflection Δ at the center, due to the load P there, is $\Delta = Pl^3/48EI$, and the constant k for the equivalent spring is

$$k = \frac{P}{\Delta} = \frac{48EI}{l^3} \tag{2-54}$$

Thus the general expression for the equivalent spring constant for a beam would be

$$k = \frac{P}{\Delta} = \frac{1}{\gamma} \frac{EI}{l^3} \tag{2-55}$$

where γ is a constant, the value of which depends on the support and loading arrangement of the beam. Note that, dimensionally, the right-hand side of Eq. 2-55 reduces to force per unit displacement.

For a uniform bar in elastic tension-compression, Fig. 2-11(c), the equivalent spring constant can be obtained from the deformation relation $\delta = Pl/AE$, where δ is the deformation due to load P, A is the cross-section area of the bar, l is the length, and E is the tension-compression modulus of elasticity for the bar material. The equivalent spring constant k is the force per unit displacement, or P/δ. Thus

$$k = \frac{P}{\delta} = \frac{AE}{l} \tag{2-56}$$

The torsional system may be similarly treated. The angle of twist θ, for a shaft subjected to torque T, is given by $\theta = Tl/GJ$, where l is the shaft length, G is the shearing modulus of elasticity, and J is the polar moment of inertia of the circular section. Then, for the arrangement shown in Fig. 2-11(d), the equivalent torsional spring constant k_T is defined by

$$k_T = \frac{T}{\theta} = \frac{GJ}{l} \tag{2-57}$$

The dimension will be that of moment per radian, and represents the moment required to produce the unit angle of twist.

In machines and structures arrangements occur that, in effect, represent various combinations of springs. It is usually desirable to determine the single spring which could replace the separate springs and would be equivalent to them, insofar as producing the same physical conditions and motion are concerned. The general method of determining the equivalent spring constant for such combinations will now be established.

The springs for Fig. 2-12(a) and (b) are said to be in parallel. (It is assumed that the connecting bar does not tip but remains horizontal.) For such a system, let Δ represent the displacement due to the load P. Note that all springs will be equally deformed, and that P will equal the sum of the forces exerted by the separate springs. Thus

$$P = k_1\Delta + k_2\Delta + \cdots + k_n\Delta \tag{2-58a}$$

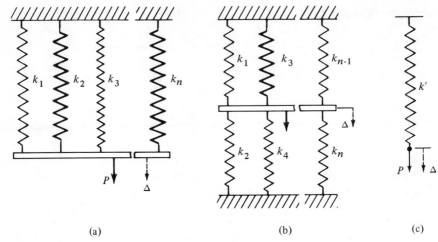

(a) (b) (c)

Figure 2-12

For the single spring of Fig. 2-12(c), which would replace the set, the equivalent spring constant k' is defined by

$$P = k'\Delta \tag{2-58b}$$

since the displacement must be the same. Then, from Eqs. 2-58a and 2-58b,

$$k'\Delta = k_1\Delta + k_2\Delta + \cdots + k_n\Delta$$

so that

$$k' = k_1 + k_2 + \cdots + k_n \quad \text{or} \quad k' = \sum_{j=1}^{n} k_j \tag{2-59}$$

Figure 2-13(a) shows an arrangement of n springs in series, that is, fastened end to end. Let Δ represent the displacement of the spring system (that is, of the end point B) caused by the load P. Also let Δ_j represent the deformation of spring j. Load P, placed at B, is transmitted through the system so that the force on each spring is P. Then

$$P = k_1\Delta_1 = k_2\Delta_2 = \cdots = k_n\Delta_n$$

Also

$$P = k'\Delta \tag{2-60}$$

where k' is for the equivalent spring shown in Fig. 2-13(b). Now

$$\Delta = \Delta_1 + \Delta_2 + \cdots + \Delta_n \tag{2-61}$$

whence, by Eq. 2-60,

$$\frac{P}{k'} = \frac{P}{k_1} + \frac{P}{k_2} + \cdots + \frac{P}{k_n}$$

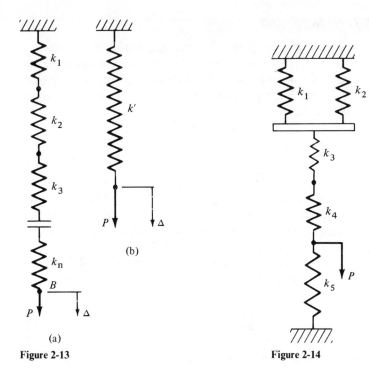

Figure 2-13

(a)

(b)

Figure 2-14

so that

$$\frac{1}{k'} = \frac{1}{k_1} + \frac{1}{k_2} + \cdots + \frac{1}{k_n} \qquad \text{or} \qquad \frac{1}{k'} = \sum_{j=1}^{n} \frac{1}{k_j} \qquad (2\text{-}62)$$

EXAMPLE 2-4   Determine the equivalent spring constant for the combination of springs shown in Fig. 2-14.

SOLUTION   This can be accomplished by expressing the equivalent spring constant for the separate parts and then combining these. The method of separating the system depends on the arrangement and also on the load position. Thus k_1 and k_2 are in parallel, but they in turn are in series with k_3 and k_4, and this subsystem is then in parallel with k_5. Proceeding in this manner, since k_1 and k_2 are in parallel, $k' = k_1 + k_2$, where k' is the equivalent spring for this part. Then k', k_3, and k_4 being in series gives $1/k'' = 1/k' + 1/k_3 + 1/k_4$, where k'' replaces the system to this point. Finally, k'' and k_5 are in parallel, so the equivalent spring constant k''' for the complete system is expressed by

$$k''' = k'' + k_5$$

$$= \frac{1}{1/(k_1 + k_2) + 1/k_3 + 1/k_4} + k_5$$

2-11. ENERGY METHOD

The differential equation of motion can be obtained by using the principle of conservation of energy rather than by Newton's second law of motion. For a free-vibration system, since no energy is removed, the total energy must remain constant. Hence if T represents the kinetic energy and V the potential energy of the system,

$$T + V = \text{constant}$$

and

$$\frac{d}{dt}(T + V) = 0 \tag{2-63}$$

In general, for small oscillations T is a function of the velocity and V is a function of the displacement. The time derivative (Eq. 2-63) should then yield the differential equation which governs the motion of the system.

This procedure may be applied to any single-degree-of-freedom undamped free system. For example, consider the mass-spring oscillator shown in Fig. 2-15. For this case, the kinetic energy is

$$T = \tfrac{1}{2}m\dot{x}^2 \tag{2-64}$$

Using the equilibrium position as the reference, the potential energy can be expressed as the integral of the mechanical work done by the forces of the system in returning from the dynamic configuration to the reference position. That is,

$$V = \int_x^0 [W - k(\Delta + x)]\, dx = -\int_x^0 kx\, dx = \frac{kx^2}{2} \tag{2-65}$$

Equation 2-63 then results in

$$\frac{d}{dt}\left(\frac{m\dot{x}^2}{2} + \frac{kx^2}{2}\right) = 0$$

$$(m\ddot{x} + kx)\dot{x} = 0$$

and

$$m\ddot{x} + kx = 0 \tag{2-66}$$

providing the trivial solution for $\dot{x} = 0$ is disregarded. This is identical to the differential equation previously obtained for this case by Newton's second law, and it has the same solution. The present method is simply a different means of deriving the differential equation of motion, which may be convenient or desirable for certain cases.

The energy method can be used for systems in which the mass of the elastic member is significant and should not be neglected. Although not precise in this instance, it gives results that are in good agreement with those for the exact analysis. The energy method for the mass-spring model shown in Fig. 2-16, and including the mass of the spring, is presented below. It is

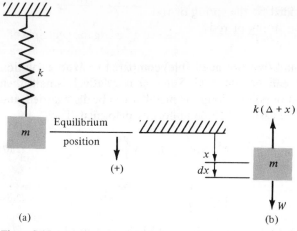

Figure 2-15

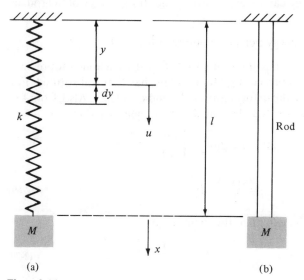

Figure 2-16

convenient here to consider the spring as an equivalent elastic rod in tension and compression.

Let

y = position coordinate of the spring or rod
ρ = weight of unit length of the spring or rod
m = mass of the spring or rod
dm = mass of the element having length dy
u = displacement of the element of the spring or rod

M = mass suspended by the spring or rod
l = length of the spring or rod
x = displacement of M

Assuming m as small (but not negligible) compared to M, the solution for the motion of M can be obtained. Since m is relatively small, then variation in the motion of points along the member can be disregarded, and the spring or rod can be assumed to be uniformly deformed. Then

$$\frac{u}{x} = \frac{y}{l} \qquad \text{whence} \qquad u = x\frac{y}{l} \tag{2-67}$$

and

$$\dot{u} = \dot{x}\frac{y}{l} \tag{2-68}$$

since y is a position coordinate and is therefore independent of time.

The potential energy can be expressed by the strain energy of the rod, as follows:

$$\text{strain energy per unit volume} = \tfrac{1}{2}\sigma\varepsilon = \tfrac{1}{2}E\varepsilon^2$$

where σ is the unit stress, ε is the unit strain, E is the tension-compression modulus, and $\sigma/\varepsilon = E$, representing Hooke's law, has been substituted. For the element of length dy the deformation is du, and $\varepsilon = du/dy$. Also, from Eq. 2-67, $du/dy = x/l$. Then, for the element the strain energy dV is

$$dV = (A\,dy)\frac{1}{2}E\left(\frac{du}{dy}\right)^2$$

$$= \frac{AE}{2}\frac{x^2}{l^2}\,dy \tag{2-69}$$

and integrating over the extent of the member gives

$$V = \frac{AE}{2}\frac{x^2}{l^2}\int_0^l dy = \frac{1}{2}\frac{AE}{l}x^2 \tag{2-70}$$

From the deformation relation $\delta = Pl/AE$ for a rod, where P is the axial load,

$$k = \frac{P}{\delta} = \frac{AE}{l}$$

and substituting into the relation for V gives

$$V = \tfrac{1}{2}kx^2 \tag{2-71}$$

Thus if the spring is assumed to be uniformly strained, the potential energy is not altered as compared to the massless-spring case.

The kinetic energy T' of the suspended mass M is

$$T' = \tfrac{1}{2}M\dot{x}^2$$

For the spring, the kinetic energy T'' can be obtained by considering an element, as follows:

$$dT'' = \frac{1}{2}dm\,\dot{u}^2 = \frac{1}{2}\frac{\rho}{g}dy\left(\dot{x}\frac{y}{l}\right)^2$$

$$T'' = \frac{1}{2}\frac{\rho}{g}\frac{\dot{x}^2}{l^2}\int_0^l y^2\,dy = \frac{1}{2}\left(\frac{1}{3}\frac{\rho l}{g}\right)\dot{x}^2 = \frac{1}{2}\frac{m}{3}\dot{x}^2$$

The total kinetic energy is then

$$T = T' + T'' = \frac{1}{2}\left(M + \frac{m}{3}\right)\dot{x}^2 \tag{2-72}$$

The time derivative of $(T + V)$ will yield the differential equation

$$\left(M + \frac{m}{3}\right)\ddot{x} + kx = 0 \tag{2-73}$$

having the solution

$$x = C\sin(\omega't + \phi) \tag{2-74}$$

where

$$\omega' = \sqrt{\frac{k}{M + m/3}} \tag{2-75}$$

The exact analysis for this case is somewhat involved.† However, in many cases the above analysis is as accurate as would be justified by the data for the physical system being investigated. This is borne out by a comparison of frequencies determined by the different methods. If m is neglected entirely and the frequency is calculated by $\omega = \sqrt{k/M}$, the result is in error by only about 5%, compared to the exact value, for m as large as $0.3M$. On the other hand, if the frequency is determined by Eq. 2-75 and compared to the exact value, the error is less than 1% for a spring mass as large as $m = M$.

EXAMPLE 2-5   An aluminum-alloy bar, 0.5 in. square, 100 in. long, and hanging vertically, suspends a weight of 5 lb at the lower end. The bar weighs 2.5 lb, and the material has a tension-compression modulus of elasticity of 10^7 psi. Obtain the natural frequency of oscillation for the arrangement (a) neglecting the weight of the bar, and (b) based on the approximation developed in Section 2-11.

† See Section 10-8.

SOLUTION

$$k = \frac{AE}{l} = \frac{(0.5)^2 \times 10^7}{100} = 2.5 \times 10^4 \text{ lb/in.}$$

a.　$\omega = \sqrt{\frac{k}{M}} = \sqrt{\frac{2.5 \times 10^4 \times 386}{5}} = 1389 \text{ rad/sec}$

$$f = \frac{\omega}{2\pi} = \frac{1389}{2\pi} = 221 \text{ Hz}$$

b.　$\omega' = \sqrt{\frac{k}{M + m/3}} = \sqrt{\frac{2.5 \times 10^4 \times 386}{5 + 2.5/3}} = 1286 \text{ rad/sec}$

$$f' = \frac{\omega'}{2\pi} = \frac{1286}{2\pi} = 205 \text{ Hz}$$

The exact results† are

$$\omega'' = 1277 \text{ rad/sec}$$

$$f'' = 203 \text{ Hz}$$

PROBLEMS

2-1. A body weighing 12 lb is suspended by a spring having a modulus of 4 lb/in. Determine the natural frequency for the system.

2-2. A 5-lb weight, attached to a spring, causes a static equilibrium elongation of 0.143 in. Determine the natural frequency of oscillation that the system will exhibit if it is displaced from its equilibrium position and then released.

2-3. A body having a mass of 9 kg is suspended by a spring having a modulus of 800 N/m. Determine the natural period of vibration for the system.

2-4. A 1.4-kg mass causes a 4-mm static equilibrium elongation of a certain spring. Determine the natural frequency of vibration for this system.

2-5. A weight of 20 lb, attached to a spring, is found to oscillate at 330 cycles/min. Determine the spring constant k.

2-6. A motor is placed on rubber mountings to isolate it from its foundation. If the dead weight of the motor compresses the mountings 3.1 mm, determine the natural frequency of the system.

2-7. Determine the mass which must be attached to a spring having a modulus of 1750 N/m in order that the resulting system will have a frequency of 21 Hz.

2-8. Determine the length of a simple pendulum so that the half period will be 0.5 sec.

2-9. A mass-spring system has a natural period of 0.152 sec. Determine the period if the spring constant is increased by 60%.

2-10. A weight W is suspended from the end of spring k, and the system has a natural frequency of 90 cycles/min. The weight W is removed and replaced by a weight

† See Example 10-6.

$W/4$ at the middle of the spring, and the upper and lower ends of the spring are then fixed. Determine the frequency of the new system.

2-11. A mass-spring system has a frequency of 2 Hz. When the mass is decreased by 0.4 kg, the frequency is changed by 25%. Determine (a) the mass m, and (b) the spring constant k for the original system.

2-12. A mass-spring system has a frequency of 5 Hz. When the spring constant is reduced by 4 lb/in., the frequency is altered by 30%. Determine (a) the spring constant k and (b) the weight W for the original system.

2-13. The cam-shaped member shown in the accompanying figure is free to rotate about the axis at O perpendicular to the view and exhibits a natural frequency of angular oscillation of 54 cycles/min with respect to this axis. The rotation axis is moved to point B, located somewhere between O and the mass center G. The frequency about the axis at B is found to be 36 cycles/min. Determine the distance OB.

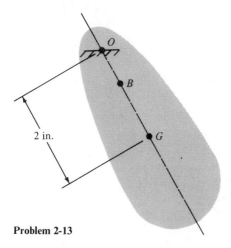

Problem 2-13

2-14. A harmonic motion is expressed by the relation $x = 0.5 \sin (10t + 4)$, where x is measured in inches, t in seconds, and the phase angle in radians. Determine (a) the circular and natural frequencies of vibration and (b) the displacement, velocity, and acceleration at $t = 0.3$ sec.

2-15. A vibratory motion has an amplitude of 0.30 in. and a period of 0.4 sec. Determine (a) the frequency of vibration, (b) the maximum velocity, and (c) the maximum acceleration.

2-16. From experimental measurements it has been found that a structure is vibrating harmonically. If the amplitude of motion was measured to be 2 cm and the period was 3 sec, determine (a) the maximum velocity and (b) the acceleration to which the structure is subjected.

2-17. A simple harmonic motion has a maximum amplitude of 50 mm and a period of 10 sec. Determine (a) the displacement, (b) the velocity, and (c) the acceleration, at the time $t = 8$ sec, if the time is zero when the motion passes through the equilibrium position.

2-18. Vibration-measuring instruments indicate that a structure is vibrating harmonically at a frequency of 4 Hz, with a maximum acceleration of 324 in./sec².

Determine (a) the displacement amplitude and (b) the maximum velocity during motion of the structure.

2-19. An accelerometer is used to measure a vibratory motion. If the frequency is found to be 20 Hz and the maximum acceleration is 12 m/s², determine (a) the period, (b) the amplitude, and (c) the maximum velocity of the motion.

2-20. If the frequency of a vibrating body is 5 Hz and its maximum velocity is 90 in./sec, determine (a) the amplitude, (b) the maximum acceleration, and (c) the period of the motion.

2-21. A slender uniform bar pivots in a vertical plane about axis A located distance a from the upper end. For small angular vibration, obtain the differential equation of motion and express the frequency.

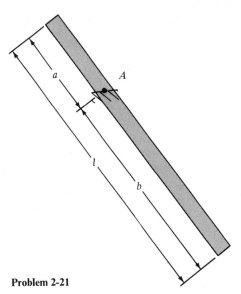

Problem 2-21

2-22. A rigid bar AB is weightless and pivots at A as shown in the accompanying figure. Obtain (a) the differential equation of motion and (b) the expression for the natural frequency of angular vibration of the system.

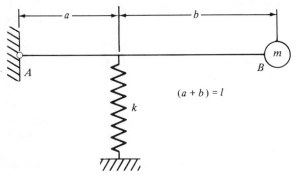

$(a + b) = l$

Problem 2-22

2-23. A rigid bar AB is weightless and pivots at A as shown in the accompanying figure. Determine (a) the differential equation for the vertical motion of m and (b) the expression for the natural frequency of oscillation of m.

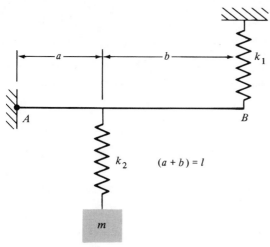

Problem 2-23

2-24. A rigid member ABC is weightless and pivots at B as shown in the accompanying figure. Obtain (a) the differential equation of motion and (b) the expression for the natural frequency of angular vibration of the system. (c) Indicate the condition for stability of the vibration.

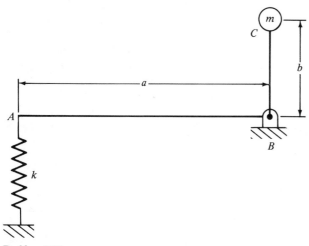

Problem 2-24

2-25. Determine (a) the differential equation of motion and (b) the expression for the natural frequency of vibration of the system shown herewith.

Problem 2-25

2-26. Obtain (a) the differential equation of motion and (b) the expression for the frequency of vibration of the system shown in the accompanying sketch.

Problem 2-26

2-27. The cylinder m is suspended by a spring and partially immersed in a liquid, as shown in the sketch. When displaced vertically, the cylinder oscillates with a small displacement, under the influence of the spring and the displaced liquid. Assume the density of the liquid to be ρ lb/in.^{3}. Determine (a) the differential equation of motion and (b) the expression for the frequency of vibration.

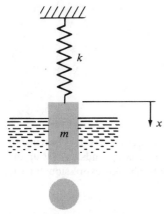

Problem 2-27

2-28. Obtain (a) the differential equation of motion and (b) the expression for the natural frequency of oscillation of the fluid in the U tube shown herewith. Assume the fluid to have a density ρ lb/in.^{3}, and let l represent the length of the liquid in the tube.

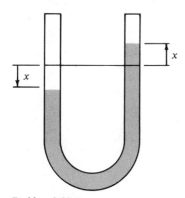

Problem 2-28

2-29. The uniform slender bar AB in the accompanying sketch is pivoted at A. (a) Write the differential equation for the angular vibration of the bar. (b) Obtain the expression for the period of oscillation.

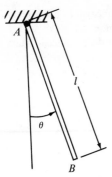

Problem 2-29

2-30. The disk shown has a mass moment of inertia I about its axis. It is fastened to shafts of different diameters, and the shafts are fixed at the ends. (a) Set up the differential equation of angular motion. (b) Obtain the expression for the frequency of angular vibration.

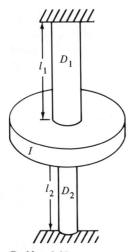

Problem 2-30

2-31. The disk shown has a mass moment of inertia I. It is fastened to the stepped shaft, as indicated. Obtain (a) the differential equation for the angular motion of the disk and (b) the expression for the period of vibration.

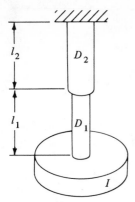

Problem 2-31

2-32. A cord AB is stretched with a tension Q and contains a small mass m, located as shown. (a) Determine the differential equation for a small vertical motion of the mass. (b) Obtain the expression for the period of vibration. Assume that the tension is not changed by the displacement.

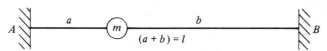

Problem 2-32

2-33. Member AB in the accompanying diagram is a weightless cantilever beam, fixed at A and fastened to a spring k and a mass m at B. (a) Determine the differential equation for a small vertical motion of m. (b) Obtain the expression for the frequency of oscillation.

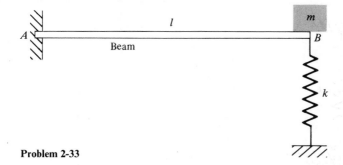

Problem 2-33

2-34. Member BC in the accompanying diagram is a uniform weightless bar. The mass m is free to move vertically in frictionless slides at F and G. The bar has a cross-sectional area A. Obtain (a) the differential equation for a small vertical motion of m, and (b) the expression for the period of the movement.

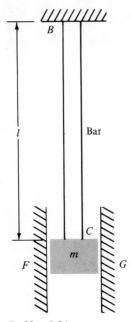

Problem 2-34

2-35. The mass m_1 is attached to spring k and is hanging at rest. Mass m_2 is then released from height h and falls on mass m_1 and adheres to it without rebounding. Obtain the differential equation that governs the motion of the system and express the natural frequency.

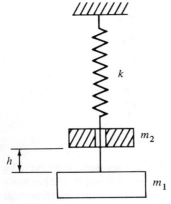

Problem 2-35

2-36. The slender bar shown pivots at *B* and is fastened to spring *k* at *A*. The bar is in equilibrium in a horizontal position. Determine the differential equation that governs the free vibrational motion and write the relation for the frequency.

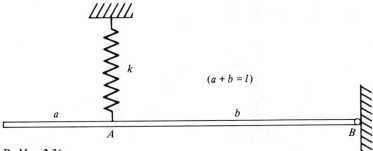

$(a + b = l)$

Problem 2-36

2-37. A slender bar with mass center *G* lies in grooves on two wheels which rotate in opposite directions, as shown. The coefficient of kinetic friction between the grooves and the bar is μ. The bar will exhibit a horizontal vibrational motion. Obtain the differential equation for this motion and express the natural frequency.

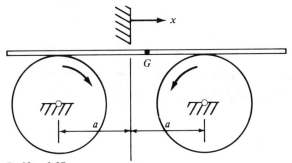

Problem 2-37

2-38. The body of the water tower shown weighs 130 000 lb and is supported on four 20-ft steel pipes, symmetrically placed as legs. The pipe section has a moment of inertia of 7 in.^{4} and the steel has a modulus of elasticity of 30×10^6 lb/in.^{2}. The legs are firmly fixed at the lower end and are considered to pivot at their upper ends against the body. The tower body can thus exhibit horizontal motion as the legs bend. Determine the differential equation of motion and determine the frequency.

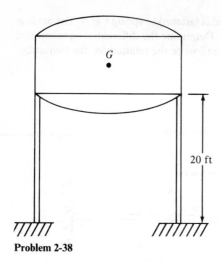

Problem 2-38

2-39. A uniform cylinder rolls without slipping on a horizontal surface and is constrained by the spring as shown. Obtain the differential equation for small oscillations of the rolling motion and express the frequency.

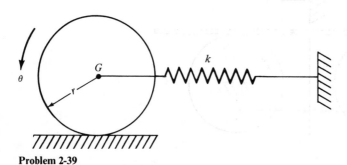

Problem 2-39

2-40. The half cylinder shown has a uniform length and rolls without slipping on the horizontal surface. The mass center is at G located distance $b = 4r/3\pi$ from the cylindrical axis at O. Obtain the differential equation for small oscillatory rolling motion and write the relation for the frequency.

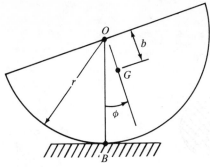

Problem 2-40

2-41. A heavy steel shoe is used in an industrial rolling operation. The face of the shoe is circular. It is desired to determine the moment of inertia of the shoe mass with respect to its mass-center axis at *G*. This is done from the small rocking motion of the shoe on a horizontal surface. (The shoe rolls without slipping.) For such small motion the angular frequency is measured as *f*. Determine the moment of inertia about the mass-center axis in terms of *f*, etc.

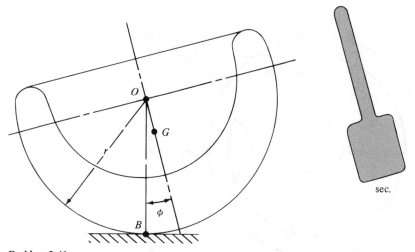

sec.

Problem 2-41

2-42. It is desired to determine the moment of inertia of the mass of a large airplane landing-gear wheel. The wheel assembly and support is fixed to the end of a 0.1-in. diameter 6-ft long steel wire, for which the shearing modulus of elasticity is 12×10^6 lb/in.^{2}. When tested in small torsional motion, the wheel assembly and support have an angular period of 90 sec. The support part alone shows a period of 5 sec. Determine the inertia of the wheel assembly about its rotational axis.

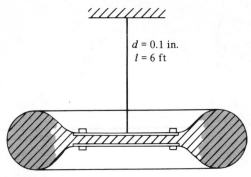

Problem 2-42

2-43. In order to determine the mass moment of inertia of a wheel, it is supported on a knife-edge, as shown and permitted to oscillate with small angular motion. The period is measured as 1.14 sec. Determine the moment of inertia of the wheel about its mass-center axis at C. The wheel has a mass of 34 kg.

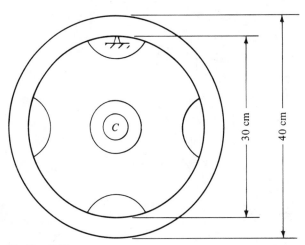

Problem 2-43

2-44. The connecting rod shown has a mass of 2.5 kg and its mass center is located at G. When placed on a knife-edge, as indicated, and allowed to oscillate through a small angle, the frequency is found to be 54 cycles/min. Determine the mass moment of inertia about the axis at G.

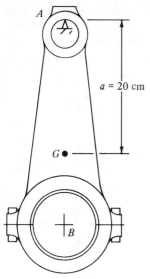

$a = 20$ cm

Problem 2-44

2-45. Gears A and B are meshed together, and each gear is fastened to a shaft having torsional stiffness K. The moments of inertia of the gears are I_A and I_B, respectively. The ratio of the number of teeth in B to that in A is n. When the gears are displaced torsionally, the system vibrates with small angular motion. Write the differential equations that govern the motion and express the frequency.

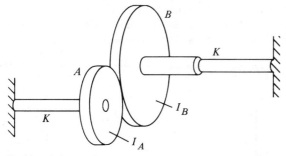

Problem 2-45

2-46. A proving ring is used for checking the accuracy and calibration of a testing machine. It consists of a steel ring which is deformed by the application of load P, as indicated. The load can then be determined from the change in the vertical dimension AD of the ring. This change is measured by adjusting the micrometer CD until the curved anvil at C just touches the lower end B of the steel reed AB. Delicate contact is determined by an electrical circuit and buzzer—the reed is flicked or vibrated and the adjustment is made until the slight buzzer sound indicates the touching condition. The steel-reed dimensions are indicated in the figure, and the mass of *B* is 9.408 grams; also for steel the modulus of elasticity is 2×10^{11} N/m^2 or pascals. Determine the time between electrical contacts of the curved end of the reed and the curved anvil.

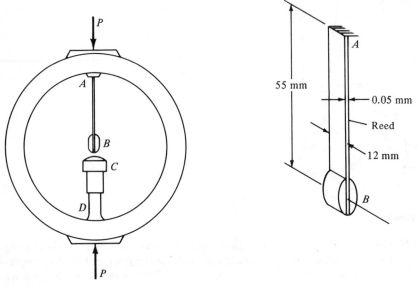

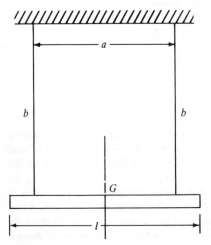

Problem 2-46

2-47. A uniform bar is suspended in a horizontal position by light flexible cables of equal length *b*. When twisted about the vertical centerline through mass center *G* and released, the bar exhibits small angular oscillations. Obtain the differential equation for this motion and express the frequency.

Problem 2-47

2-48. Two tanks, having cross-sectional areas A_1 and A_2, are connected by a pipe having area A_3 and length b, as shown. When at rest, the fluid level in the two tanks has a height h. If the fluid level is made to differ momentarily in the two tanks, small oscillations of the fluid between the tanks will result. Using the energy method, obtain (a) the differential equation of motion and (b) the expression for the frequency of oscillation. Assume the fluid to have a density ρ lb/in.^{3}.

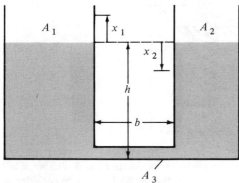

Problem 2-48

2-49. A uniform cylinder, of radius r and weight W, rolls, without slipping, on a cylindrical surface of radius R. (a) Determine the differential equation for a small motion about the equilibrium point B. (b) Obtain the expression for the frequency of oscillation about this point.

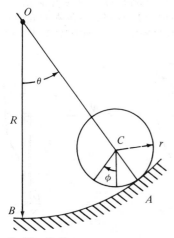

Problem 2-49

2-50. Mass m in the accompanying diagram moves in frictionless horizontal slides. When the spring k_1 is vertical, it is stretched with tension Q, and the spring k_2 is unstrained. Determine (a) the differential equation of motion for m, and (b) the expression for the frequency of vibration. Assume x/b to be small, so that the change in tension of the spring k_1 can be neglected.

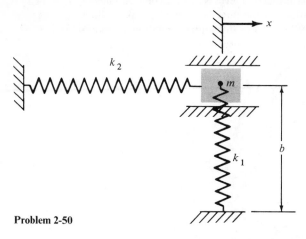

Problem 2-50

2-51. For the case of Prob. 2-50, assume x/b to be moderate (that is, neither small nor large), so that the change in tension of the spring k_1 must be considered. Then determine the differential equation of motion.

2-52. A one-story building frame is composed of columns AB and CD, and rigid girder BC. The girder has a weight W, including the roof, and so on, which it supports. The columns are to be considered weightless. The columns are fixed against rotation at the lower ends; E and I represent the modulus of elasticity and moment of inertia (of the section), respectively, of a column. Determine the natural frequency of horizontal vibration of the girder (a) assuming the upper end of the columns to be pivoted as shown, and (b) considering that the upper end of columns are fixed against rotation at their connection with the girder, as indicated.

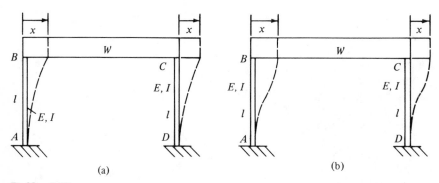

(a)   (b)

Problem 2-52

2-53. Using the energy method, solve Prob. 2-52(b), considering that the weight w of the columns is not negligible.

chapter 3

Damped Free Vibrations for
Single-Degree-of-Freedom
Systems

3-1. INTRODUCTION

The vibrations considered in the preceding chapters were self-sustaining and would not increase, diminish, or change in character with time. That is, there was no source which would excite the system and hence increase the amplitude of the vibration, nor was there any form of resistance that would dissipate energy and reduce the oscillation in any way. A consideration of practical cases, however, would reveal that this condition is not realistic, since all vibrations gradually lose amplitude and eventually cease altogether, unless they are maintained by some external source. Since the amplitude of a free vibration slowly dies away, something must cause energy to be removed from the system. The vibration is said to be *damped*, and the means of energy removal is called a *damper*. There are three main forms of damping. These will now be identified and discussed briefly.

Viscous damping is fairly common, occurring, for example, when bodies move in or through fluids, at low velocities. The resisting force F_d produced upon the body is proportional to the first power of the velocity of the motion. Thus

$$F_d = -c\dot{x}$$

where the *damping constant c* is the constant resistance developed per unit velocity. It has the units of pound second per inch (lb sec/in.), so that F_d has the dimension of pounds. In the SI system, the units of c are newton seconds per meter ($N \cdot s/m = kg/s$), and F_d is in newtons (N). The negative sign is used, since the damping force opposes the direction of motion. This mathematical model of damping is a good approximation for cases in which bodies slide on lubricated surfaces, when bodies move in air, oil, or other fluids, and for simple shock absorbers and hydraulic dashpots, providing the speed is not too great. At high speeds such resistance may be proportional to the square or a higher power of the velocity.

Coulomb or dry-friction damping is encountered when bodies slide on dry surfaces. This type of resistance is approximately constant, providing the surfaces are uniform and that the difference between the starting and moving conditions is negligible. The resisting force depends on the kind of materials and the nature of the sliding surfaces, and also on the force normal to the surface. This is expressed as

$$F_d = \mu N$$

where μ is the coefficient of kinetic friction for the materials and N is the normal force. The value of μ is determined experimentally. In the final stages of motion this form of damping tends to predominate, since it is constant and the other types become negligible for small velocity and displacement.

Hysteresis damping, which is also called solid or structural damping, is due to internal friction of the material. The deformation of a member is an accumulation of the internal displacements of the material. Such displacements are accompanied by frictional resistance, and the energy so absorbed is dissipated in the form of heat. This type of resistance is approximately proportional to the displacement amplitude and is independent of the frequency.

3-2. FREE VIBRATIONS WITH VISCOUS DAMPING; ANALYSIS AND SOLUTION

Consider the system shown in Fig. 3-1(a), which includes a viscous damping element represented symbolically by the dashpot shown, in addition to the mass and elastic elements. Since the displacement x referred to the equilibrium level is taken positive downward, for convenience, the time derivatives $\dot{x}$ and $\ddot{x}$ are also positive in that sense. Then, for a positive velocity $\dot{x}$, the damping force $c\dot{x}$ will be in the indicated upward direction in the dynamic free-body diagram shown in Fig. 3-1(b). It should be noted that this force will be automatically and properly reversed when $\dot{x}$ becomes negative. Newton's second law yields

$$m\ddot{x} = -kx - c\dot{x} \tag{3-1}$$

which can be rearranged as

$$\ddot{x} + \frac{c}{m}\dot{x} + \frac{k}{m}x = 0 \tag{3-2}$$

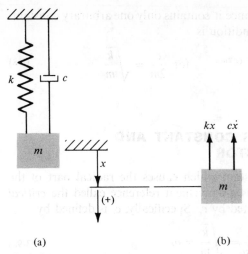

(a)   (b)

Figure 3-1

This is a homogeneous linear differential equation with constant coefficients. The presence of both the first and second derivatives suggests assuming the solution as

$$x = Ce^{st} \tag{3-3}$$

since it returns to the same form in all derivatives and hence would tend to satisfy the differential equation of motion. Substituting this relation into Eq. 3-2 gives

$$\left(s^2 + \frac{c}{m}s + \frac{k}{m}\right)Ce^{st} = 0 \tag{3-4}$$

Since neither C nor e^{st} can vanish except for the trivial case, it is necessary that

$$s^2 + \frac{c}{m}s + \frac{k}{m} = 0 \tag{3-5}$$

This is known as the auxiliary equation, the solution of which determines the conditions for which Eq. 3-3 will represent a solution. Equation 3-5 yields two roots, namely,

$$s = -\frac{c}{2m} \pm \sqrt{\left(\frac{c}{2m}\right)^2 - \frac{k}{m}} \tag{3-6}$$

The general solution is then

$$x = C_1 e^{[-(c/2m) + \sqrt{(c/2m)^2 - (k/m)}]t} + C_2 e^{[-(c/2m) - \sqrt{(c/2m)^2 - (k/m)}]t} \tag{3-7}$$

providing $c/2m \neq \sqrt{k/m}$. If $c/2m = \sqrt{k/m}$, Eq. 3-7 reduces to

$$x = (C_1 + C_2)e^{-(c/2m)t} = C_3 e^{-(c/2m)t}$$

which is not a general solution since it contains only one arbitrary constant. The general solution for this condition is

$$x = (A + Bt)e^{-(c/2m)t} \qquad \text{for } \frac{c}{2m} = \sqrt{\frac{k}{m}} \tag{3-8}$$

and will be derived later.

3-3. CRITICAL DAMPING CONSTANT AND THE DAMPING FACTOR

The value of the damping constant which causes the radical part of the exponent of Eq. 3-7 to vanish is a convenient reference called the *critical damping constant* and is designated by c_c. Specifically, c_c is defined by

$$\frac{c_c}{2m} = \sqrt{\frac{k}{m}} = \omega \tag{3-9a}$$

or

$$c_c = 2\sqrt{mk} = 2m\omega \tag{3-9b}$$

For a damped system the ratio of the damping constant to the critical value is a dimensionless parameter which represents a meaningful measure of the amount of damping present in the system. This ratio ζ is called the *damping factor* and is defined by

$$\zeta = \frac{c}{c_c} \tag{3-10}$$

whence

$$\frac{c}{2m} = \frac{c}{c_c} \cdot \frac{c_c}{2m} = \zeta\omega \tag{3-11}$$

Making use of these symbols, Eq. 3-7 may be written as

$$x = C_1 e^{(-\zeta + \sqrt{\zeta^2 - 1})\omega t} + C_2 e^{(-\zeta - \sqrt{\zeta^2 - 1})\omega t} \qquad \text{for } \zeta \neq 1 \tag{3-12}$$

The two cases represented by this solution, and also the solution for critical damping, will be investigated in the sections that follow.

3-4. FURTHER EXAMINATION OF SOLUTION

Case for Which $\zeta > 1$ or $c/2m > \sqrt{k/m}$

In this instance since $\sqrt{\zeta^2 - 1} < \zeta$, both exponents of Eq. 3-12 are real and negative, and the solution can be left in the form shown; namely,

$$x = C_1 e^{(-\zeta + \sqrt{\zeta^2 - 1})\omega t} + C_2 e^{(-\zeta - \sqrt{\zeta^2 - 1})\omega t} \qquad \text{for } \zeta > 1 \tag{3-13}$$

This represents the sum of two decaying exponentials. Hence the motion is nonperiodic, or *aperiodic*. A more detailed interpretation of the motion for this case will be made subsequently.

Case for Which $\zeta < 1$ or $c/2m < \sqrt{k/m}$

For this condition, $(\zeta^2 - 1)$ is negative, and the exponential multipliers of ωt in Eq. 3-12 are conjugate complex numbers. It is therefore desirable to write the solution as

$$x = C_1 e^{(-\zeta + i\sqrt{1-\zeta^2})\omega t} + C_2 e^{(-\zeta - i\sqrt{1-\zeta^2})\omega t}$$
$$= e^{-\zeta \omega t}(C_1 e^{i\sqrt{1-\zeta^2}\,\omega t} + C_2 e^{-i\sqrt{1-\zeta^2}\,\omega t}) \tag{3-14}$$

which, by Euler's relation (Eq. 2-28), and in the manner of the second method of Sec. 2-6, can be written as

$$x = e^{-\zeta \omega t}(C' \sin \sqrt{1 - \zeta^2}\,\omega t + C'' \cos \sqrt{1 - \zeta^2}\,\omega t)$$
$$= X e^{-\zeta \omega t} \sin (\sqrt{1 - \zeta^2}\,\omega t + \phi)$$
$$= X e^{-\zeta \omega t} \sin (\omega_d t + \phi) \qquad \text{for } \zeta < 1 \tag{3-15}$$

where

$$\omega_d = \sqrt{1 - \zeta^2}\,\omega \tag{3-16}$$

represents the *damped circular frequency*. Both X and ϕ are determined by initial motion conditions. The first part, $X e^{-\zeta \omega t}$, represents a decaying amplitude for the trigonometric function $\sin \omega_d t$. Thus the motion is oscillatory but with an amplitude that reduces with time. A rather detailed analysis of this motion will be made later.

Case for Which $\zeta = 1$ or $c/2m = \sqrt{k/m}$

In order to obtain the general solution, Eq. 3-15, for which $\zeta < 1$, may be used and ζ be made to approach unity.† Now as $\zeta \to 1$, $\omega_d \to 0$, so that for finite time, $\sin \omega_d t \to \omega_d t$ and $\cos \omega_d t \to 1$. Then the first form listed for Eq. 3-15 yields

$$x = e^{-\omega t}(C'\omega_d t + C'')$$
$$= (A + Bt)e^{-\omega t} \qquad \text{for } \zeta = 1 \tag{3-17}$$

which agrees with the previously listed solution, Eq. 3-8. This relation also is a nonperiodic function. The form of the motion will be explained in some detail subsequently.

---

† A somewhat more sophisticated method of obtaining the solution for this case can be made by using operator notation and procedures. See R. E. Gaskell, *Engineering Mathematics* (New York: Dryden Press, 1958), pp. 12–15.

3-5. MOTION FOR ABOVE-CRITICAL DAMPING

For $\zeta > 1$, the damping is above critical. This is also called supercritical damping, or the system is said to be overdamped. The solution expressed as Eq. 3-13 applies. For convenience, this can be written as

$$x = C_1 e^{-\alpha t} + C_2 e^{-\beta t} \tag{3-18}$$

where

$$\alpha = (\zeta - \sqrt{\zeta^2 - 1})\omega > 0$$
$$\beta = (\zeta + \sqrt{\zeta^2 - 1})\omega > 0 \tag{3-19}$$

and

$$\beta \gg \alpha$$

Both parts of Eq. 3-18 represent decaying exponentials. However, the second part drops much more rapidly than does the first part since β is considerably greater than α. These characteristics are shown in Fig. 3-2. It should be noted that C_1 or C_2 (or both) may be either positive or negative. Five basic types of motion are possible here. In justifying these, consider first the case for which C_1 is positive and C_2 is negative, but with the absolute value of C_2 greater than that of C_1. The two parts, $C_1 e^{-\alpha t}$ and $C_2 e^{-\beta t}$, of the curve will then be as shown separately in Fig. 3-3(a). It will be observed that the negative part reduces much more rapidly than does the positive part, since $\beta \gg \alpha$. By adding, the two parts are combined into the single x curve shown in Fig. 3-3(b). At the initial point 1, x will be negative. At point 2, the positive and negative parts have the same magnitude, so $x = 0$. Beyond this point, the value of the positive part is greater, so x is positive (as shown at point 3) and reaches a maximum at point 4. Then both parts reduce, so that x becomes smaller (as at point 5) and finally approaches zero. Thus motion of the type shown in Fig. 3-3(b) can occur. This curve, Fig. 3-3(b), can also be used to justify the four remaining types of motion. A time shift may be made so that point 2 becomes the initial point, with the initial displacement and velocity as defined there, resulting in the motion curve shown in Fig. 3-3(c). Similarly, time shifts of the curve may be made so that points 3, 4, and 5 become the initial points, resulting in the motion curves shown in Fig. 3-3(d), (e), and (f), respectively. Finally, note that the mirror image (about the time axis) of each of these curves is proper, that is, where the displacement and motion are of opposite direction to that shown in Fig. 3-3(b) through (f).

For the foregoing, the initial conditions are expressed by

$$\left.\begin{array}{l} x_0 = C_1 + C_2 \\ \dot{x}_0 = -(\alpha C_1 + \beta C_2) \end{array}\right\} \quad \text{for } t = 0 \tag{3-20}$$

From these, the values of C_1 and C_2 can be obtained.

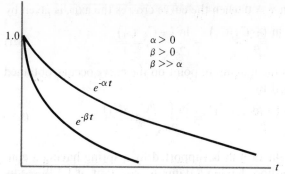

$$\alpha > 0$$
$$\beta > 0$$
$$\beta \gg \alpha$$

Figure 3-2

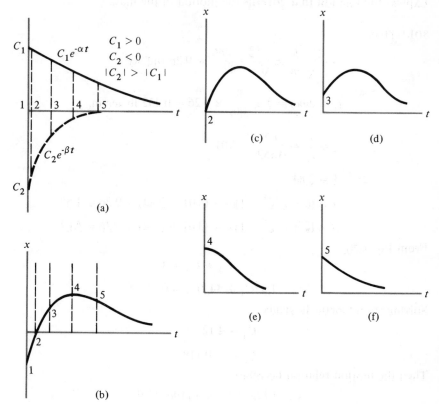

Figure 3-3

The time t', for which $x = 0$ when the curve crosses the axis, is given by

$$t' = \frac{\ln(-C_2/C_1)}{\beta - \alpha} = \frac{\ln(-C_2/C_1)}{2\omega\sqrt{\zeta^2 - 1}} \qquad (3\text{-}21)$$

The time t'', at which the maximum point on the curve occurs, obtained by setting $\dot{x} = 0$, is defined by

$$t'' = \frac{\ln(-\beta C_2/\alpha C_1)}{\beta - \alpha} = \frac{\ln(-\beta C_2/\alpha C_1)}{2\omega\sqrt{\zeta^2 - 1}} \qquad (3\text{-}22)$$

EXAMPLE 3-1 A weight of 9 lb is supported by a spring having a constant of 2 lb/in. and a dashpot having a damping constant of 1.3 lb sec/in. The mass has an initial displacement of 4 in. and an initial velocity of zero. Express the relation that governs the motion of the mass.

SOLUTION

$$\omega = \sqrt{\frac{k}{m}} = \sqrt{\frac{2 \times 386}{9}} = 9.26 \text{ rad/sec}$$

$$c_c = 2m\omega = 2 \times \frac{9}{386} \times 9.26 = 0.432 \text{ lb sec/in.}$$

$$\zeta = \frac{c}{c_c} = \frac{1.3}{0.432} = 3.01$$

$$\sqrt{\zeta^2 - 1} = 2.84$$

$$\alpha = (\zeta - \sqrt{\zeta^2 - 1})\omega = (3.01 - 2.84) \times 9.26 = 1.57$$

$$\beta = (\zeta + \sqrt{\zeta^2 - 1})\omega = (3.01 + 2.84) \times 9.26 = 54.2$$

From Eq. 3-20,

$$C_1 + C_2 = 4$$

$$1.57C_1 + 54.2C_2 = 0$$

Solving simultaneously yields

$$C_1 = 4.12$$

$$C_2 = -0.119$$

Then the motion relation becomes

$$x = 4.12e^{-1.57t} - 0.119e^{-54.2t}$$

3-6. CRITICALLY DAMPED MOTION

For the case of critical damping, corresponding to $\zeta = 1$, the motion is governed by Eq. 3-17. This relation is a product of the linear function $(A + Bt)$ and the decaying exponential $e^{-\omega t}$. Separate curves for these parts

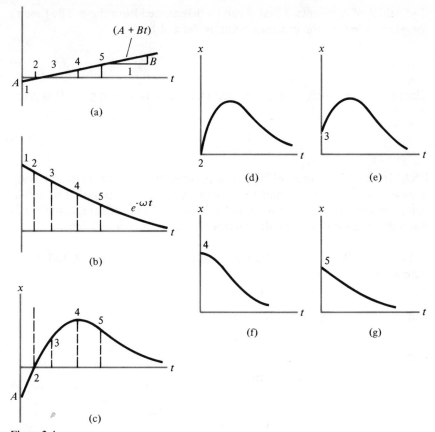

Figure 3-4

are shown in Fig. 3-4(a) and (b), where A is taken as negative and B as positive. The single curve shown in Fig. 3-4(c) and representing the complete function, Eq. 3-17, is then the product of values from Fig. 3-4(a) and (b). In this manner the value of x at points 1 through 5 can be determined, and the shape of the motion curve of Fig. 3-4(c) can be justified. Then, by shifting the time so that point 2 becomes the initial point, the curve shown in Fig. 3-4(d) can be verified. Similar shifts to points 3, 4, and 5 justify the curves exhibited in Fig. 3-4(e), (f), and (g). Thus five types of aperiodic motion exist. The mirror image, with respect to the time axis, of each curve is also proper here. Note that the curves are similar in form to those for the case of above-critical damping.

In the foregoing it should be noted from Eq. 3-17 that as t becomes large, x approaches zero. The initial conditions are given by

$$\left. \begin{array}{l} x_0 = A \\ \dot{x}_0 = B - A\omega \end{array} \right\} \quad \text{for } t = 0 \qquad (3\text{-}23)$$

The values of constants A and B can be determined from these. The time t', for which $x = 0$ at the crossing point, is defined by

$$t' = -\frac{A}{B} \qquad (3\text{-}24)$$

The time t'', for which x maximum occurs, obtained by setting $\dot{x} = 0$, is given by

$$t'' = \frac{1}{\omega} - \frac{A}{B} \qquad (3\text{-}25)$$

EXAMPLE 3-2   A mass of 20 kg is supported by an elastic member having a modulus of 320 N/m, and the system is critically damped. For initial conditions of zero displacement and velocity of 3 m/s, determine the maximum displacement and the displacement at $t = 0.5$, 1, and 2 sec.

SOLUTION   Based on the initial conditions, the constants A and B are determined as

$$A = 0 \qquad B = \dot{x}_0$$

and the solution (Eq. 3-17) becomes

$$x = \dot{x}_0 t e^{-\omega t}$$

with

$$\dot{x}_0 = 3 \text{ m/s}$$

$$\omega = \sqrt{\frac{k}{m}} = \sqrt{\frac{320}{20}} = 4 \text{ rad/s}$$

The time at which the maximum displacement occurs is

$$t'' = \frac{1}{\omega} - \frac{A}{B} = \frac{1}{\omega} = \frac{1}{4} = 0.25 \text{ s}$$

and

$$x_{\max} = \dot{x}_0 t e^{-\omega t} = 3 \times 0.25 e^{-4 \times 0.25} = 0.2759 \text{ m} = 27.59 \text{ cm}$$

$$x_{t=0.5} = 3 \times 0.5 e^{-4 \times 0.5} = 0.2030 \text{ m} = 20.30 \text{ cm}$$

$$x_{t=1} = 3 \times 1 e^{-4 \times 1} = 0.054\,95 \text{ m} = 5.495 \text{ cm}$$

$$x_{t=2} = 3 \times 2 e^{-4 \times 2} = 0.002\,013 \text{ m} = 0.2013 \text{ cm}$$

3-7. MOTION FOR BELOW-CRITICAL DAMPING

Damping corresponding to $\zeta < 1$ is referred to as subcritical or below-critical damping, and the system is said to be *underdamped*. For this condition, the solution is specified by Eq. 3-15. The motion is of harmonic form,

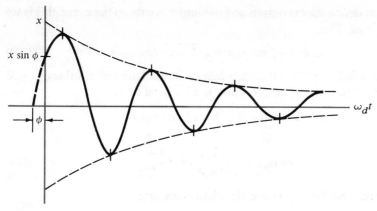

Figure 3-5

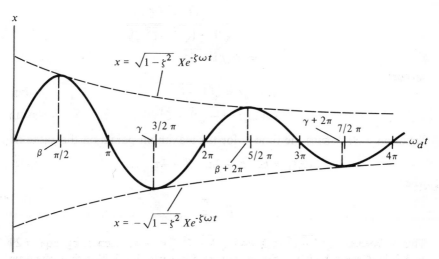

Figure 3-6

with a decaying amplitude. Actually, except for small values of ζ, the motion differs from harmonic. This difference becomes more pronounced as ζ increases. The general form of the motion is shown in Fig. 3-5.

In order to examine this more closely, ϕ can be taken as zero without loss of generality. The solution then is

$$x = Xe^{-\zeta\omega t}\sin\omega_d t \tag{3-26}$$

which is represented in Fig. 3-6. By taking the time derivative,

$$\dot{x} = Xe^{-\zeta\omega t}(-\zeta\omega\sin\omega_d t + \sqrt{1-\zeta^2}\,\omega\cos\omega_d t) \tag{3-27}$$

In order to define the maximum and minimum points on the curve, this is set equal to zero. Thus

$$X\omega e^{-\zeta\omega t}(-\zeta \sin \omega_d t + \sqrt{1-\zeta^2} \cos \omega_d t) = 0 \qquad (3\text{-}28)$$

This is satisfied by $e^{-\zeta\omega t} = 0$, which defines the final minimum value of $x = 0$ when t becomes large. The condition is also satisfied by

$$(-\zeta \sin \omega_d t + \sqrt{1-\zeta^2} \cos \omega_d t) = 0$$

whence

$$\tan \omega_d t = \frac{\sin \omega_d t}{\cos \omega_d t} = \frac{\sqrt{1-\zeta^2}}{\zeta} \qquad (3\text{-}29)$$

From which, by the trigonometric relation for sine,

$$\sin \omega_d t = \pm \frac{\tan \omega_d t}{\sqrt{1 + \tan^2 \omega_d t}}$$

$$= \pm \frac{\sqrt{1-\zeta^2}/\zeta}{\sqrt{1 + (\sqrt{1-\zeta^2}/\zeta)^2}}$$

$$= \pm \sqrt{1-\zeta^2}$$

so that

$$\sin \omega_d t = \sqrt{1-\zeta^2} \qquad (3\text{-}30)$$

$$\sin \omega_d t = -\sqrt{1-\zeta^2} \qquad (3\text{-}31)$$

Equation 3-30 defines the time t_1 for the maximum points on the curve as

$$\omega_d t_1 = \beta, \quad \beta + 2\pi, \quad \beta + 4\pi, \quad \ldots \qquad (3\text{-}32)$$

where

$$\beta < \frac{\pi}{2}$$

This is because $\sqrt{1-\zeta^2} < 1$ and $\sqrt{1-\zeta^2}/\zeta < +\infty$; hence, by Eqs. 3-29 and 3-30, β must be in the first quadrant. Similarly, Eq. 3-31 defines the time t_2 for the minimum points on the curve as

$$\omega_d t_2 = \gamma, \quad \gamma + 2\pi, \quad \gamma + 4\pi, \quad \ldots \qquad (3\text{-}33)$$

where

$$\gamma < \tfrac{3}{2}\pi$$

Here, since the sine is negative but the tangent is positive, Eqs. 3-29 and 3-31 specify γ to be a third-quadrant value. It should also be noted that $\gamma = \beta + \pi$.

The curve will cross the $\omega_d t$ axis at points for which $\sin \omega_d t = 0$. Such crossings will occur at

$$\omega_d t_3 = \pi, \quad 2\pi, \quad 3\pi, \quad \ldots \qquad (3\text{-}34)$$

Examination of Eqs. 3-32 through 3-34 reveals that in every instance the time between an $x = 0$ point and the next x_{max} point is less than the time between this x_{max} point and the following $x = 0$ point. Similarly, the time between an $x = 0$ point and the next x_{min} point is less than the time between this x_{min} point and the following $x = 0$ point. The described unequal time intervals occur because the damping aids the spring in stopping the outward motion but opposes the spring during the inward motion.

The foregoing is made clearer by the curve for Eq. 3-26, shown in Fig. 3-6. It should be noted that the period τ, as defined by the time between successive points of the same kind (for example, maximum points), is given by

$$\omega_d \tau = 2\pi$$

$$\tau = \frac{2\pi}{\omega_d} = \frac{2\pi}{\sqrt{1 - \zeta^2}\,\omega} \tag{3-35}$$

From Eqs. 3-30 and 3-26 the curve which goes through the maximum points is found to have the equation

$$x = \sqrt{1 - \zeta^2}\, Xe^{-\zeta\omega t} \tag{3-36}$$

Likewise, by substituting Eq. 3-31 into Eq. 3-26, the curve through the minimum points is expressed by

$$x = -\sqrt{1 - \zeta^2}\, Xe^{-\zeta\omega t} \tag{3-37}$$

EXAMPLE 3-3 A mass of 3.174 kg is supported by a spring having a modulus of 700 N/m and a dashpot having a damping constant of 14.18 N · s/m. Write the equation that governs the motion of the mass.

SOLUTION

$$\omega = \sqrt{\frac{k}{m}} = \sqrt{\frac{700}{3.174}} = 14.85 \text{ rad/sec}$$

$$c_c = 2m\omega = 2 \times 3.174 \times 14.85 = 94.27 \text{ N · s/m}$$

$$\zeta = \frac{14.18}{94.27} = 0.15$$

$$\sqrt{1 - \zeta^2} = \sqrt{1 - (0.15)^2} = 0.989$$

$$\omega_d = 0.989 \times 14.85 = 14.7$$

$$\zeta\omega = 0.15 \times 14.85 = 2.23$$

Then the equation of motion is

$$x = Xe^{-2.23t} \sin (14.7t + \phi)$$

3-8. EFFECT OF DAMPING ON FREE MOTION

In order to gain an understanding of the effect of damping, it is desirable to study the family of motion curves resulting from different values of the damping factor ζ. This is shown in Fig. 3-7, where Eqs. 3-26, 3-13, and 3-17 have been plotted against ωt so that a comparison can be made. The initial conditions were the same for each curve; namely, $x_0 = 0$ and $\dot{x}_0 = 1$. The relations plotted then were

$$x = \frac{1}{\omega}\sin \omega t \qquad \text{for } \zeta = 0$$

$$x = \frac{e^{-\zeta \omega t}}{\omega\sqrt{1 - \zeta^2}}\sin \sqrt{1 - \zeta^2}\,\omega t \qquad \text{for } \zeta < 1$$

$$x = te^{-\omega t} \qquad \text{for } \zeta = 1$$

and

$$x = \frac{e^{(-\zeta + \sqrt{\zeta^2 - 1})\omega t} - e^{(-\zeta - \sqrt{\zeta^2 - 1})\omega t}}{2\omega\sqrt{\zeta^2 - 1}} \qquad \text{for } \zeta > 1$$

(3-38)

Damping has the effect of reducing the amplitude of the oscillatory motion. It also lessens the time for the first peak or maximum point to occur. By Eq. 3-30, this is given by

$$\omega t_1 = \frac{\text{arc sin }\sqrt{1 - \zeta^2}}{\sqrt{1 - \zeta^2}} \tag{3-39}$$

Writing the series for arc sin results in

$$\omega t_1 = \frac{1}{\sqrt{1 - \zeta^2}}\left[(1 - \zeta^2)^{1/2} + \frac{(1 - \zeta^2)^{3/2}}{2 \cdot 3} + \frac{1}{2}\cdot\frac{3}{4}\frac{(1 - \zeta^2)^{5/2}}{5}\right.$$

$$\left. + \frac{1}{2}\cdot\frac{3}{4}\cdot\frac{5}{6}\frac{(1 - \zeta^2)^{7/2}}{7} + \cdots\right]$$

$$= \left[1 + \frac{(1 - \zeta^2)}{2 \cdot 3} + \frac{1}{2}\cdot\frac{3}{4}\frac{(1 - \zeta^2)^2}{5} + \frac{1}{2}\cdot\frac{3}{4}\cdot\frac{5}{6}\frac{(1 - \zeta^2)^3}{7} + \cdots\right]$$

$$\text{for } (1 - \zeta^2) < 1 \tag{3-40}$$

From Eq. 3-39 it is seen that for $\zeta = 0$, the value of ωt_1 will be $\pi/2$. Then, beginning with small ζ, Eq. 3-40 shows that as ζ increases, the value of ωt_1 decreases. Thus, increasing the damping has the effect of reducing the time at which the first maximum displacement occurs.

On the other hand, the first crossing of the ωt axis takes place at

$$\omega t_2 = \frac{\pi}{\sqrt{1 - \zeta^2}} \tag{3-41}$$

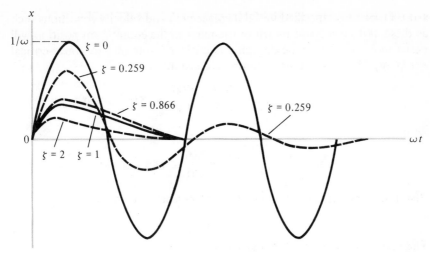

Figure 3-7

It is evident that ωt_2 will be smallest (and equal to π) for $\zeta = 0$, and as ζ increases ωt_2 will become larger. Thus damping has the effect of increasing the time for the motion to return to the neutral position.

Likewise, the period of the damped motion is greater than that of the undamped motion, and this effect is intensified by an increase in damping. This is observed from the expression for the damped period, or

$$\omega\tau = \frac{2\pi}{\sqrt{1-\zeta^2}} \tag{3-42}$$

From this, $\omega\tau = 2\pi$ for $\zeta = 0$, becoming larger as ζ increases. Even though in the early stages, the peaks for the damped motion occur to the left of those having less or no damping, eventually they will occur to the right because of the greater period accompanying the larger damping condition.

For very small damping the motion curve is quite close to that for the undamped case. As damping increases the effects described above become more and more pronounced. As ζ approaches the value of 1, the first peak is considerably to the left and below that for no damping, and the first crossing is far to the right. When critical damping is reached, the first peak is moved farther to the left (also down), and there is no crossing. These two effects are magnified for above-critical damping. In this case, note that the return is very slow, so that the curve crosses the curve shown for critical damping.

EXAMPLE 3-4   A damped quick-return mechanism is to be designed for a certain machine operation. Due to impact, the moving part has an initial velocity of 22 fps. The part weighs 250 lb and is restrained by a spring having a stiffness of 1500 lb/ft. Determine the damping constant required and the maximum displacement attained by the part.

SOLUTION   For specified initial displacement and velocity conditions such as these, the subsequent return of the mass to the equilibrium position will occur most quickly if the damping is critical (compared to above-critical damping). Hence the desired damping constant is

$$c = c_c = 2\sqrt{mk} = 2\sqrt{\frac{250}{386} \times \frac{1500}{12}} = 18.0 \text{ lb sec/in.}$$

Also,

$$\omega = \sqrt{\frac{k}{m}} = \sqrt{\frac{1500}{12} \times \frac{386}{250}} = 13.9 \text{ rad/sec}$$

The initial displacement is zero, and the motion relation is then

$$x = \dot{x}_0 t e^{-\omega t}$$

The maximum displacement occurs at

$$t'' = \frac{1}{\omega} = \frac{1}{13.9} = 0.0719 \text{ sec}$$

and

$$\omega t'' = 1$$

Then

$$x = (22 \times 12) \times 0.0719 \times e^{-1} = 6.98 \text{ in.}$$

3-9. LOGARITHMIC DECREMENT

For a damped system the rate of decay of the motion is conveniently expressed by the ratio of any two successive amplitudes. If x_j and x_{j+1} represent the amplitudes for the jth and $(j + 1)$th cycles, then by Eq. 3-36,

$$x_j = \sqrt{1 - \zeta^2}\, X e^{-\zeta \omega t_j}$$
$$x_{j+1} = \sqrt{1 - \zeta^2}\, X e^{-\zeta \omega (t_j + \tau)} \qquad (3\text{-}43)$$

whence

$$\frac{x_j}{x_{j+1}} = \frac{\sqrt{1 - \zeta^2}\, X e^{-\zeta \omega t_j}}{\sqrt{1 - \zeta^2}\, X e^{-\zeta \omega (t_j + \tau)}} = e^{\zeta \omega \tau} = \text{constant} \qquad (3\text{-}44)$$

Note that this ratio is the same for any successive amplitudes. The measure of decay commonly used is the natural logarithm of the foregoing ratio. This is called the *logarithmic decrement* and is designated by δ. Thus

$$\delta = \ln \frac{x_j}{x_{j+1}} \qquad (3\text{-}45)$$

$$= \ln e^{\zeta \omega \tau} = \zeta \omega \tau = \frac{2\pi\zeta}{\sqrt{1 - \zeta^2}} \qquad (3\text{-}46)$$

For small damping this becomes

$$\delta \approx 2\pi\zeta \qquad (3\text{-}47)$$

The logarithmic decrement can also be calculated from the ratio of amplitudes several cycles apart. Thus if x_n is the amplitude n cycles after amplitude x_0, then

$$\frac{x_0}{x_n} = \frac{x_0}{x_1} \cdot \frac{x_1}{x_2} \cdot \frac{x_2}{x_3} \cdots \frac{x_{n-1}}{x_n}$$

$$= \left(\frac{x_j}{x_{j+1}}\right)^n \qquad (3\text{-}48)$$

The natural log of this is

$$\ln\left(\frac{x_0}{x_n}\right) = n \ln\left(\frac{x_j}{x_{j+1}}\right) = n\delta$$

whence

$$\delta = \frac{1}{n} \ln\left(\frac{x_0}{x_n}\right) \qquad (3\text{-}49)$$

so that δ can be obtained from the amplitude loss occurring over several cycles. This relation may be rearranged as

$$n = \frac{1}{\delta} \ln\left(\frac{x_0}{x_n}\right)$$

$$= \frac{\sqrt{1-\zeta^2}}{2\pi\zeta} \ln\left(\frac{x_0}{x_n}\right) \qquad (3\text{-}50)$$

This is convenient for determining the number of cycles necessary for a given system to reach a specified reduction in amplitude. It should be noted that n is then not necessarily an integer. The corresponding time Δt to attain a certain amplitude loss is also useful. Since the damped period is $\tau = 2\pi/\omega_d = 2\pi/\sqrt{1-\zeta^2}\,\omega$,

$$\Delta t = n\tau = \frac{\sqrt{1-\zeta^2}}{2\pi\zeta} \left[\ln\left(\frac{x_0}{x_n}\right)\right] \frac{2\pi}{\sqrt{1-\zeta^2}\,\omega}$$

$$= \frac{1}{\zeta\omega} \ln\left(\frac{x_0}{x_n}\right) \qquad (3\text{-}51)$$

This can be employed for determinations such as the half-life or the time required for a 50% reduction in amplitude to take place.

The foregoing can be used as the basis for the experimental determination of the system damping. This is particularly valuable where the damping mechanism and the amount of damping, as expressed by ζ or c, cannot be determined directly. However, by measuring the decay of the motion, the damping can be obtained. Equation 3-46 can be rearranged as

$$\zeta = \frac{\delta}{\sqrt{(2\pi)^2 + \delta^2}} \tag{3-52}$$

By experimental observation of the change of amplitude and the number of cycles, δ may be calculated from Eq. 3-49. Then ζ can be obtained from Eq. 3-52. If k and m are known, or can be determined from static deflection measurements, then the value of the damping constant c can be computed.

EXAMPLE 3-5   A weight of 11.42 lb is suspended by a spring having a modulus of 5 lb/in. When displaced and permitted to oscillate freely, it is found that the amplitude diminishes from 0.750 in. to 0.658 in. in exactly 8 cycles. Determine the value of the damping constant.

SOLUTION   By Eq. 3-49,

$$\delta = \frac{1}{8} \ln \frac{0.750}{0.658} = 0.0164$$

From Eq. 3-52,

$$\zeta = \frac{0.0164}{\sqrt{(2\pi)^2 + (0.0164)^2}} = 0.002\,61$$

$$c_c = 2\sqrt{mk} = 2\sqrt{\frac{11.42 \times 5}{386}} = 0.769$$

$$c = \zeta c_c = 0.002\,61 \times 0.769 = 0.0020 \text{ lb sec/in.}$$

EXAMPLE 3-6   A suspension shock-absorber system is to be designed for a vehicle. The damping is to be less than critical and the initial amplitude is to be reduced to 1/3 in the first half cycle. Thus the displacement time curve will be as shown in Fig. 3-8, wherein $x_{1.5} = x_1/3$. (The system avoids severe snubbing of the motion, and small movement is tolerated in the second half cycle and beyond.) Other specifications include $m = 450$ kg and that the damped period is to be 1 sec. The required damping constant and stiffness constant are to be determined. Also, if clearance is 30 cm, obtain the minimum initial velocity that will result in "bottoming" (striking a rubber base) of the system.

SOLUTION

$$x_{1.5} = \tfrac{1}{3}x_1 \quad \text{and} \quad x_2 = \tfrac{1}{3}x_{1.5} = \tfrac{1}{3}(\tfrac{1}{3}x_1) = \tfrac{1}{9}x_1$$

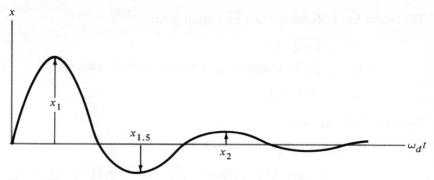

Figure 3-8

From the logarithmic decrement relation,

$$\ln\left(\frac{x_1}{x_1/9}\right) = \frac{2\pi\zeta}{\sqrt{1-\zeta^2}}$$

$$\ln 9 = \frac{2\pi\zeta}{\sqrt{1-\zeta^2}}$$

$$2.19722 = \frac{2\pi\zeta}{\sqrt{1-\zeta^2}}$$

Solving gives $\zeta = 0.330097$. For the damped period,

$$\omega_d t = 2\pi$$

$$\omega_d \times 1 = 2\pi$$

$$\omega_d = 2\pi$$

$$\sqrt{1-(0.330097)^2}\,\omega = 2\pi$$

$$0.818476\omega = 2\pi$$

$$\omega = 7.67669 \text{ rad/s}$$

$$c_c = 2m\omega = 2 \times 450 \times 7.67669 = 6909.02 \text{ N} \cdot \text{s/m}$$

$$c = \zeta c_c = 0.330097 \times 6909.02 = 2280.65 \text{ N} \cdot \text{s/m}$$

$$k = m\omega^2 = 450 \times (7.67669)^2 = 26.5192 \text{ kN/m}$$

From Eq. 3-30, the displacement x_1 occurs for

$$\sin \omega_d t = \sqrt{1-\zeta^2}$$

$$\sin 2\pi t = \sqrt{1-(0.330097)^2}$$

$$t = \frac{\text{arc sin } 0.818476}{2\pi} = 0.152590 \text{ s}$$

Then from Eq. 3-36, for the first maximum point

$$x = \sqrt{1 - \zeta^2}\, Xe^{-\zeta\omega t}$$

$$0.30 = \sqrt{1 - (0.330\,097)^2}\, Xe^{-0.330\,097 \times 7.676\,69 \times 0.152\,590}$$

$$X = 0.539\,567 \text{ m}$$

From Eq. 3-27, for $t = 0$

$$\dot{x}_0 = X\omega\sqrt{1 - \zeta^2}$$

$$= 0.539\,567 \times 7.676\,69 \times \sqrt{1 - (0.330\,097)^2}$$

$$= 3.390\,20 \text{ m/s}$$

3-10. FREE VIBRATIONS WITH COULOMB DAMPING

As discussed in Section 3-1, the magnitude of Coulomb damping due to dry sliding friction is assumed to be constant. The friction force is always opposite to the direction of motion. Thus for the system shown in Fig. 3-9(a), and designating the friction damping force as F_d, the separate free-body diagrams of Fig. 3-9(b) and (c) are constructed, one for each motion direction. From these, the differential equation governing the motion can be written as

$$m\ddot{x} = -kx - F_d(\text{sgn } \dot{x}) \tag{3-53}$$

where (sgn $\dot{x}$) stands for the sign of $\dot{x}$. This single differential equation properly describes the motion. However, for clarity of solution separate differential equations—one for each direction of travel—may be written as follows:

$$m\ddot{x} = -kx + F_d \qquad \text{for } \dot{x} < 0 \tag{3-54}$$

$$m\ddot{x} = -kx - F_d \qquad \text{for } \dot{x} > 0 \tag{3-55}$$

These differential equations and their solutions are discontinuous at the end points of their motion. The case for motion to the left, and Eq. 3-54, will now be solved. Rearranging gives

$$\ddot{x} + \frac{k}{m}x = \frac{F_d}{m} \tag{3-56}$$

Due to the term on the right-hand side, the differential equation is not homogeneous, and the solution will be composed of two parts. This may be written as

$$x = x_a + x_b \tag{3-57}$$

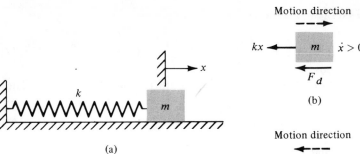

Figure 3-9

where

x = the complete solution

x_a = the complementary function (the solution of the differential equation with the right-hand side set to zero)

= $A \sin \omega t + B \cos \omega t$

x_b = the particular solution (a function which satisfies the complete differential equation and which contains no arbitrary or unknown constants)

In this case it is reasonable that the particular solution must be a constant. Thus it is assumed that

$$x_b = C \tag{3-58}$$

where C is to be determined by substitution in Eq. 3-56, resulting in

$$\frac{k}{m} C = \frac{F_d}{m}$$

whence $C = F_d/k$ and

$$x_b = \frac{F_d}{k} \tag{3-59}$$

The general solution is then

$$x = A \sin \omega t + B \cos \omega t + \frac{F_d}{k} \qquad (\dot{x} < 0) \tag{3-60}$$

for motion toward the left. For the initial motion conditions of $x = x_0$ and $\dot{x} = 0$ at $t = 0$ for the extreme position at the right, the constants are found to be

$$A = 0 \qquad \text{and} \qquad B = x_0 - \frac{F_d}{k} \tag{3-61}$$

Equation 3-60 then becomes

$$x = \left(x_0 - \frac{F_d}{k}\right)\cos \omega t + \frac{F_d}{k} \qquad (\dot{x} < 0) \qquad (3\text{-}62)$$

This holds for motion toward the left, or until $\dot{x}$ again becomes zero. Thus

$$\dot{x} = \left(x_0 - \frac{F_d}{k}\right)\omega(-\sin \omega t) = 0 \qquad (3\text{-}63)$$

whence $\omega t = \pi$ and

$$x = \left(x_0 - \frac{F_d}{k}\right)(-1) + \frac{F_d}{k}$$

$$= -\left(x_0 - \frac{2F_d}{k}\right) \qquad (3\text{-}64)$$

Thus the displacement is negative, or to the left of the neutral position, and has a magnitude $2F_d/k$ less than the initial displacement x_0.

Similar analysis of the differential equation (Eq. 3-55) and solution for motion to the right would show that when the motion has reached its limiting position at the right, the displacement would be

$$x = x_0 - \frac{4F_d}{k} \qquad (3\text{-}65)$$

Hence in each cycle the amplitude loss is $4F_d/k$. The motion is shown in Fig. 3-10. The decay is linear; that is, the curve going through maximum points on the diagram will be a straight line.

The motion is harmonic in form, consisting of one-half sine-wave parts which are offset successively up or down by F_d/k, as shown, depending on whether the motion is to the left or to the right. The frequency is the same as that for the undamped system.

The eventual cessation of the motion is of interest. For an amplitude position X_j, the spring force is kX_j. If this force is balanced by the friction force, the motion will cease. Thus for $kX_j \leq F_d$ the motion will stop. Although the exact point of coming to rest can be determined, ordinarily this is not significant.

The method presented here is a general technique used for some cases of nonlinear analysis—specifically, that of piecewise linearity.

EXAMPLE 3-7 A weight of 2 lb is arranged horizontally with a spring k, as in Fig. 3-9(a). The sliding dry-friction force is known to be 0.3 lb. The amplitude diminishes from 8.00 in. to 4.40 in. in 15 cycles. Determine the time required for this to occur.

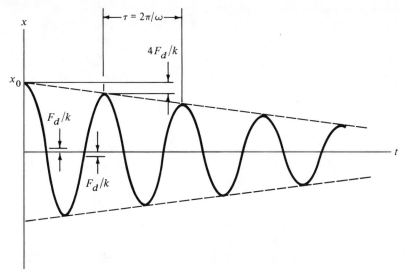

Figure 3-10

SOLUTION

$$\text{decay per cycle} = \Delta X = \frac{8.00 - 4.40}{15} = 0.24 \text{ in.}$$

$$\frac{4F_d}{k} = \Delta X$$

$$k = \frac{4F_d}{\Delta X} = \frac{4 \times 0.3}{0.24} = 5.00 \text{ lb/in.}$$

$$\omega = \sqrt{\frac{k}{m}} = \sqrt{\frac{5 \times 386}{2}} = 31.1 \text{ rad/sec}$$

$$\tau = \frac{2\pi}{\omega} = \frac{2\pi}{31.1} = 0.202 \text{ sec}$$

$$t = n \cdot \tau = 15 \times 0.202 = 3.03 \text{ sec}$$

3-11. HYSTERESIS DAMPING

As previously explained, when materials are deformed, energy is absorbed and dissipated by the material. This hysteresis effect is due to friction between internal planes which slip or slide as the deformations take place. This form of damping results in offsetting the parts of the force-displacement curve, as shown in Fig. 3-11, where P is the spring force, x is the displacement, and X is the amplitude. All materials exhibit this phenomenon with

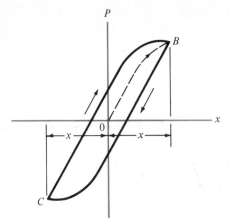

Figure 3-11

rubberlike materials showing a large loop and metals such as steel display-ing a very small enclosure.

If ΔU represents the energy loss per cycle, then

$$\Delta U = \int P \, dx = A_{P,x} \tag{3-66}$$

That is, the energy loss per cycle is represented by the area within the loop. It has been found experimentally that the energy loss is independent of the frequency but is proportional to the (approximate) square of the amplitude. It is also considered to be directly related to the stiffness of the member. Thus the energy loss per cycle can be expressed as

$$\Delta U = \pi k b X^2 \tag{3-67}$$

where b is a dimensionless solid damping constant for the material. The factor kb then relates the energy loss to the size and shape of the member as well as to the material characteristics. Thus b is considered to be property of the material alone. Including the factor π was originally done so that the form would be similar to the relation for energy dissipation for maintained harmonic motion with viscous damping.† The exponent of X is from 2.0 to 2.3 for certain steels, and it may be taken as 2.0 for many materials, including steel, unless extreme accuracy is required.

In reality Eq. 3-67 becomes the definition for the hysteresis or solid damping constant b.

In many cases ΔU is small and the motion will be very nearly harmonic in form. The loss of amplitude per cycle may be determined from a consider-ation of the energy. Referring to Fig. 3-12, the loss for a quarter cycle is assumed to be $\frac{1}{4}(\pi k b X_j^2)$, where X_j is the amplitude of that part. The energy

† See Eq. 4-116.

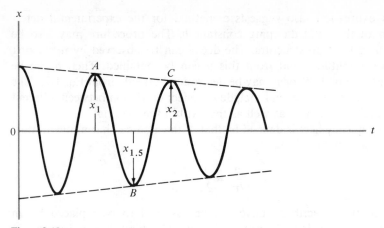

Figure 3-12

equation for the half cycle from A to B is then

$$\frac{kX_1^2}{2} - \frac{\pi k b X_1^2}{4} - \frac{\pi k b X_{1.5}^2}{4} = \frac{kX_{1.5}^2}{2} \tag{3-68}$$

whence

$$(2 - \pi b)X_1^2 = (2 + \pi b)X_{1.5}^2$$

$$\frac{X_1}{X_{1.5}} = \sqrt{\frac{2 + \pi b}{2 - \pi b}} \tag{3-69}$$

Similarly, for the next half cycle from B to C,

$$\frac{X_{1.5}}{X_2} = \sqrt{\frac{2 + \pi b}{2 - \pi b}} \tag{3-70}$$

Multiplying Eqs. 3-69 and 3-70 gives

$$\frac{X_1}{X_2} = \frac{2 + \pi b}{2 - \pi b} = \text{constant}$$

$$= 1 + \frac{2\pi b}{2 - \pi b} \tag{3-71}$$

$$\approx 1 + \pi b \tag{3-72}$$

since b is very small.

Since the ratio of the successive amplitudes is constant, the decay is exponential. The log decrement is defined by

$$\delta = \ln\left(\frac{X_1}{X_2}\right) \tag{3-73}$$

$$\approx \ln(1 + \pi b) \tag{3-74}$$

$$\approx \pi b \tag{3-75}$$

This expression also suggests a method for the experimental determination of the solid damping constant b. The procedure may also be applied to a built-up structure. The decay can be observed by measuring successive amplitudes and from this δ can be obtained. Then b can be calculated by Eq. 3-75, or b may be determined directly from Eq. 3-72. The value of b so obtained would include the effect of friction between adjacent parts of the arrangement as well as that within the material.

The frequency here is defined by that of the assumed harmonic motion, or

$$f = \frac{\omega}{2\pi} = \frac{1}{2\pi}\sqrt{\frac{k}{m}} \tag{3-76}$$

The motion described above can be assumed to be replaced by an equivalent viscously damped motion that would exhibit the same characteristics. The corresponding equivalent viscous damping factor ζ_e and constant c_e are defined by equating the approximate relations for δ for the two cases. Thus

$$2\pi\zeta_e \approx \delta \approx \pi b$$

whence

$$\zeta_e = \frac{b}{2} \tag{3-77}$$

and

$$c_e = c_c\zeta_e = 2\sqrt{mk}\,\frac{b}{2} = b\sqrt{mk} = \frac{bk}{\omega} \tag{3-78}$$

These equivalent values would then be used in Eq. 3-15 for viscous damping, resulting in

$$x = Xe^{-\zeta_e\omega t}\sin(\omega t + \phi)$$
$$= Xe^{-(b/2)\omega t}\sin(\omega t + \phi) \tag{3-79}$$

wherein ω has replaced ω_d, since ζ is small. This expression defines a decaying motion which is "equivalent" to that for the solid damping case.

EXAMPLE 3-8   A small structure is found to behave as a single-degree-of-freedom system. From static tests the spring constant is determined to be 10 lb/in., and the effective suspended mass has a weight of 2 lb. Vibration tests show the ratio of successive amplitudes to be 1.064. Calculate the value of the solid damping constant and the equivalent viscous damping constant. Also determine the energy absorbed per cycle for an amplitude of 0.5 in.

SOLUTION

$$\frac{X_1}{X_2} = 1 + \pi b$$

$$1.064 = 1 + \pi b \quad \text{and} \quad b = \frac{0.064}{\pi} = 0.0204$$

$$c_e = b\sqrt{mk} = 0.0204 \sqrt{\frac{2 \times 10}{386}} = 0.004\,64 \text{ lb sec/in.}$$

$$\Delta U = \pi kbX^2 = \pi \times 10 \times 0.0204 \times (0.5)^2 = 0.160 \text{ in. lb}$$

PROBLEMS

3-1. A certain shock absorber exhibits viscous damping characteristics. The resistance developed by the absorber is measured as 34.7 lb for the constant velocity of 7.41 ft/sec developed during the test. Determine the viscous damping constant.

3-2. A weight of 13.5 lb attached to a spring produces a static deformation of 0.450 in. Determine the value of the critical damping constant for the system.

3-3. When tested, a certain dashpot shows viscous damping characteristics. A constant force of 165 N produces a velocity of 1.43 m/s. Determine the value of the damping constant c.

3-4. A body has a mass of 6.80 kg. When the body is suspended from a spring, a deformation of 9.525 mm results. Determine the value of the critical damping constant for the system.

3-5. A mass of 4.5 kg is suspended by a spring having a modulus of 1400 N/m. A dashpot having a viscous damping constant of 50 N·s/m is connected to the system. (a) Determine the damping factor. (b) Obtain the natural frequency and the damped frequency for the system.

3-6. A weight of 12 lb is suspended by a spring having a modulus of 9.6 lb/in. A dashpot having a viscous damping constant of 0.218 lb sec/in. is connected to the system. (a) Determine the damping factor. (b) Obtain the natural frequency and the damped frequency for the system.

3-7. A mass-spring system has an undamped frequency of 8 Hz. The suspended mass is 0.20 lb sec²/in. A dashpot for which the viscous damping constant is 2.4 lb sec/in. is connected to the system. (a) Determine the damped frequency for the arrangement. (b) Determine the period of the damped motion.

3-8. A mass-spring system has an undamped frequency of 5 Hz. The suspended mass is 30 kg. A dashpot for which the viscous damping constant is 280 N·s/m is connected to the system. Determine the damped frequency and the damped period for the arrangement.

3-9. A mass-spring system has a natural frequency of oscillation measured as f_1. After connecting a dashpot to the system, the frequency is lowered to f_2. Obtain an expression for the evaluation of the damping coefficient c.

3-10. Rigid bar AB in the accompanying figure is weightless and pivots at A. (a) Write the differential equation of motion for the system. (b) Obtain expressions for the critical damping constant and the damped frequency of vibration (assuming that $\zeta < 1$).

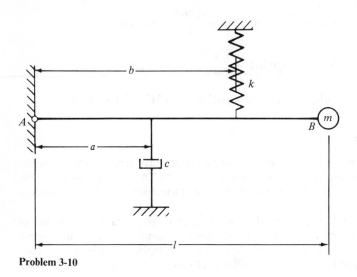

Problem 3-10

3-11. A member AB is a weightless cantilever beam fixed at A and fastened to a dashpot c and mass m at B. (a) Determine the differential equation of motion for the system shown. (b) Obtain relations for the critical damping constant and the period of the damped oscillation (assuming that $\zeta < 1$).

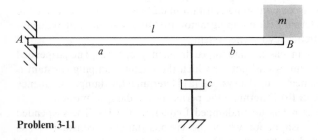

Problem 3-11

3-12. The uniform slender bar AC of mass m is pivoted at C and connected to a spring at A and a viscous dashpot at B, as shown. The bar is in equilibrium in a horizontal position. (a) Obtain the differential equation of motion for the system. (b) Determine the relation for the critical damping constant. (c) Assuming the damping is less than critical, express the period of the damped oscillation.

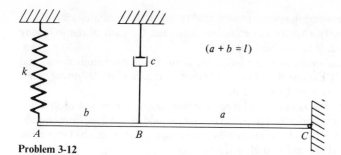

$(a + b = l)$

A b B a C

Problem 3-12

3-13. (a) Set up the differential equation of motion for the system shown. (b) Obtain the expression for the frequency of damped vibration (assuming that $\zeta < 1$).

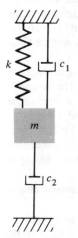

Problem 3-13

3-14. The torsional system shown is subjected to damping. (a) Write the differential equation for the motion. (b) Assuming that c_T is unknown but that k_{T1}, k_{T2}, and I are known, and that the frequency f_d of the damped vibration has been measured, obtain an expression for the evaluation of c_T.

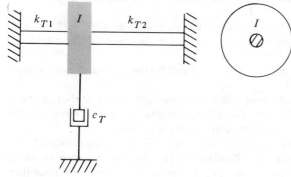

Problem 3-14

3-15. A damped mass-spring system is subjected to the initial conditions $x = 0$ and $\dot{x} = \dot{x}_0$ (at $t = 0$). Express the equation of motion for each of the following cases: (a) $\zeta = 2$, (b) $\zeta = 1$, and (c) $\zeta = 0.1$.

3-16. A damped mass-spring system is subjected to the initial conditions $x = x_0$ and $\dot{x} = 0$ (at $t = 0$). Express the equation of motion for each of the following cases: (a) $\zeta = 2$, (b) $\zeta = 1$, and (c) $\zeta = 0.1$.

3-17. A weight of 9.65 lb is suspended from a spring having a constant of 30 lb/in., and the system is critically damped. The mass is held at rest 2 in. from equilibrium and then released. (a) Determine the damping constant. (b) Obtain the displacement at the end of 0.01, 0.1, and 1 sec.

3-18. A mass of 9 kg is suspended from a spring having a modulus of 8100 N/m. The system is also critically damped. The initial displacement is zero and the initial velocity is 60 cm/s. Determine the maximum displacement for the system.

3-19. For a damped system, $k = 7875$ N/m, $m = 8.75$ kg, and the damping factor $\zeta = 0.1$. The initial displacement is 10 cm and the initial velocity is zero. (a) Determine the damping constant. (b) Obtain the amplitude one-half and one cycle after the initial condition.

3-20. A damped system is composed of a body for which $W = 38.6$ lb, spring constant $k = 90$ lb/in., and damping constant $c = 13.416$ lb sec/in. The system is given an initial displacement of 2 in. and released from rest. (a) Determine the damping factor. (b) Obtain the displacement at $t = 0.01$, 0.1, and 1 sec.

3-21. A suspended mass of 4.5 kg causes a static displacement of 1.00 cm for a spring. The system is also connected to a dashpot having a viscous damping constant of 35 N · s/m. Determine (a) the damping factor, (b) the undamped frequency, and (c) the damped frequency.

3-22. Determine (a) the logarithmic decrement and (b) the ratio of successive amplitudes for the data of Prob. 3-21.

3-23. A static displacement of 0.375 in. of a spring is caused by a suspended weight of 12 lb. The system also includes a dashpot having a viscous damping constant of 0.15 lb sec/in. Determine (a) the damping factor, (b) the undamped frequency, and (c) the damped frequency.

3-24. For the data of Prob. 3-23, determine (a) the logarithmic decrement and (b) the ratio of successive amplitudes of free vibration.

3-25. For a viscously damped system, a certain amplitude is measured as 74.8% of the immediately preceding amplitude. Determine the damping factor for the system.

3-26. The physical constants for a damped system are $k = 45$ lb/in., $W = 19.3$ lb, and $c = 0.057$ lb sec/in. If the initial amplitude is 3 in., determine the amplitude 12 cycles later.

3-27. For Prob. 3-26 determine (a) the half-life and (b) the number of cycles occurring in the half-life.

3-28. For a certain viscously damped system, measurements show that the amplitude reduction is 80% in 15 cycles. The critical damping constant is known to be 70 N · s/m. Determine the damping constant for the system.

3-29. A system is composed of a body for which $W = 7.72$ lb, spring constant $k = 5$ lb/in., and damping constant $c = 0.125$ lb sec/in. If the system is given an initial displacement and released from rest, determine the percent overshoot (the displacement attained past the equilibrium position).

3-30. A system composed of a mass of 5 kg and an elastic member having a modulus of 45 N/m is less than critically damped. When the mass is given an initial displacement and released from rest, the overshoot (the displacement attained past the equilibrium position) is 25%. Determine the damping factor and the damping constant.

3-31. A mass-spring system is critically damped. The spring constant is 3 lb/in. and the body weighs 11.58 lb. (a) The system is given an initial displacement $x_0 = 2$ in. and initial velocity $\dot{x}_0 = -30$ in./sec. Determine (a) the overshoot. (b) If $x_0 = 2$ in. and $\dot{x}_0 = -20$ in./sec, obtain the overshoot.

3-32. A large gun with supporting base weighs 2500 lb, and has a recoil system composed of spring $k = 32\,000$ lb/ft and a viscous shock absorber (dashpot) for which damping is critical. The recoil distance is 3 ft. Determine (a) the initial recoil velocity, (b) the time to return to a position 0.1 in. from the initial position, and (c) displacement at $t = 0.5$ sec.

3-33. An artillery piece and integral supporting base weigh 1800 lb and have a recoil spring for which $k = 24\,000$ lb/ft. When fired, the system recoils 40 in. A dashpot having a critical damping constant is then engaged on the return stroke. Determine (a) the initial velocity of recoil, (b) the damping constant, and (c) the position of the gun and base at 0.5 sec of the return stroke.

3-34. For the case of Coulomb damping, by equating the loss of potential energy for a half cycle to the work done by the constant friction force, show that the loss of amplitude is $2F_d/k$. Designate the initial amplitude as x_0 and a half cycle later as $x_{0.5}$.

3-35. A mass-spring system is subjected to an undetermined damping condition. The vibrating body weighs 47 lb and the spring modulus is 30 lb/in. The amplitudes of successive cycles were measured as 1.64, 1.59, 1.54, 1.49, 1.44, ..., in. (a) Define the type and magnitude of the damping force and (b) determine the frequency of the damped oscillation.

3-36. A system composed of mass of 57.61 kg and a spring having a modulus of 9100 N/m is subjected to a Coulomb damping force of 6.825 N. If the initial amplitude is 4 cm, determine (a) the amplitude at the end of 8 cycles and (b) the frequency of the damped oscillation.

3-37. A simple structure exhibits hysteresis damping. The period of the vibration is 0.30 sec and the amplitude at the ninth cycle is 0.9309 times that at the first cycle. Determine the hysteresis damping constant b and write the equation of motion for this case.

chapter

4

Harmonically Forced
Vibrations for Single-Degree-
of-Freedom Systems

4-1. INTRODUCTION

A mass-elastic system is often subjected to an external force of some nature. This impressed force is usually time dependent and is called the forcing or driving function. The force may be harmonic, nonharmonic but periodic (as in the case of a square or triangular wave), nonperiodic but having a defined form, or random. A noncyclic force may be of long or short duration, and it often approaches a condition of impact, that is, a force of large magnitude applied for a small time interval. In all of these cases, it is desired to determine the *dynamic response* of the system, that is, to ascertain the motion or change in motion caused by the force.

In this chapter the response of a single-degree-of-freedom system to a harmonic forcing condition will be investigated. The most general form for a harmonic force would be $P = P_0 \sin (\omega_f t + \chi)$, wherein P_0 is a constant representing the amplitude of force P, ω_f is the circular frequency for the harmonic variation, and χ is the phase angle. The value of χ depends on the initial conditions for the force, and generality is not sacrificed by considering the force to be defined by $P = P_0 \sin \omega_f t$ or by $P = P_0 \cos \omega_f t$.

Although most systems will contain damping, the case of forced motion without damping will be considered separately here. This is done mainly to promote better understanding.

4-2. FORCED UNDAMPED VIBRATIONS

The model for an undamped system subjected to the harmonic force $P = P_0 \sin \omega_f t$ is represented in Fig. 4-1(a). The positive direction for force as well as displacement, and so on, is taken to be downward. The corre-

90

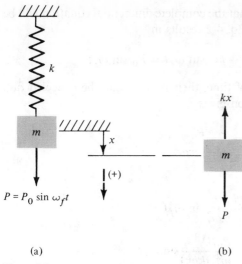

(a)   (b)

Figure 4-1

sponding dynamic free-body diagram, for an assumed plus x position, is shown in Fig. 4-1(b). From this, the equation of motion is

$$m\ddot{x} = -kx + P \tag{4-1}$$

whence

$$m\ddot{x} + kx = P_0 \sin \omega_f t \tag{4-2}$$

The solution to this will be

$$x = x_a + x_b \tag{4-3}$$

where

x_a = the complementary function (the solution to the
    differential equation with the right-hand side set to zero)
    = $A \sin \omega t + B \cos \omega t$ (4-4)
x_b = the particular solution (a solution which satisfies
    the complete differential equation)

The complementary function is the free-vibration component, and the particular solution represents the forced-vibration part of the motion. The complete motion consists of the sum of these two parts. The free-vibrational portion of the motion was covered in Chapter 2. The forced-vibration part will now be determined.

The need for the particular solution x_b is evident since the x_a part will reduce to zero when substituted into the left side of the differential equation. Since the x_b part must then reduce to a $\sin \omega_f t$ form, it is reasonable to assume the particular solution as

$$x_b = C \sin \omega_f t \tag{4-5}$$

where C is to be determined so that the complete differential equation will be satisfied. Substituting this into Eq. 4-2 results in

$$-m\omega_f^2 C \sin \omega_f t + kC \sin \omega_f t = P_0 \sin \omega_f t$$

Since this holds for all values of time, then $\sin \omega_f t$ may be canceled out. Rearranging the resulting relation gives

$$C = \frac{P_0}{k - m\omega_f^2} \tag{4-6}$$

Thus

$$x_b = \frac{P_0}{k - m\omega_f^2} \sin \omega_f t$$

$$= \frac{P_0/k}{1 - \omega_f^2/(k/m)} \sin \omega_f t$$

$$= \frac{X_0}{1 - r^2} \sin \omega_f t$$

$$= X \sin \omega_f t \tag{4-7}$$

where

$$X = \frac{X_0}{1 - r^2}$$
$$= \text{the amplitude of } x_b \tag{4-8}$$

$X_0 = \dfrac{P_0}{k}$ = the static displacement of the spring which would be caused by a constant force of magnitude P_0 (this is a fictitious displacement reference)

$r = \dfrac{\omega_f}{\omega}$ = the frequency ratio

$\qquad$ = the ratio of forced frequency to the free-vibration frequency

Three cases are of interest here, depending on whether r is less than, greater than, or equal to 1. If $r < 1$, then $1 - r^2$ is positive, and the forced motion is defined by Eq. 4-7 without change. In this case, x_b will be in phase with the force, as shown in Fig. 4-2. Here the amplitude of motion is defined by

$$X = \frac{X_0}{1 - r^2} \qquad r < 1 \tag{4-9}$$

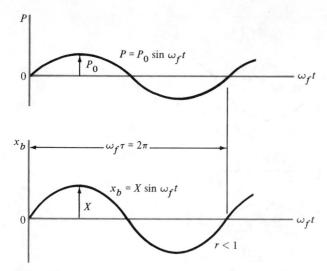

Figure 4-2

When $r > 1$, since $1 - r^2$ is then negative, it is desirable to write the solution as

$$x_b = \frac{X_0}{r^2 - 1}(-\sin \omega_f t) \qquad r > 1$$

$$= X(-\sin \omega_f t)$$

(4-10)

The amplitude of motion is positive and is defined by

$$X = \frac{X_0}{r^2 - 1} \qquad r > 1$$

(4-11)

The motion is shown in Fig. 4-3. Note that the motion is of opposite phase to the force.

If the forcing function had been taken as $P = P_0 \cos \omega_f t$, the particular solution would have been

$$x_b = \frac{X_0}{1 - r^2} \cos \omega_f t \qquad r \neq 1$$

(4-12)

with X_0 and r defined as above.

Finally, of concern is the case wherein $r = 1$. The amplitude, defined by $X = \pm X_0/(1 - r^2)$, then becomes infinite. This condition, for which the forced frequency ω_f is equal to the natural frequency ω of the system, is called *resonance*, and the resulting amplitude is referred to as the *resonant amplitude*. When $r = 1$, the solution expressed by Eq. 4-7 does not define the variation of displacement x_b with time, for this solution would not then be valid. It will subsequently be shown, in Section 4-6, that for this case the

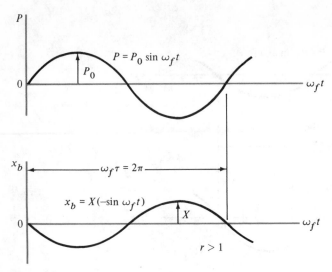

Figure 4-3

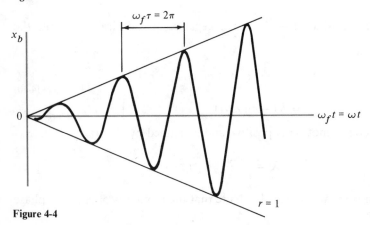

Figure 4-4

solution is

$$x_b = -\frac{X_0 \omega_f t}{2} \cos \omega_f t \qquad r = 1 \qquad (4\text{-}13)$$

The forced motion represented by this function is shown in Fig. 4-4. The solution here may also be written in the form

$$x_b = \left(\frac{X_0 \omega_f t}{2}\right) \sin\left(\omega_f t - \frac{\pi}{2}\right) \qquad (4\text{-}14)$$

The motion is of harmonic form but has an amplitude that increases linearly with time. Thus the amplitude does not become large instantaneously but requires finite time to build up. Note that the motion lags the force by 90 degrees.

It should be reemphasized that the foregoing discussions have been of the forced-motion part *only* of the complete motion as defined by Eq. 4-3. The free-vibration portion of the motion was covered in Chapter 2.

4-3. FORCED-AMPLITUDE AND MAGNIFICATION FACTOR

Frequently the most important consideration is the amplitude of the forced motion. This can be studied by observing the manner in which the amplitude X varies with respect to the frequency ratio r. For this purpose X_0 may be used as a reference, and the magnification factor MF is defined as

$$MF = \frac{X}{X_0} \qquad (4\text{-}15)$$

This is simply the ratio of the forced amplitude to the arbitrary reference X_0. Note, from Eqs. 4-9 and 4-11, that

$$MF = \frac{X}{X_0} = \frac{1}{1 - r^2} \qquad r < 1 \qquad (4\text{-}16)$$

$$MF = \frac{X}{X_0} = \frac{1}{r^2 - 1} \qquad r > 1 \qquad (4\text{-}17)$$

The influence of frequency on forced amplitude can be studied by plotting MF against r, as shown in Fig. 4-5. This indicates that the MF is greater than 1 in the range from $r = 0$ to $r = 1$, approaching infinity as r approaches unity. As previously explained, the resonant amplitude is not reached instantly but requires time to build up. However, the eventual large amplitude resulting from resonance is of great concern, since it may result in destruction of the system. As r becomes large, the MF and amplitude become small.

Since $r = \omega_f / \omega$ and $\omega = \sqrt{k/m}$, then r can be changed by altering ω_f or m or k. In studying Fig. 4-5 it is proper to contemplate the effect on the amplitude if ω_f or m is changed; however, if k is altered, it must be kept in mind that not only does the MF change but the reference X_0 also changes, since it depends on k. The effect of varying k will be discussed in detail later, when forced damped motion is investigated.

EXAMPLE 4-1 A mass-spring system is excited by a force $P = 4 \sin \omega_f t$. The spring has a modulus of 20 lb/in., and the mass has a weight of 23.83 lb. (a) Obtain the magnification factor for ω_f values of 3.6, 16.2, and 54 rad/sec. (b) If ω_f is equal to the natural frequency, obtain the amplitude at the end of 10 cycles, as well as the time required to reach this stage.

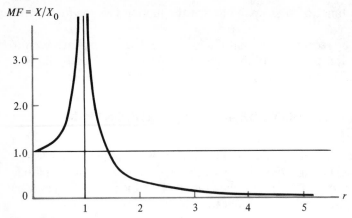

Figure 4-5

SOLUTION

$$\omega = \sqrt{\frac{k}{m}} = \sqrt{\frac{20 \times 386}{23.83}} = 18.0 \text{ rad/sec}$$

a. $r_1 = \dfrac{3.6}{18.0} = 0.2 \qquad MF = \dfrac{1}{1 - r^2} = \dfrac{1}{1 - (0.2)^2} = 1.04$

$r_2 = \dfrac{16.2}{18} = 0.9 \qquad MF = \dfrac{1}{1 - (0.9)^2} = 5.26$

$r_3 = \dfrac{54}{18} = 3.0 \qquad MF = \dfrac{1}{3^2 - 1} = 0.125$

b. $x = -\dfrac{X_0 \omega_f t}{2} \cos \omega_f t$

$X_0 = \dfrac{P_0}{k} = \dfrac{4}{20} = 0.2 \text{ in.}$

$\omega_f t = 2\pi \times 10 = 20\pi \text{ rad}$

$x = \left| -\dfrac{0.2 \times 20\pi}{2} \times \cos 20\pi \right| = 6.28 \text{ in.}$

$t = \dfrac{20\pi}{\omega_f} = \dfrac{20\pi}{\omega} = \dfrac{20\pi}{18} = 3.49 \text{ sec}$

4-4. THE COMPLETE SOLUTION AND MOTION

In the interpretation of the solution here, only the forced motion x_b has been considered. The total motion is defined by

$$x = x_a + x_b$$

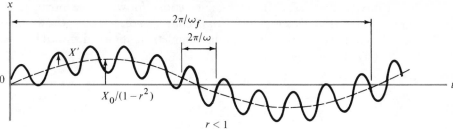

Figure 4-6. For $r < 1(\omega_f < \omega)$.

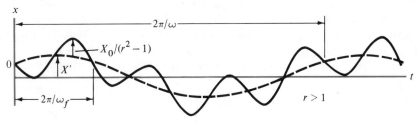

Figure 4-7. For $r > 1(\omega_f > \omega)$.

so that

$$x = X' \sin (\omega t + \phi) + \frac{X_0}{1 - r^2} \sin \omega_f t \qquad r < 1 \qquad (4\text{-}18)$$

$$x = X' \sin (\omega t + \phi) - \frac{X_0}{r^2 - 1} \sin \omega_f t \qquad r > 1 \qquad (4\text{-}19)$$

In these cases the complete motion is the sum of two sinusoidal curves of different frequency. For Eq. 4-18, shown in Fig. 4-6, where the forced frequency is smaller than the natural frequency, the forced-motion part conveniently becomes the axis for plotting the free-vibration part. In Fig. 4-7, for Eq. 4-19, since the forced frequency is greater than the natural frequency, the free-vibration part serves as the axis for the forced portion.

For the case of resonance the complete solution would be

$$x = X' \sin (\omega t + \phi) - \frac{X_0 \omega_f t}{2} \cos \omega_f t \qquad \omega_f = \omega \qquad (4\text{-}20)$$

This is the sum of a sine wave of constant amplitude and a similar curve having an increasing amplitude. The two parts have the same frequency. In the early stages the first part may be significant, but eventually the latter (forced-motion) part becomes predominant.

4-5. FREE VIBRATION AFTER REMOVAL OF EXCITATION

If a system has been vibrating under the action of a harmonic force and this driving condition ceases, a free vibration of some nature must continue. In certain instances the final free motion may have undesirable features.

Consider the case, in the preceding sections, for which the forcing function was $P = P_0 \sin \omega_f t$. For convenience take $x = 0$ and $\dot{x} = X_0 \omega_f / (1 - r^2)$ at $t = 0$. Then the complete solutions, identified by Eqs. 4-18 and 4-19, become

$$x = X_1 \sin \omega_f t \qquad r < 1 \tag{4-21}$$

where

$$X_1 = \frac{X_0}{1 - r^2}$$

This represents forced motion only. It is shown in Fig. 4-8(a). The motion is harmonic, with a circular frequency ω_f. If the force P ceases at any moment, such as for time t_n, the motion that follows must also be harmonic, but it will have a circular frequency ω. The frequencies ω_f and ω will be assumed here to have different values. The displacement and velocity occurring at time t_n would be defined by

$$x_n = X_1 \sin \omega_f t_n \qquad \dot{x}_n = X_1 \omega_f \cos \omega_f t_n \tag{4-22}$$

The values of x_n and $\dot{x}_n$ determined from these may be used as the initial values of displacement and velocity for the free-vibration motion that follows after the force has stopped. For this purpose the time may be shifted so that it is now measured from the moment P ceases. This is shown in Fig. 4-8(b), where the new time is designated by t_2. The free-vibration motion may then be expressed as

$$x = \frac{\dot{x}_n}{\omega} \sin \omega t_2 + x_n \cos \omega t_2 \tag{4-23}$$

Thus for any case in which t_n is specified, Eqs. 4-22 and 4-23 may be used to determine the free vibration which continues after the forcing condition has vanished.

Certain cases of special significance will now be considered. First, if the force ceases at point η' in Fig. 4-8(a) when the displacement is maximum, then from Eq. 4-22, $x_n = X_1$ and $\dot{x}_n = 0$, and the continuing motion, as expressed by Eq. 4-23, is

$$x = X_1 \cos \omega t_2$$

Thus the continuing free vibration has the same amplitude as that of the preceding forced motion, but the frequency is different. Examples of this case are shown in Fig. 4-9.

Next, consider the case in which the force P stops at point η'' in Fig. 4-8(a) when the displacement is zero, so that from Eq. 4-22, $x_n = 0$ and $\dot{x}_n = X_1 \omega_f$. The continuing free motion, defined by Eq. 4-23, is now

$$x = \frac{X_1 \omega_f}{\omega} \sin \omega t_2 = X_2 \sin \omega t_2$$

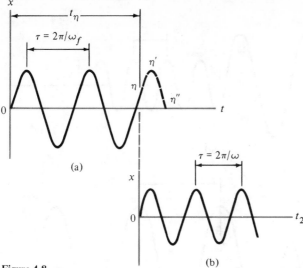

(a)

(b)

Figure 4-8

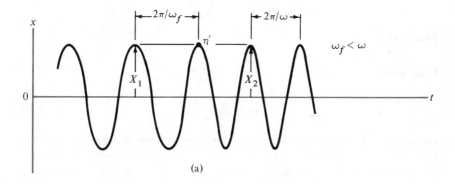

(a)

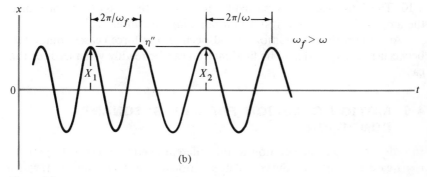

(b)

Figure 4-9

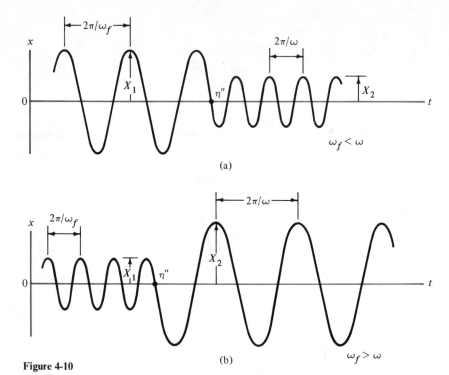

Figure 4-10

from which

$$X_2 = X_1 \frac{\omega_f}{\omega} \tag{4-24}$$

represents the amplitude of the free-vibration motion that occurs after the force has ceased. For $\omega_f \neq \omega$, a change in amplitude takes place. If $\omega_f < \omega$, there will be a drop in amplitude; if $\omega_f > \omega$, there will be a gain. The frequency will also differ for the two parts. These two cases are shown in Fig. 4-10. The latter case is of particular interest. If the mass lacks clearance, the amplitude gain could result in a bumping condition.

An intermediate condition would occur if the force ceases somewhere between points η' and η''. The final motion can be readily visualized for this case.

4-6. MOTION SOLUTION FOR THE RESONANT CONDITION

In order to obtain the solution to the differential equation (Eq. 4-2) for the $r = 1$ case, first consider the complete solution for the condition $r < 1$, which can be written as

$$x = A \sin \omega t + B \cos \omega t + \frac{P_0/m}{\omega^2 - \omega_f^2} \sin \omega_f t \qquad (4\text{-}25)$$

Designating the initial values as $x = x_0$ and $\dot{x} = \dot{x}_0$ at $t = 0$ and substituting gives

$$A = \frac{\dot{x}_0}{\omega} - \frac{(P_0/m)(\omega_f/\omega)}{\omega^2 - \omega_f^2} \qquad \text{and} \qquad B = x_0 \qquad (4\text{-}26)$$

whence Eq. 4-25 can be written as

$$x = \frac{\dot{x}_0}{\omega} \sin \omega t + x_0 \cos \omega t + \frac{P_0/m}{\omega^2 - \omega_f^2} \left(\sin \omega_f t - \frac{\omega_f}{\omega} \sin \omega t \right) \qquad (4\text{-}27)$$

For the initial conditions $x_0 = 0 = \dot{x}_0$, this becomes

$$x = \frac{P_0/m}{\omega^2 - \omega_f^2} \left(\sin \omega_f t - \frac{\omega_f}{\omega} \sin \omega t \right) \qquad (4\text{-}28)$$

representing the response of the system to the forcing function $P = P_0 \sin \omega_f t$. This particular solution satisfies the differential equation (Eq. 4-2) and also the initial conditions of zero displacement and velocity.

Now consider the case for which ω_f is almost equal to, but slightly less than, ω and set

$$\omega - \omega_f = 2\Delta \qquad (4\text{-}29)$$

where Δ is a very small positive quantity. Note that

$$\omega + \omega_f \approx 2\omega_f \qquad (4\text{-}30)$$

Multiplying these together results in

$$\omega^2 - \omega_f^2 \approx 4\omega_f \Delta \qquad (4\text{-}31)$$

Noting that $\omega_f/\omega \approx 1$, and by using a trigonometric identity, Eq. 4-28 can be written in the form

$$x = \frac{P_0/m}{\omega^2 - \omega_f^2} 2 \sin \left(\frac{\omega_f - \omega}{2} \right) t \cdot \cos \left(\frac{\omega_f + \omega}{2} \right) t \qquad (4\text{-}32)$$

Then substituting in Eqs. 4-29, 4-30, and 4-31 gives

$$x = \frac{2P_0/m}{4\omega_f \Delta} \sin \left(-\Delta t \right) \cos \omega_f t$$

$$= -\frac{P_0/m}{2\omega_f \Delta} \sin \Delta t \cos \omega_f t \qquad (4\text{-}33)$$

Now let $\omega_f \to \omega$, so that $\Delta \to 0$ and $\sin \Delta t \to \Delta t$. Then

$$x = -\frac{(P_0/k)\cdot(k/m)}{2\omega_f\Delta}\Delta t \cos \omega_f t$$

$$= -\frac{X_0\omega^2}{2\omega_f}t \cos \omega_f t$$

$$= -\frac{X_0\omega_f}{2}t \cos \omega_f t \qquad \omega_f = \omega \tag{4-34}$$

representing the particular solution for this case. Thus

$$x_b = \frac{-X_0\omega_f}{2}t \cos \omega_f t \qquad r = 1 \tag{4-35}$$

which is the same as Eq. 4-13. This solution will, of course, satisfy the differential equation (Eq. 4-2), provided that the condition $\omega_f = \omega$ is also included.

4-7. BEATING

When the forced frequency is close to but slightly different from the natural frequency, a vibration takes place in which the amplitude builds up and then dies out, repeating this process continuously. This phenomenon is known as *beating*, and it is often observed in machines, structures, and elsewhere. In effect, the two parts of the motion are sometimes in phase so that they add together, and then they gradually get out of phase so that the two motion parts are opposite and thus cancel; and this process continues on and on.

In order to explain this type of motion, consider Eq. 4-33, where ω_f was slightly less than ω. Rearrange this relation as follows:

$$x = \left(\frac{-P_0/m}{2\omega_f\Delta}\cos \omega_f t\right)\sin \Delta t \tag{4-36}$$

Since ω_f is much larger than Δ, the $\cos \omega_f t$ curve will go through several cycles while the $\sin \Delta t$ wave is going through a single cycle. The part inside the parentheses represents a cyclically varying amplitude of the sine curve. Stated differently, the $\sin \Delta t$ curve becomes an envelope of the expression within the parentheses. The resulting motion curve is shown in Fig. 4-11. Observe the manner in which the amplitude builds up and dies down continuously. The number of cycles of the $\cos \omega_f t$ curve occurring in 1 cycle of the $\sin \Delta t$ curve is not necessarily an integral number; that is, ω_f may or may not be an integral multiplier of Δ. The beat period τ_b, as defined by the time between points of zero motion or the time between maximum motions, would be

$$\tau_b = \frac{\pi}{\Delta} = \frac{2\pi}{\omega - \omega_f} \tag{4-37}$$

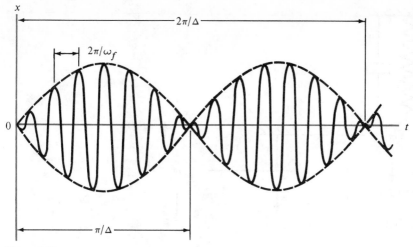

Figure 4-11

Note that $\omega - \omega_f$ represents the relative circular frequency of the motion parts, and hence verifies this relation for the beat period.

This analysis and interpretation would apply to the addition of any two sinusoidal curves of close, but unequal, frequencies and of approximately equal amplitudes.

4-8. FORCED VIBRATIONS WITH VISCOUS DAMPING

A viscously damped system subjected to a harmonic force is shown in Fig. 4-12(a), with the displacement x and the force P being positive downward, as indicated. Based on the free-body diagram of Fig. 4-12(b), for an assumed plus x position and positive $\dot{x}$, the equation of motion is

$$m\ddot{x} = -kx - c\dot{x} + P$$

which can be rearranged as

$$m\ddot{x} + c\dot{x} + kx = P_0 \sin \omega_f t \qquad (4\text{-}38)$$

The solution to this differential equation is

$$x = x_a + x_b \qquad (4\text{-}39)$$

where x_a is the complementary function and x_b is the particular solution.

The complementary function is defined by Eq. 3-13, Eq. 3-15, or Eq. 3-17, depending on the system damping relative to the critical value. Assuming the damping to be small (in particular, less than critical), the complementary function is given by Eq. 3-15 as

$$x_a = X' e^{-\zeta \omega t} \sin (\omega_d t + \phi) \qquad \zeta < 1 \qquad (4\text{-}40)$$

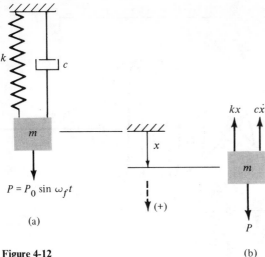

Figure 4-12   (b)

This represents a free oscillatory part of the motion, which decays with time and eventually approaches zero; hence it is called the *transient* solution. It will be present only in the initial stages of motion or as a decaying free vibration which follows after the cessation of the forcing condition.

The particular solution represents a part of the motion that will occur continuously while the forcing condition is present, and hence is called the *steady state*.

The particular solution must reduce to a sin $\omega_f t$ form. Since $\dot{x}$ (as well as x and $\ddot{x}$) is present in the differential equation, it is reasonable to assume that

$$x_b = M \sin \omega_f t + N \cos \omega_f t \tag{4-41}$$

This is general, in that it assumes a phase difference for the motion relative to the force. The constant coefficients M and N are to be determined so that the differential equation will be satisfied. Thus substituting the particular solution into the differential equation results in

$$-m\omega_f^2(M \sin \omega_f t + N \cos \omega_f t) + c\omega_f(M \cos \omega_f t - N \sin \omega_f t)$$
$$+ k(M \sin \omega_f t + N \cos \omega_f t) = P_0 \sin \omega_f t \tag{4-42}$$

Equating the coefficients of the sine terms on the two sides of this relation, and similarly equating those for the cosine terms, gives

$$(k - m\omega_f^2)M - c\omega_f N = P_0$$
$$c\omega_f M + (k - m\omega_f^2)N = 0 \tag{4-43}$$

The values M and N can be obtained by using determinants and Cramer's rule, resulting in

$$M = \frac{(k - m\omega_f^2)P_0}{(k - m\omega_f^2)^2 + (c\omega_f)^2}$$

$$N = \frac{-c\omega_f P_0}{(k - m\omega_f^2)^2 + (c\omega_f)^2} \tag{4-44}$$

Trigonometric substitutions enable Eq. 4-41 to be written in the form

$$x_b = \sqrt{M^2 + N^2} \sin(\omega_f t - \psi) \tag{4-45}$$

where

$$\tan \psi = \frac{-N}{M}$$

Substituting Eq. 4-44 into Eq. 4-45 yields

$$x_b = \frac{P_0}{\sqrt{(k - m\omega_f^2)^2 + (c\omega_f)^2}} \sin(\omega_f t - \psi) \tag{4-46}$$

where

$$\tan \psi = \frac{c\omega_f}{k - m\omega_f^2}$$

The following substitutions are useful here:

$$\frac{1}{k}(k - m\omega_f^2) = 1 - \frac{\omega_f^2}{k/m} = 1 - \frac{\omega_f^2}{\omega^2} = 1 - r^2$$

$$\frac{1}{k}(c\omega_f) = \left(\frac{2c}{2\sqrt{mk}}\right)\left(\frac{\omega_f}{\sqrt{k/m}}\right) = 2\left(\frac{c}{c_c}\right)\left(\frac{\omega_f}{\omega}\right) = 2\zeta r \tag{4-47}$$

Equation 4-46 can then be written as

$$x_b = X \sin(\omega_f t - \psi) \tag{4-48}$$

where

$$X = \frac{P_0}{\sqrt{(k - m\omega_f^2)^2 + (c\omega_f)^2}} = \frac{P_0/k}{\sqrt{[(k - m\omega_f^2)/k]^2 + (c\omega_f/k)^2}}$$

$$= \frac{X_0}{\sqrt{(1 - r^2)^2 + (2\zeta r)^2}} \tag{4-49}$$

and

$$\tan \psi = \frac{c\omega_f}{k - m\omega_f^2} = \frac{c\omega_f/k}{(k - m\omega_f^2)/k} = \frac{2\zeta r}{1 - r^2} \tag{4-50}$$

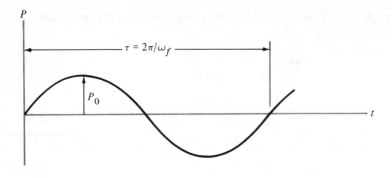

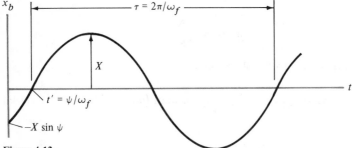

Figure 4-13

The particular solution (Eq. 4-48) represents a steady-state motion of amplitude X. This motion has the same frequency as the forcing condition, but it lags behind the force P by the phase angle ψ, or by the time t' defined by

$$t' = \frac{\psi}{\omega_f} \tag{4-51}$$

A plot of the force and motion is shown in Fig. 4-13.

Both the steady-state amplitude X and the phase angle ψ are dependent on the damping factor ζ and the frequency ratio r. This will be discussed in some detail in subsequent sections.

The complete motion is, of course, defined by x. Writing the complete solution results in

$$x = X'e^{-\zeta\omega t} \sin(\omega_d t + \phi) + X \sin(\omega_f t - \psi) \qquad \zeta < 1 \tag{4-52}$$

Note that X and ψ depend on the forcing condition as well as the physical constants of the system. The constants X' and ϕ can be determined from the initial displacement and velocity. An example of the complete motion is plotted in Fig. 4-14, where the forced frequency has been assumed to be less than the natural damped frequency.

If the force ceases after steady-state motion has been attained, a free damped vibration will follow. Whether an immediate gain, a loss, or no

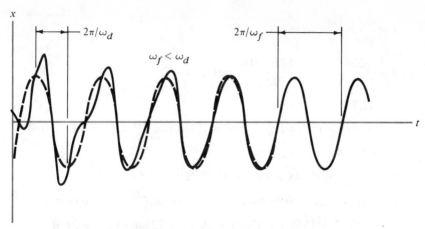

Figure 4-14

change in amplitude occurs, depends on when the force stops. In this connection, the analysis of Section 4-5 can be applied, except for the influence of damping. Small damping will have little effect on the initial amplitude of this final free vibration, but it will cause the subsequent motion to decay.

EXAMPLE 4-2 A body having a mass of 44 kg moves along a straight line according to the relation

$$x = 0.05e^{-0.72t} \sin (5.96t + 29°) + 0.08 \sin (3.6t - \lambda)$$

where x is the displacement in meters measured from a fixed reference, and t is time in seconds. Obtain the physical constants and parameters of the system, including any appropriate frequencies, and so on. Also obtain the initial velocity and displacement.

SOLUTION The first part is a transient, showing characteristic decay; hence the system is damped. The second part is a sustained function and thus is the steady state due to a forcing condition.

$$\zeta\omega = 0.72 \quad \text{and} \quad \omega_d = \sqrt{1 - \zeta^2}\,\omega = 5.96 \text{ rad/s}$$

Solving these gives

$\zeta = 0.12$

$\omega = 6$ rad/s

$k = m\omega^2 = 44 \times 6^2 = 1584$ N/m

$c_c = 2m\omega = 2 \times 44 \times 6 = 528$ N · s/m

$c = \zeta c_c = 0.12 \times 528 = 63.36$ N · s/m

$$r = \frac{\omega_f}{\omega} = \frac{3.6}{6} = 0.6$$

$$\tan \lambda = \frac{2\zeta r}{1 - r^2} = \frac{2 \times 0.12 \times 0.6}{1 - (0.6)^2} = 0.225$$

$$\lambda = 12.7°$$

$$X_0 = X\sqrt{(1 - r^2)^2 + (2\zeta r)^2} = 0.08\sqrt{[1 - (0.6)^2]^2 + (2 \times 0.12 \times 0.6)^2}$$
$$= 0.052\,48 \text{ m} = 5.248 \text{ cm}$$

$$P_0 = X_0 k = 0.052\,48 \times 1584 = 83.13 \text{ N}$$

$$x_0 = 0.05 \sin 29° + 0.08 \sin (-12.7°) = 0.006\,653 \text{ m} = 6.653 \text{ mm}$$

$$\dot{x} = 0.05 e^{-0.72t}[5.96 \cos (5.96t + 29°) - 0.72 \sin (5.96t + 29°)]$$
$$+ 0.288 \cos (3.6t - 12.7°)$$

$$\dot{x}_0 = 0.05(5.96 \cos 29° - 0.72 \sin 29°) + 0.288 \cos 12.7° = 0.5241 \text{ m/s}$$

4-9. ALTERNATIVE-SOLUTION METHOD, BY COMPLEX ALGEBRA

A simple method for obtaining the particular solution of the differential equation for the case of a harmonic forcing function employs complex notation. Before proceeding, it is desirable to introduce an additional complex relation, as follows:

$$\frac{1}{a + ib} = \frac{1}{a + ib} \times \frac{a - ib}{a - ib} = \frac{a - ib}{a^2 + b^2}$$

$$= \frac{1}{\sqrt{a^2 + b^2}} \left(\frac{a}{\sqrt{a^2 + b^2}} - \frac{ib}{\sqrt{a^2 + b^2}} \right)$$

$$= \frac{1}{\sqrt{a^2 + b^2}} (\cos \alpha - i \sin \alpha)$$

$$= \frac{1}{\sqrt{a^2 + b^2}} e^{-i\alpha} \qquad (4\text{-}53)$$

where

$$\cos \alpha = \frac{a}{\sqrt{a^2 + b^2}} \qquad \sin \alpha = \frac{b}{\sqrt{a^2 + b^2}}$$

so that

$$\tan \alpha = \frac{b}{a} \qquad (4\text{-}54)$$

It is also necessary to express the forcing function in complex form. This can be done (see Third Method of Section 2-6) in the following manner:

$$P = P_0 \cos \omega_f t = \mathscr{R}[P_0(\cos \omega_f t + i \sin \omega_f t)]$$
$$= \mathscr{R}P_0 e^{i\omega_f t} \tag{4-55}$$

It is considered that the force P, so defined, acts on the system of Fig. 4-12(a), for which the differential equation is then

$$m\ddot{x} + c\dot{x} + kx = P \tag{4-56}$$

This is the same as Eq. 4-38, except that P is defined by the harmonic cosine function instead of sine. This is not significant but is merely convenient. The particular solution is then expressed as

$$x_b = X \cos (\omega_f t - \psi) = \mathscr{R}X[\cos (\omega_f t - \psi) + i \sin (\omega_f t - \psi)]$$
$$= \mathscr{R}\bar{X}e^{i\omega_f t} \tag{4-57}$$

where the complex amplitude $\bar{X}$ is defined by

$$\bar{X} = Xe^{-i\psi} \tag{4-58}$$

Substituting the complete function (that is, both the real and the imaginary parts) for Eqs. 4-55 and 4-57 into Eq. 4-56 results in

$$(-m\omega_f^2 + ic\omega_f + k)\bar{X}e^{i\omega_f t} = P_0 e^{i\omega_f t} \tag{4-59}$$

Since this must be independent of t,

$$[(k - m\omega_f^2) + i(c\omega_f)]\bar{X} = P_0$$

$$\bar{X} = \frac{P_0}{(k - m\omega_f^2) + i(c\omega_f)} \tag{4-60}$$

which defines the complex amplitude $\bar{X}$ of the forced motion. Substituting Eq. 4-58 on the left-hand side and replacing the right-hand side according to Eq. 4-53 gives

$$Xe^{-i\psi} = \frac{P_0}{\sqrt{(k - m\omega_f^2)^2 + (c\omega_f)^2}} e^{-i\alpha} \tag{4-61}$$

whence

$$X = \frac{P_0}{\sqrt{(k - m\omega_f^2)^2 + (c\omega_f)^2}} \tag{4-62}$$

and

$$\psi = \alpha$$

so that, by Eq. 4-54,

$$\tan \psi = \frac{c\omega_f}{k - m\omega_f^2} \tag{4-63}$$

By Eq. 4-57, the particular solution then is

$$x_b = \mathscr{R} \bar{X} e^{i\omega_f t} = \mathscr{R} X e^{i(\omega_f t - \psi)}$$

$$= X \cos (\omega_f t - \psi) \tag{4-64}$$

where X and ψ are given by Eqs. 4-62 and 4-63. These correspond to the relations previously obtained for the case of a harmonically forced system with viscous damping.

4-10. MAGNIFICATION FACTOR AND STEADY-STATE AMPLITUDE

The amplitude X of the steady-state motion is generally an important consideration. From Eq. 4-49, the magnification factor MF is defined by

$$MF = \frac{X}{X_0} = \frac{1}{\sqrt{(1 - r^2)^2 + (2\zeta r)^2}} \tag{4-65}$$

Plotting MF against r results in a family of curves which are dependent on ζ, as shown in Fig. 4-15. In obtaining maximum and minimum points, the derivative $d(MF)/dr$ may be set to zero. This results in

$$\frac{r(1 - r^2 - 2\zeta^2)}{[(1 - r^2)^2 + (2\zeta r)^2]^{3/2}} = 0 \tag{4-66}$$

which is satisfied by the following:

1. For $r = 0$. This defines the initial point on the curves. This will be a minimum point, provided that $\zeta < 0.707$. For $\zeta \geq 0.707$, it will be a maximum point.
2. For $r = \infty$. This defines the final minimum point on each curve.
3. For $(1 - r^2 - 2\zeta^2) = 0$, which yields

$$r = \sqrt{1 - 2\zeta^2} \qquad \text{for } \zeta \leq 0.707 \tag{4-67}$$

This last expression defines the maximum point in the resonant region. Since

$$r = \sqrt{1 - 2\zeta^2} < 1 \tag{4-68}$$

the peak of the curve occurs to the left of the resonant value of $r = 1$. The maximum amplitude can be determined by substituting $r = \sqrt{1 - 2\zeta^2}$ into the amplitude expression (Eq. 4-65), resulting in

$$\frac{X_{max}}{X_0} = \frac{1}{2\zeta\sqrt{1 - \zeta^2}} \tag{4-69}$$

$$\approx \frac{1}{2\zeta} \qquad \text{for } \zeta \ll 1 \tag{4-70}$$

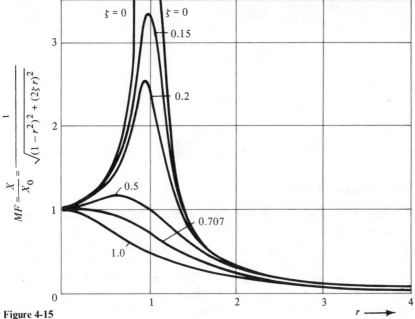

$$MF = \frac{X}{X_0} = \frac{1}{\sqrt{(1-r^2)^2 + (2\zeta r)^2}}$$

Figure 4-15

$r \longrightarrow$

The damping serves to limit the maximum amplitude here to a finite value. For $\zeta \geq 0.707$, the maximum point occurs at $r = 0$, and the curve drops continuously as r increases.

The family of curves indicates that a reduction in MF—and hence in amplitude—is attained only in the region where r is large due to a high forcing frequency relative to the natural frequency of the system.

The effect of varying r on the maximum displacement amplitude can be observed from Fig. 4-15. However, such a consideration should be limited to varying r by changing ω_f only. The conditions resulting from changing r by altering k and m will be discussed later.

EXAMPLE 4-3 The physical constants for a damped system are $W = 19.3$ lb, $k = 45$ lb/in., and $c = 0.60$ lb sec/in. The system is subject to a harmonic force having a maximum value of 25 lb. Determine the resonant amplitude and the maximum amplitude for the steady-state motion.

SOLUTION

$$\omega = \sqrt{\frac{k}{m}} = \sqrt{\frac{45 \times 386}{19.3}} = 30 \text{ rad/sec}$$

$$c_c = 2m\omega = 2 \times \frac{19.3}{386} \times 30 = 3.0 \text{ lb sec/in.}$$

$$\zeta = \frac{c}{c_c} = \frac{0.60}{3.0} = 0.20$$

$$X_0 = \frac{P_0}{k} = \frac{25}{45} = 0.556 \text{ in.}$$

$$X_{res} = \frac{X_0}{2\zeta} = \frac{0.556}{2 \times 0.2} = 1.39 \text{ in.}$$

$$X_{max} = \frac{X_0}{2\zeta\sqrt{1 - \zeta^2}} = \frac{0.556}{2 \times 0.2\sqrt{1 - (0.2)^2}} = \frac{1.39}{0.980} = 1.42 \text{ in.}$$

4-11. PHASE ANGLE

The value of the phase angle, as defined by Eq. 4-50, depends on the damping factor ζ and the frequency ratio r. This can be studied by plotting ψ against r for various values of ζ. The resulting family of curves is shown in Fig. 4-16. For no damping, $\psi = 0$ from $r = 0$ to $r < 1$, $\psi = 90$ degrees for $r = 1$, and $\psi = 180$ degrees for $r > 1$. This agrees with the analysis and discussion of Section 4-2. For small values of ζ, these same conditions are approximated; that is, the curve approaches the curve for the zero-damping case. All curves go through the point of $\psi = 90$ degrees for $r = 1$. Note that for $\zeta = 0.707$ the ψ curve is approximately linear from $r = 0$ through $r = 1$.

4-12. INFLUENCE OF MASS AND ELASTICITY ON AMPLITUDE

In determining the effect of varying r on the steady-state amplitude, recall that $r = \omega_f \sqrt{m/k}$, and hence r can be varied by changing k or m as well as ω_f. However, if either k or m is changed, this will alter ζ (as $\zeta = c/2\sqrt{mk}$) and distort the interpretation of Fig. 4-15, since a different ζ curve would then have to be used. In addition, altering k will change the reference value X_0. If it is desired to study the effect of varying k, the amplitude relation (Eq. 4-49) can be written in the form

$$X = \frac{P_0}{\sqrt{(k - m\omega_f^2)^2 + (c\omega_f)^2}} \tag{4-71}$$

and X can then be plotted against k for several values of the damping constant c. The resulting family of curves is shown in Fig. 4-17. It should be noted that P_0, m, and ω_f are constant in this consideration. Maximum and minimum points on the curves can be obtained by setting $dX/dk = 0$. From this, it is found that the maximum point occurs for $k = m\omega_f^2$ and is defined by

$$X_{max} = \frac{P_0}{c\omega_f} \tag{4-72}$$

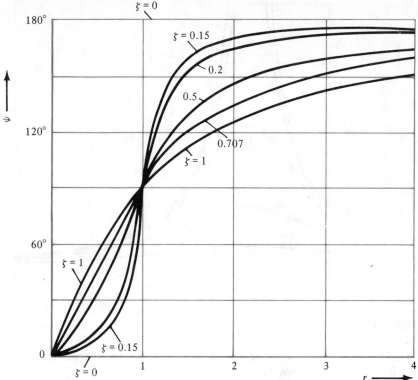

Figure 4-16

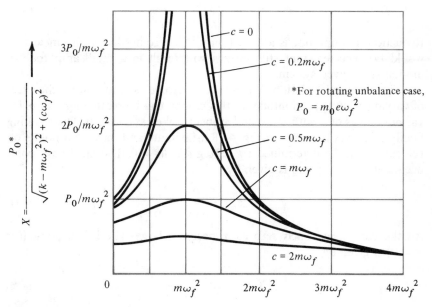

Figure 4-17

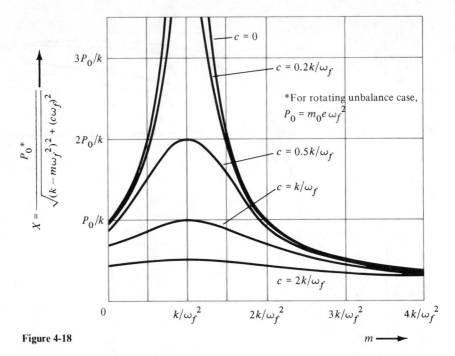

Figure 4-18

In addition, all the curves approach zero as k becomes large. The initial point (for $k = 0$) is given by

$$X = \frac{P_0}{\sqrt{(m\omega_f^2)^2 + (c\omega_f)^2}} \qquad (4\text{-}73)$$

Reduction in amplitude is attained only as k becomes large. This means, as would be expected, that stiff springs will result in a small amplitude of motion for a given system.

The amplitude relation (Eq. 4-71) can also be used to observe the effect of varying m on the amplitude. In this case X can be plotted against m for various values of c, with P_0, k, and ω_f being taken as constant. The resulting family of curves is shown in Fig. 4-18. Maximum and minimum points on the curves can be determined by setting $dX/dm = 0$. The maximum point occurs at $m = k/\omega_f^2$ and is given by

$$X_{max} = \frac{P_0}{c\omega_f} \qquad (4\text{-}74)$$

All the curves approach zero as m becomes large. The initial point (for $m = 0$) is given by

$$X = \frac{P_0}{\sqrt{k^2 + (c\omega_f)^2}} \qquad (4\text{-}75)$$

As might be anticipated, large values of m result in a reduction in the amplitude.

EXAMPLE 4-4   A machine weighing 19.3 lb is subjected to a harmonic force having a maximum value of 12 lb and a frequency of 300 cycles/min. The clearance for the vibrational movement of the machine is 1 in. Design a lightly damped elastic-support system for the machine, so that the machine does not collide in its movement.

SOLUTION   In order to properly limit the movement of the machine, the allowable movement is set at one-half the actual clearance. Thus

$$X = 0.5 \text{ in.}$$

Also,

$$\omega_f = \frac{300}{60} \times 2\pi = 10\pi \text{ rad/sec}$$

For small damping, from Fig. 4-17, the value of c is selected as

$$c = 0.1 m \omega_f = 0.1 \times \frac{19.3}{386} \times 10\pi = 0.05\pi \text{ lb sec/in.}$$

$$= 0.157\,08 \text{ lb sec/in.}$$

Then from Eq. 4-71,

$$X = \frac{P_0}{\sqrt{(k - m\omega_f^2)^2 + (c\omega_f)^2}}$$

$$0.5 = \frac{12}{\sqrt{[k - (19.3/386)(10\pi)^2]^2 + (0.05\pi \times 10\pi)^2}}$$

Expanding and rearranging gives

$$k^2 - 98.6960k + 1883.58 = 0$$

which has the single positive root

$$k = 25.861 \text{ lb/in.}$$

and the design is composed of an elastic support and damping device having the values of k and c determined.

EXAMPLE 4-5   A damped system is driven by the force $P = 0.54 \sin 12t$, where P is in newtons and t is in seconds. The system has a mass of 0.1 kg, and the damping constant is 0.24 N · s/m. (a) Obtain the steady-state amplitude for spring-constant k values of 2, 25, and 90 N/m. (b) Determine the spring constant that will produce the maximum amplitude, and calculate this amplitude.

SOLUTION

$$X = \frac{P_0}{\sqrt{(k - m\omega_f^2)^2 + (c\omega_f)^2}} = \frac{0.54}{\sqrt{[k - 0.1 \times (12)^2]^2 + (0.24 \times 12)^2}}$$

$$= \frac{0.54}{\sqrt{(k - 14.4)^2 + (2.88)^2}}$$

a. For $k = 2$, $X = 0.042\,42$ m $= 4.242$ cm
   For $k = 25$, $X = 4.916$ cm
   For $k = 90$, $X = 0.7138$ cm

b. $X_{max} = \dfrac{P_0}{c\omega_f} = \dfrac{0.54}{0.24 \times 12} = 0.1875$ m $= 18.75$ cm

   for $k = m\omega_f^2 = 0.1 \times (12)^2 = 14.4$ N/m

4-13. ROTATING UNBALANCE

A common source of forced vibration is caused by the rotation of a small eccentric mass such as that represented by m_0 in Fig. 4-19(a). This condition results from a setscrew or a key on a rotating shaft, crankshaft rotation, and many other simple but unavoidable situations. Rotating unbalance is inherent in rotating parts, because it is virtually impossible to place the axis of the mass center on the axis of rotation.

For the system shown, the total mass is m and the eccentric mass is m_0, so the mass of the machine body is $(m - m_0)$. The length of the eccentric arm, or the eccentricity of m_0, is represented by e. If the arm rotates with an angular velocity ω_f rad/sec, then the angular position of the arm is defined by $\omega_f t$ with respect to the indicated horizontal reference, where t is time, in seconds. The free-body diagram for this system is shown in Fig. 4-19(b), positive x having been taken as upward. The horizontal motion of $(m - m_0)$ is considered to be prevented by guides. The vertical displacement of m_0 is $(x + e \sin \omega_f t)$. From Eq. 1-8, the differential equation of motion can then be written as

$$(m - m_0)\frac{d^2x}{dt^2} + m_0 \frac{d^2}{dt^2}(x + e \sin \omega_f t) = -kx - c\frac{dx}{dt} \tag{4-76}$$

which can be rearranged in the form

$$m\frac{d^2x}{dt^2} + c\frac{dx}{dt} + kx = m_0 e\omega_f^2 \sin \omega_f t \tag{4-77}$$

Examination of this and comparison to the differential equation (Eq. 4-38) for motion forced by $P = P_0 \sin \omega_f t$ enable the steady-state solution to be set down, from Eq. 4-48, as

$$x = X \sin (\omega_f t - \psi) \tag{4-78}$$

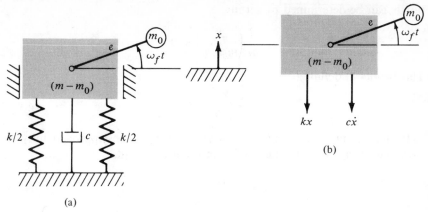

Figure 4-19

where

$$X = \frac{m_0 e \omega_f^2}{\sqrt{(k - m\omega_f^2)^2 + (c\omega_f)^2}} \tag{4-79}$$

$$= \frac{(m_0 e/m)\omega_f^2(m/k)}{\sqrt{[(k - m\omega_f^2)/k]^2 + (c\omega_f/k)^2}} = \frac{m_0 e}{m} \frac{r^2}{\sqrt{(1 - r^2)^2 + (2\zeta r)^2}} \tag{4-80}$$

and

$$\frac{X}{m_0 e/m} = \frac{r^2}{\sqrt{(1 - r^2)^2 + (2\zeta r)^2}} \tag{4-81}$$

Also

$$\tan \psi = \frac{2\zeta r}{1 - r^2} \tag{4-82}$$

Here $\omega = \sqrt{k/m}$ represents the natural circular frequency of the undamped system (including the mass m_0), but x defines the forced motion of the main mass $(m - m_0)$. It should be noted that for this case ψ will be represented physically by the angle of the eccentric arm relative to the horizontal reference of $\omega_f t$. Thus for a value of ψ determined by Eq. 4-82, the arm would be at this angle when the main body is at its neutral position, moving upward. (Since the motion lags the forcing condition, the arm then leads the motion by the angle ψ determined.) The steady-state amplitude is generally significant, and this can be studied by plotting

$$\frac{X}{m_0 e/m}$$

against the frequency ratio r for various values of the damping factor ζ, resulting in the family of curves shown in Fig. 4-20. Maximum and minimum points can be determined by setting

$$\frac{d}{dr}\left(\frac{X}{m_0\,e/m}\right) = 0$$

This results in defining

$$\frac{X}{m_0\,e/m} = 0$$

as the initial minimum point of all curves. Also, all curves approach unity as r becomes large. Finally, the maximum point occurs at

$$r = \frac{1}{\sqrt{1 - 2\zeta^2}} > 1 \tag{4-83}$$

Accordingly, the peaks occur to the right of the resonance value of $r = 1$. For $\zeta = 0.707$ the curve rises through its entirety, with the maximum equal to 1 as r approaches infinity.

Figure 4-20 is adequate, provided that the variation in r is limited to changing ω_f. Note that small amplitude occurs only at low operating frequencies, as would be expected.

Since ζ (as well as r) is dependent on k and m, Fig. 4-20 does not properly show the effect of varying k or m. Also, the reference $m_0\,e/m$ is affected by altering m. The effect of varying k or m can be observed by writing the amplitude relation here in the form given by Eq. 4-79. The amplitude X can then be plotted against either k or m for various values of the damping constant c. The resulting families of curves will be identical to those of Figs. 4-17 and 4-18, provided P_0 is replaced by $m_0\,e\omega_f^2$.

In all the preceding discussion, it should be noted that amplitude is dependent on the quantity $m_0\,e$ and that if either m_0 or e is small, the amplitude will become small. This merely emphasizes the importance of reducing the eccentric condition insofar as may be possible.

EXAMPLE 4-6   A machine with a rotating shaft has a total weight of 200 lb and is supported by springs. The damping constant for the system is found to be 3 lb sec/in. The resonant speed is determined experimentally to be 1200 rpm, and the corresponding amplitude of the main mass of the machine is 0.50 in. Determine the amplitude for a speed of 2400 rpm. Also determine the fixed value the amplitude will eventually approach at high speed.

SOLUTION   Since the machine oscillates when in operation, the rotating part must contain an eccentric mass. At resonance, Eq. 4-79 becomes

$$X = \frac{m_0\,e\omega_f}{c}$$

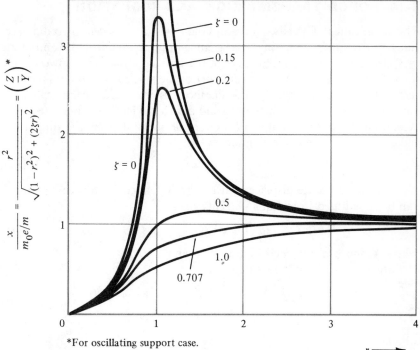

$$\frac{x}{m_0 e/m} = \frac{r^2}{\sqrt{(1-r^2)^2 + (2\zeta r)^2}} = \left(\frac{z}{Y}\right)^*$$

$\zeta = 0$

0.15

0.2

$\zeta = 0$

0.5

1.0

0.707

*For oscillating support case.

Figure 4-20

so that

$$m_0 e = \frac{Xc}{\omega_f} = \frac{0.5 \times 3}{40\pi} = 0.011\,93 \text{ lb sec}^2$$

$$k = m\omega_f^2 = \frac{200}{386} \times (40\pi)^2 = 8180 \text{ lb/in.}$$

Then, at 2400 rpm,

$$X = \frac{m_0 e\omega_f^2}{\sqrt{(k - m\omega_f^2)^2 + (c\omega_f)^2}}$$

$$= \frac{0.011\,93 \times (80\pi)^2}{\sqrt{[8180 - \frac{200}{386} \times (80\pi)^2]^2 + (3 \times 80\pi)^2}}$$

$$= 0.0306 \text{ in.}$$

For further increase in speed, the amplitude approaches the value defined by

$$X = \frac{m_0 e}{m} = \frac{0.011\,93 \times 386}{200} = 0.0230 \text{ in.}$$

4-14. FORCE TRANSMISSION AND ISOLATION

The significance of steady-state amplitude was discussed in preceding sections. The manner of altering the system in order to minimize this amplitude was also explained. Amplitude is not the only consideration, however; the force that is transmitted to the support of the system is also an important factor. For a harmonically forced damped system, the force carried by the support must be due to the spring and dashpot which are connected to it. Accordingly, the *dynamic* force F exerted on the system by the support, as indicated in the free-body diagram of Fig. 4-21, is

$$F = kx + c\dot{x} \tag{4-84}$$

Since the steady-state displacement has been defined for this case, x and $\dot{x}$ can be substituted from Eq. 4-48. This gives

$$F = kX \sin (\omega_f t - \psi) + c\omega_f X \cos (\omega_f t - \psi) \tag{4-85}$$

where X and ψ are defined by Eqs. 4-49 and 4-50.

Equation 4-85 can be written as

$$
\begin{aligned}
F &= \sqrt{(kX)^2 + (c\omega_f X)^2} \sin (\omega_f t - \psi - \beta) \\
&= \sqrt{k^2 + (c\omega_f)^2} \, X \sin (\omega_f t - \gamma)
\end{aligned}
\tag{4-86}
$$

where

$$\gamma = \psi + \beta \quad \text{and} \quad \tan \beta = -\frac{c\omega_f}{k} = -2\zeta r \tag{4-87}$$

The maximum force F_T, or force amplitude, transmitted will be

$$
\begin{aligned}
F_T &= X\sqrt{k^2 + (c\omega_f)^2} \\
&= P_0 \frac{\sqrt{k^2 + (c\omega_f)^2}}{\sqrt{(k - m\omega_f^2)^2 + (c\omega_f)^2}}
\end{aligned}
\tag{4-88}
$$

Dividing the numerator and denominator by k enables this expression to be written as

$$F_T = P_0 \frac{\sqrt{1 + (2\zeta r)^2}}{\sqrt{(1 - r^2)^2 + (2\zeta r)^2}} \tag{4-89}$$

The ratio F_T/P_0 is defined as the transmissibility TR. Thus

$$TR = \frac{F_T}{P_0} = \frac{\sqrt{1 + (2\zeta r)^2}}{\sqrt{(1 - r^2)^2 + (2\zeta r)^2}} \tag{4-90}$$

The manner in which the transmitted force is influenced by the physical parameters of the system can be shown by plotting TR against r for various values of the damping factor ζ, as represented in Fig. 4-22. All curves start at $TR = 1$ for $r = 0$. There is a common point, or crossover, at $r = \sqrt{2}$. While

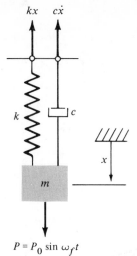

$P = P_0 \sin \omega_f t$

Figure 4-21

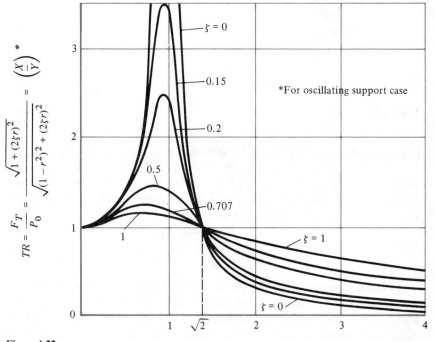

$$TR = \frac{F_T}{P_0} = \frac{\sqrt{1 + (2\zeta r)^2}}{\sqrt{(1 - r^2)^2 + (2\zeta r)^2}} = \left(\frac{X}{Y}\right)^{*}$$

$\zeta = 0$

0.15

*For oscillating support case

0.2

0.5

0.707

1

$\zeta = 1$

$\zeta = 0$

Figure 4-22

$r \longrightarrow$

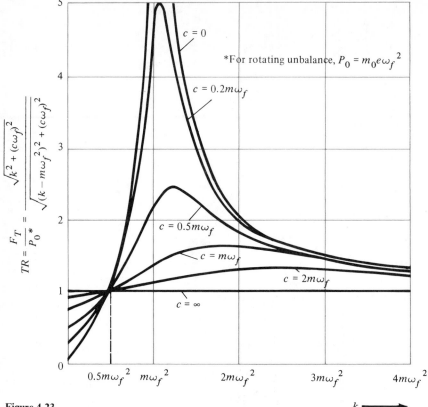

Figure 4-23

$k \longrightarrow$

damping cuts down the peak force transmitted in the resonant region, it also results in greater force transmission for $r > \sqrt{2}$. This latter effect is opposed to the influence which the damping increase has in reducing the displacement amplitude for large values of r, as indicated by Fig. 4-15. It can be shown that the peak of the curve occurs at

$$r = \frac{\sqrt{-1 + (1 + 8\zeta^2)^{1/2}}}{2\zeta} < 1 \qquad (4\text{-}91)$$

Consideration of Fig. 4-22 should be limited to conditions for which P_0 is constant and r is varied by changing ω_f only. If it is desired to observe the effect of varying the spring constant k, Eq. 4-88 can be used to plot TR against k for various values of the damping constant c. The transmissibility is then expressed as

$$TR = \frac{F_T}{P_0} = \frac{\sqrt{k^2 + (c\omega_f)^2}}{\sqrt{(k - m\omega_f^2)^2 + (c\omega_f)^2}} \qquad (4\text{-}92)$$

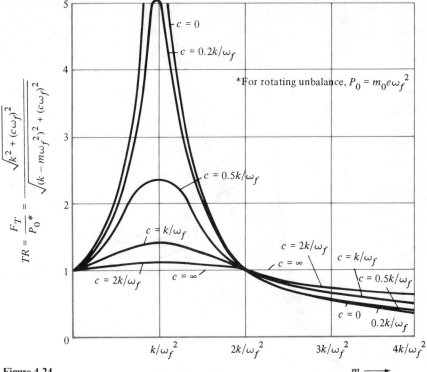

Figure 4-24

$$TR = \frac{F_T}{P_0^*} = \frac{\sqrt{k^2 + (c\omega_f)^2}}{\sqrt{(k - m\omega_f^2)^2 + (c\omega_f)^2}}$$

*For rotating unbalance, $P_0 = m_0 e \omega_f^2$

The resulting family of curves is shown in Fig. 4-23. The initial value of TR and the crossover point can be readily determined, as well as the trend of the curves in approaching $TR = 1$ for large values of k. The peak value occurs at $k = m\omega_f^2/2 + \sqrt{(m\omega_f^2/2)^2 + (c\omega_f)^2}$. This will be greater than $m\omega_f^2$ if $c \neq 0$. Thus the peak occurs to the right of the $k = m\omega_f^2$ value. An appreciable reduction in the transmitted force is achieved only in the region for which $k < m\omega_f^2/2$ by light damping in addition to a small spring constant.

The effect of varying the mass can be also determined, from Eq. 4-92, by plotting TR against m for various values of c, resulting in the family of curves shown in Fig. 4-24. Here, the peak value occurs at $m = k/\omega_f^2$ and is given by

$$(TR)_{\max} = \frac{\sqrt{k^2 + (c\omega_f)^2}}{c\omega_f}$$

A reduction in the transmitted force occurs only in the region where m is large. This might not be expected. In this connection it is important to note that an increase in the mass m will, however, also result in increasing the static force carried by the support.

The force transmitted for the case of a rotating eccentric mass should also be investigated. The relation for F_T can be obtained by substituting

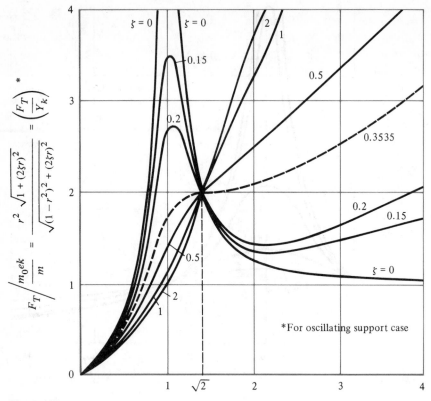

Figure 4-25

$r \longrightarrow$

$m_0 e\omega_f^2$ for P_0 in Eqs. 4-88 and 4-89. Then

$$F_T = m_0 e\omega_f^2 \frac{\sqrt{k^2 + (c\omega_f)^2}}{\sqrt{(k - m\omega_f^2)^2 + (c\omega_f)^2}} = m_0 e\omega_f^2 \frac{\sqrt{1 + (2\zeta r)^2}}{\sqrt{(1 - r^2)^2 + (2\zeta r)^2}} \quad (4\text{-}93)$$

Multiplying the numerator and denominator by k/m and rearranging gives

$$F_T = \frac{m_0 ek}{m} \cdot \frac{r^2\sqrt{1 + (2\zeta r)^2}}{\sqrt{(1 - r^2)^2 + (2\zeta r)^2}} \quad (4\text{-}94)$$

whence

$$\frac{F_T}{m_0 ek/m} = \frac{r^2\sqrt{1 + (2\zeta r)^2}}{\sqrt{(1 - r^2)^2 + (2\zeta r)^2}} \quad (4\text{-}95)$$

The effect of varying ω_f on the transmitted force can be shown by plotting $F_T/(m_0 ek/m)$ against r for various values of the damping factor ζ. In so doing, k and m are taken as constant. The reference $m_0 ek/m$ is then fixed. The resulting family of curves is shown in Fig. 4-25. Damping serves to limit the transmitted force in the region of resonance. A crossover point occurs at $r = \sqrt{2}$, for which $F_T/(m_0 ek/m)$ has a value of 2. For no damping, the curve approaches a value of 1 as r approaches infinity. When damping is present, the force becomes very large as r increases, and the greater the damping, the more rapidly this occurs. Even for small damping, the increase in the transmitted force is significant. Since frequency ratios of 10 or more are common in practice, the seriousness of damping is evident.

The maximum and minimum points for the family of curves here can be determined by setting

$$\frac{d}{dr}\left(\frac{F_T}{m_0\,ek/m}\right) = 0$$

The resulting expression is satisfied by the following conditions:

1. For $r = 0$. This defines the initial point of $F_T/(m_0\,ek/m) = 0$ for all curves.
2. By the roots of the relation

$$2\zeta^2 r^6 + (16\zeta^4 - 8\zeta^2)r^4 + (8\zeta^2 - 1)r^2 + 1 = 0$$

If $0 < \zeta < \sqrt{2}/4$, there are two positive real roots of this relation. One of these will be between $r = 0$ and $r = \sqrt{2}$, and will define a maximum point on the curve. The other will be $r > \sqrt{2}$ and will define a minimum point on the curve. If $\zeta > \sqrt{2}/4$, there is no maximum point on the curve.

If it is desired to determine the effect of varying k and m on the transmitted force, this can be done by using Eq. 4-93 and arranging it as

$$\frac{F_T}{m_0\,e\omega_f^2} = \frac{\sqrt{k^2 + (c\omega_f)^2}}{\sqrt{(k - m\omega_f^2)^2 + (c\omega_f)^2}} \tag{4-96}$$

Since the forcing frequency ω_f, the mass m_0, and the eccentricity e are to be held constant, this case becomes identical to those shown in Figs. 4-23 and 4-24, and no further analysis is needed here.

EXAMPLE 4-7 For the rotating eccentric of Example 4-6, calculate the maximum dynamic force transmitted to the foundation at the resonant speed. Also obtain F_T for the crossover point of Fig. 4-25.

SOLUTION   At resonance, Eq. 4-93 reduces to

$$F_T = \frac{m_0\,e\omega_f}{c}\sqrt{k^2 + (c\omega_f)^2} = X_{res}\sqrt{k^2 + (c\omega_f)^2}$$

$$= 0.5\sqrt{(8180)^2 + (3 \times 40\pi)^2} = 4094\text{ lb}$$

For the crossover,

$$F_T = \frac{2m_0\,e}{m}k = 2 \times 0.0230 \times 8180 = 376\text{ lb}$$

EXAMPLE 4-8   A harmonic force having a maximum value of 20 N and a frequency of 180 cycles/min acts on a machine having a mass of 25 kg. Design the machine support, consisting of a linearly elastic member and viscous damping device, so that 10% of the dynamic force is carried by the base of the support.

SOLUTION   Based on the curves of Fig. 4-23, a small damping constant would be desirable. From this, $c \approx 0.02 m\omega_f$ is chosen. Then

$$\omega_f = \frac{180}{60} \times 2\pi = 6\pi \text{ rad/s}$$

$$c \approx 0.02 \times 25 \times 6\pi = 9.425 \text{ N} \cdot \text{s/m}$$

and $c = 10 \text{ N} \cdot \text{s/m}$ is selected.

Next, k can be determined from Eq. 4-92, as follows:

$$TR = \frac{F_T}{P_0} = \frac{\sqrt{k^2 + (c\omega_f)^2}}{\sqrt{(k - m\omega_f^2)^2 + (c\omega_f)^2}}$$

$$0.1 = \frac{\sqrt{k^2 + (10 \times 6\pi)^2}}{\sqrt{[k - 25 \times (6\pi)^2]^2 + (10 \times 6\pi)^2}}$$

Expanding and rearranging results in the equation

$$k^2 + 179.447k - 761\,453 = 0$$

having the single positive root

$$k = 787.49 \text{ N/m}$$

Thus the design consists of the values of c and k determined.

4-15. OSCILLATING SUPPORT

A common source of forced motion is the oscillation of the system support. This condition is often due to isolating equipment from shock and vibration, as, for example, in aircraft, missiles, ships, and ground vehicles. A support condition of this type is represented in Fig. 4-26(a). The absolute displacement of the mass m is designated by x, and the absolute displacement of the support is defined by y. Consider that the support is subjected to the harmonic movement $y = Y \sin \omega_f t$. From the dynamic free-body diagram shown in Fig. 4-26(b), in which x is assumed to be greater than y, and $\dot{x}$ to be larger than $\dot{y}$, the differential equation for the motion of m is

$$m\ddot{x} = -k(x - y) - c(\dot{x} - \dot{y}) \tag{4-97}$$

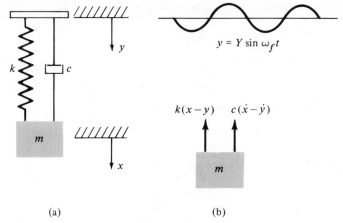

(a)   (b)

Figure 4-26

This may be arranged in the form appropriate for the solution representing the absolute motion of m. Thus

$$m\ddot{x} + c\dot{x} + kx = ky + c\dot{y}$$

$$= kY \sin \omega_f t + c\omega_f Y \cos \omega_f t$$

$$= Y\sqrt{k^2 + (c\omega_f)^2} \sin (\omega_f t - \beta) \qquad (4\text{-}98)$$

where

$$\tan \beta = -\frac{c\omega_f}{k} = -2\zeta r \qquad (4\text{-}99)$$

The forcing condition is represented by the right-hand side of Eq. 4-98, wherein the force amplitude would be $Y\sqrt{k^2 + (c\omega_f)^2}$. From Eqs. 4-48 through 4-50, the steady-state solution is

$$x = \frac{Y\sqrt{k^2 + (c\omega_f)^2}}{\sqrt{(k - m\omega_f^2)^2 + (c\omega_f)^2}} \sin (\omega_f t - \gamma)$$

$$= \frac{Y\sqrt{1 + (2\zeta r)^2}}{\sqrt{(1 - r^2)^2 + (2\zeta r)^2}} \sin (\omega_f t - \gamma) \qquad (4\text{-}100)$$

$$= X \sin (\omega_f t - \gamma)$$

where

$$X = \text{amplitude of } x$$

$$= Y\frac{\sqrt{1 + (2\zeta r)^2}}{\sqrt{(1 - r^2)^2 + (2\zeta r)^2}} \qquad (4\text{-}101)$$

$$\gamma = \beta + \psi \qquad \text{and} \qquad \tan \psi = \frac{2\zeta r}{1 - r^2} \qquad (4\text{-}102)$$

The force carried by the support can be readily determined. This force is defined by

$$F = k(x - y) + c(\dot{x} - \dot{y}) \tag{4-103}$$

which by Eq. 4-97 can be written as

$$F = -m\ddot{x} \tag{4-104}$$

Then substituting the solution (Eq. 4-100) for x gives

$$
\begin{aligned}
F &= m\omega_f^2 X \sin(\omega_f t - \gamma) \\
&= m\omega_f^2 Y \frac{\sqrt{1 + (2\zeta r)^2}}{\sqrt{(1 - r^2)^2 + (2\zeta r)^2}} \sin(\omega_f t - \gamma) \\
&= Yk \frac{\omega_f^2}{k/m} \frac{\sqrt{1 + (2\zeta r)^2}}{\sqrt{(1 - r^2)^2 + (2\zeta r)^2}} \sin(\omega_f t - \gamma) \\
&= Yk \frac{r^2\sqrt{1 + (2\zeta r)^2}}{\sqrt{(1 - r^2)^2 + (2\zeta r)^2}} \sin(\omega_f t - \gamma)
\end{aligned}
\tag{4-105}
$$

$$= F_T \sin(\omega_f t - \gamma) \tag{4-106}$$

where

$$F_T = Yk \frac{r^2\sqrt{1 + (2\zeta r)^2}}{\sqrt{(1 - r^2)^2 + (2\zeta r)^2}} \tag{4-107}$$

Here F_T represents the amplitude, or maximum value, of the transmitted force. Note that the transmitted force F is in phase with the motion of m.

The displacement amplitude can be studied from Eq. 4-101, arranged as

$$\frac{X}{Y} = \frac{\sqrt{1 + (2\zeta r)^2}}{\sqrt{(1 - r^2)^2 + (2\zeta r)^2}} \tag{4-108}$$

This is the same as the transmissibility expression (Eq. 4-90) plotted in Fig. 4-22.

The transmitted force can be interpreted by arranging Eq. 4-107 as

$$\frac{F_T}{Yk} = \frac{r^2\sqrt{1 + (2\zeta r)^2}}{\sqrt{(1 - r^2)^2 + (2\zeta r)^2}} \tag{4-109}$$

The right-hand side is the same form as Eq. 4-95, plotted in Fig. 4-25.

It is observed that as r becomes large the displacement amplitude is small but the force amplitude is large. This is due to the high velocity involved, causing most of the force to be transmitted by the damping mechanism.

Relative Motion

The relative motion of the mass may also be determined here. Referring to Fig. 4-26, note that $x - y = z$ is the motion of mass m relative to the support or base. Substituting this and the time derivatives $\dot{x} - \dot{y} = \dot{z}$ and $\ddot{x} - \ddot{y} = \ddot{z}$ into the differential equation (Eq. 4-97) and rearranging gives

$$m(\ddot{z} + \ddot{y}) = -kz - c\dot{z}$$

$$m\ddot{z} + c\dot{z} + kz = -m\ddot{y} \tag{4-110}$$

$$= m\omega_f^2 Y \sin \omega_f t$$

The right-hand side becomes the forcing condition specified by the motion of the support and the left side defines the motion of mass m relative to the support. The steady-state solution is

$$z = \frac{m\omega_f^2 Y/k}{\sqrt{(1-r^2)^2 + (2\zeta r)^2}} \sin (\omega_f t - \psi)$$

$$= \frac{r^2}{\sqrt{(1-r^2)^2 + (2\zeta r)^2}} Y \sin (\omega_f t - \psi) \tag{4-111}$$

$$= Z \sin (\omega_f t - \psi)$$

where

$$Z = \text{amplitude of } z$$

$$= Y \frac{r^2}{\sqrt{(1-r^2)^2 + (2\zeta r)^2}} \tag{4-112}$$

and

$$\tan \psi = \frac{2\zeta r}{1-r^2}$$

The relative displacement amplitude can be studied from Eq. 4-112, rearranged as

$$\frac{Z}{Y} = \frac{r^2}{\sqrt{(1-r^2)^2 + (2\zeta r)^2}} \tag{4-113}$$

The right-hand side is the same as that for Eq. 4-81 and is plotted in Fig. 4-20.

EXAMPLE 4-9 A flexible system, with a moving base similar to that shown in Fig. 4-26, is composed of a body having a mass of 225 kg and a spring having a modulus of 35 000 N/m. However, the damping is unknown. When the support oscillates with an amplitude of 0.280 cm at the natural frequency of the supported system, the mass exhibits an amplitude of 0.794 cm. Determine the damping constant for the system and the dynamic-force amplitude carried by the support. Also obtain the amplitude of the displacement of the mass relative to the support.

SOLUTION   At resonance, Eq. 4-101 reduces to

$$X = \frac{Y\sqrt{1 + (2\zeta)^2}}{2\zeta}$$

whence

$$0.007\,94 = \frac{0.0028\sqrt{1 + (2\zeta)^2}}{2\zeta}$$

and

$$\zeta = 0.188$$

$$c_c = 2\sqrt{mk} = 2\sqrt{225 \times 35\,000} = 5612 \text{ N} \cdot \text{s/m}$$

$$c = \zeta c_c = 0.188 \times 5612 = 1055 \text{ N} \cdot \text{s/m}$$

From Eq. 4-107,

$$F_T = Yk\frac{\sqrt{1 + (2\zeta)^2}}{2\zeta} = kX = 35\,000 \times 0.007\,94 = 277.9 \text{ N}$$

Also, from Eq. 4-112,

$$Z = \frac{Y}{2\zeta} = \frac{0.0028}{2 \times 0.188} = 0.007\,45 \text{ m} = 0.745 \text{ cm}$$

Note that $Z \neq X - Y$

$$\neq 0.794 - 0.280$$

This is due to the phase difference of x, y, and z.

4-16. ENERGY CONSIDERATIONS FOR FORCED MOTION

For any system containing damping, energy will be dissipated because the damping will do negative work during each cycle. In the case of forced motion, energy will be introduced by means of the positive work done by the driving force during each cycle. For the maintained condition of steady-state motion, these energies must be equal. A consideration of the energy values involved is often helpful in analysis. Accordingly, expressions will be obtained for energy input and energy dissipation for typical conditions.

Energy Input for Harmonic Force

For steady-state motion produced by a forcing condition, the energy put into the system is due to the work done by the force. The expression for $\Delta U'$, the energy input per cycle, can be readily obtained. For the harmonic force $P = P_0 \sin \omega_f t$, the steady-state motion for the case of viscous damping is

$x = X \sin (\omega_f t - \psi)$, so that

$$dx = X\omega_f \cos (\omega_f t - \psi) \, dt$$

Then, from the free-body diagram of Fig. 4-12(b), the energy input per cycle is expressed by

$$\Delta U' = \int P \, dx$$

$$= \int_0^\tau P_0 \sin \omega_f t X\omega_f \cos (\omega_f t - \psi) \, dt$$

$$= P_0 X\omega_f \int_0^{2\pi/\omega_f} \sin \omega_f t (\cos \omega_f t \cos \psi + \sin \omega_f t \sin \psi) \, dt$$

$$= P_0 X\omega_f \int_0^{2\pi/\omega_f} [\cos \psi (\sin \omega_f t \cos \omega_f t) + \sin \psi (\sin^2 \omega_f t)] \, dt$$

$$= P_0 X\omega_f \left[\cos \psi \frac{\sin^2 \omega_f t}{2\omega_f} + \sin \psi \left(\frac{t}{2} - \frac{\sin \omega_f t \cos \omega_f t}{2\omega_f} \right) \right]_0^{2\pi/\omega_f}$$

$$= \pi P_0 X \sin \psi \tag{4-114}$$

Note that the maximum energy input will occur when $\psi = 90$ degrees for the resonant condition, resulting in

$$\Delta U'_{\text{res}} = \pi P_0 X \tag{4-115}$$

Energy Dissipated by Viscous Damping

For steady-state motion of a damped system, energy is dissipated by the damping. The energy $\Delta U''$ dissipated per cycle is equal to the work done during a cycle by the damping. For the viscous case, the damping force is $F_d = -c\dot{x}$. The steady-state relation, $x = X \sin (\omega_f t - \psi)$, for the displacement yields

$$dx = X\omega_f \cos (\omega_f t - \psi) \, dt \qquad \text{and} \qquad \dot{x} = X\omega_f \cos (\omega_f t - \psi)$$

The work is then expressed as follows:

$$\Delta U'' = \int (-F_d) \, dx$$

$$= \int (c\dot{x}) \, dx = c(\omega_f X)^2 \int_0^{2\pi/\omega_f} \cos^2 (\omega_f t - \psi) \, dt$$

$$= c(\omega_f X)^2 \left[\frac{t}{2} + \frac{\sin (\omega_f t - \psi) \cos (\omega_f t - \psi)}{2\omega_f} \right]_0^{2\pi/\omega_f}$$

$$= \pi c\omega_f X^2 \tag{4-116}$$

EXAMPLE 4-10   Since the steady state is a maintained motion, the energy input must equal the energy dissipated per cycle. Verify this for the case of a harmonically forced system with viscous damping.

SOLUTION   Set

a.
$$\Delta U' = \Delta U''$$

Then

$$\pi P_0 X \sin \psi = \pi c \omega_f X^2$$

b.
$$X = \frac{P_0}{c\omega_f} \sin \psi$$

Now

$$\tan \psi = \frac{c\omega_f}{k - m\omega_f^2}$$

so that

c.
$$\sin \psi = \frac{c\omega_f}{\sqrt{(k - m\omega_f^2)^2 + (c\omega_f)^2}}$$

Substituting this into part (b) gives

$$X = \frac{P_0}{c\omega_f} \frac{c\omega_f}{\sqrt{(k - m\omega_f^2)^2 + (c\omega_f)^2}} = \frac{P_0}{\sqrt{(k - m\omega_f^2)^2 + (c\omega_f)^2}}$$

Since this is the correct expression for the amplitude, the relation, part (a), between the energies is verified.

The foregoing principle may be used for other types of damping to determine the amplitude of forced motion for such cases.

4-17. FORCED MOTION WITH COULOMB DAMPING

The arrangement and free-body diagram are shown in Fig. 4-27 for a vibratory system subjected to a harmonic forcing condition and Coulomb damping. An exact analysis of this case is too involved to be presented here; however, an approximate analysis that has experimental validity can be made. In Section 3-10 it was shown that, for a free-vibration system with Coulomb damping, the half-cycle parts of the motion curve were sinusoidal. It is therefore reasonable to assume that if the Coulomb friction is not large, the steady-state motion will be approximately harmonic and of the same frequency as that of the forcing function. It can then be presumed that the motion will be adequately defined by the expression $x = X \sin(\omega_f t - \psi)$ for viscously damped steady-state motion, provided that an equivalent viscous damping constant is introduced. This can be done by relating the energy absorption per cycle for the two cases. The energy dissipation $\Delta U''_\mu$

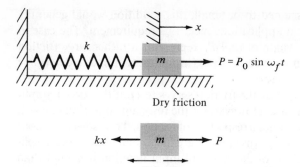

Figure 4-27

per cycle for Coulomb damping would be

$$\Delta U''_\mu = 4FX \tag{4-117}$$

since F is a constant force. Equating this to Eq. 4-116 and designating the equivalent viscous damping constant by c_e results in

$$\Delta U'' = \Delta U''_\mu$$

$$\pi c_e \omega_f X^2 = 4FX$$

$$c_e = \frac{4F}{\pi \omega_f X} \tag{4-118}$$

This may be substituted into the amplitude relation for the viscous case, as follows:

$$X = \frac{X_0}{\sqrt{(1 - r^2)^2 + (2\zeta r)^2}}$$

$$= \frac{X_0}{\sqrt{(1 - r^2)^2 + (c\omega_f/k)^2}} = \frac{X_0}{\sqrt{(1 - r^2)^2 + (4F/\pi Xk)^2}} \tag{4-119}$$

Solving this for the amplitude X gives

$$X^2 \left[(1 - r^2)^2 + \left(\frac{4F}{\pi Xk} \right)^2 \right] = X_0^2$$

$$X = \frac{\sqrt{X_0^2 - (4F/\pi k)^2}}{1 - r^2}$$

$$= X_0 \frac{\sqrt{1 - (4F/\pi P_0)^2}}{1 - r^2} \tag{4-120}$$

This expression for X has a real value, provided that

$$4F < \pi P_0 \quad \text{or} \quad F < \frac{\pi}{4} P_0 \tag{4-121}$$

However, since F was assumed to be small, this condition would generally be fulfilled. Most practical applications meet this requirement. The case in which F approaches the value of $(\pi/4)P_0$ represents a rather large friction force. If $F \geq (\pi/4)P_0$, then an exact analysis† would be required to determine the amplitude, and other factors.

Equation 4-120 shows that the friction serves to limit the forced amplitude for the nonresonant case. However, for the resonant condition of $r = 1$, the amplitude becomes infinite, irrespective of friction. This is because both the energy input and the energy dissipation are linear functions of the amplitude. Furthermore, the slope of the resonant-input function is greater than that of the energy-dissipation function, since $\pi P_0 > 4F$. This is evident from Fig. 4-28. Thus for resonance the energy input is greater than the energy dissipated by friction; there is no intersection of the curves, and the amplitude is not limited. For the nonresonant condition the energy input is $\Delta U' = \pi P_0 X \sin \psi$, so that the input curve is brought into coincidence with the dissipation curve by $\sin \psi$, and the amplitude is limited. Thus the factor that restricts the amplitude is the phase of the motion.

EXAMPLE 4-11   A mass-spring system is subject to Coulomb damping. The dry-friction force is 11 N, the spring constant is 1800 N/m, and the body has a mass of 9 kg. When subjected to a harmonic force having a frequency of 81 cycles/min, the mass develops a sustained oscillation with an amplitude of 3.712 cm. Determine the maximum value of the impressed force.

SOLUTION

$$\omega = \sqrt{\frac{k}{m}} = \sqrt{\frac{1800}{9}} = 14.14 \text{ rad/sec}$$

$$r = \frac{\omega_f}{\omega} = \frac{81 \times 2\pi/60}{14.14} = \frac{8.48}{14.14} = 0.60$$

$$X = \frac{X_0\sqrt{1 - (4F/\pi P_0)^2}}{1 - r^2}$$

$$0.037\,12 = \frac{(P_0/1800)\sqrt{1 - [(4 \times 11)^2/\pi P_0]^2}}{1 - (0.6)^2}$$

$$P_0 = 45.0 \text{ N}$$

4-18. FORCED MOTION WITH HYSTERESIS DAMPING

For free vibration with hysteresis damping, the motion was determined to be approximately harmonic. If the system is forced harmonically, it may be assumed that the steady-state motion will be harmonic. The steady-state

---

† See J. P. Den Hartog, *Trans. ASME*, 53, 1931, p. APM-107.

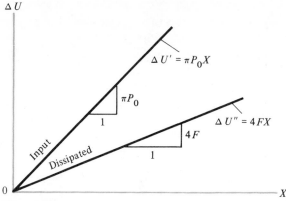

Figure 4-28

amplitude can then be determined by defining an equivalent viscous damping constant based on equating the energies. Thus, assuming the energy dissipated by hysteresis damping to be defined, from Eq. 3-67, as $\Delta U_h'' = \pi b k X^2$ results in

$$\Delta U'' = \Delta U_h''$$

$$\pi c_e \omega_f X^2 = \pi b k X^2$$

$$c_e = \frac{bk}{\omega_f} \quad \text{or} \quad b = \frac{c_e \omega_f}{k} = 2\zeta_e r \tag{4-122}$$

Substituting this into the amplitude relation for the viscous case gives

$$X = \frac{X_0}{\sqrt{(1 - r^2)^2 + b^2}} \tag{4-123}$$

It should be kept in mind that b is generally a small quantity. Two cases are then of interest, depending on whether r is equal to unity or not. If $r \neq 1$, the small value for b can be neglected, and Eq. 4-123 will become

$$X = \frac{X_0}{1 - r^2} \quad \text{for } r < 1$$

$$\tag{4-124}$$

$$X = \frac{X_0}{r^2 - 1} \quad \text{for } r > 1$$

In the first case, the motion is in phase with the force; in the second instance, the motion is of opposite phase to the force. In both cases, the amplitude is limited by the frequency ratio r rather than the damping. In this connection it will be noted that the expressions are the same as those for undamped forced motion.

For the condition of resonance the hysteresis cannot be neglected, and

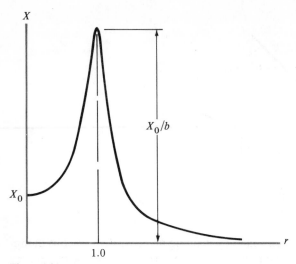

Figure 4-29

Eq. 4-123 will reduce to

$$X = \frac{X_0}{b} = \frac{P_0}{bk} \quad \text{for } r = 1 \tag{4-125}$$

Since X_0 is large compared to b, then X will be large but finite. The hysteresis damping thus serves to limit the amplitude in the region of resonance but has little effect elsewhere. The preceding conditions are represented in Fig. 4-29.

The foregoing assumes that the hysteresis loss varies as the square of the amplitude. If the exponent of the hysteresis-energy expression differs from the value of 2, a more accurate determination can be made of the amplitude for the condition of resonance. In agreement with Eq. 3-67, the hysteresis energy can be expressed as

$$\Delta U_h'' = \pi \beta k X^\gamma \tag{4-126}$$

where b has been replaced by β, as this coefficient is no longer dimensionless. The units for β will be inches$^{(2-\gamma)}$, or meters$^{(2-\gamma)}$ in the SI system. The energy input at resonance, for a harmonic forcing condition, is given by Eq. 4-115. Equating these energies results in

$$\pi \beta k X^\gamma = \pi P_0 X$$

$$X = \left(\frac{P_0}{\beta k}\right)^{1/(\gamma-1)} = \left(\frac{X_0}{\beta}\right)^{1/(\gamma-1)} \tag{4-127}$$

This agrees with Eq. 4-125 for the case of $\gamma = 2$ and $\beta = b$.

EXAMPLE 4-12   A certain mass-spring exhibits hysteresis damping characteristics only. When excited by a harmonic force, it is found that the energy input at resonance is 115.8 in. lb/cycle. The spring constant is 25 lb/in., the hysteresis coefficient is $\beta = 0.030$ in.$^{(2-\gamma)}$, and the exponent is $\gamma = 2.4$. Determine the amplitude of the resonant vibration and of the impressed force.

SOLUTION

$$\Delta U' = \Delta U''_h$$

$$= \pi \beta k X^\gamma$$

$$X = \left(\frac{\Delta U'}{\pi \beta k}\right)^{1/\gamma} = \left(\frac{115.8}{\pi \times 0.030 \times 25}\right)^{1/2.4} = 5.07 \text{ in.}$$

$$\pi P_0 X = \Delta U'$$

$$P_0 = \frac{\Delta U'}{\pi X} = \frac{115.8}{\pi \times 5.07} = 7.28 \text{ lb}$$

4-19. SELF-EXCITED VIBRATION AND INSTABILITY

In the preceding sections, the forcing condition considered was external to the system and independent of the motion. This may be verified by assuming the system to be clamped so that it cannot move. The force will continue and hence is independent of the motion. On the other hand, a forcing condition can be a function of the displacement, velocity, or acceleration of the system. The motion is then said to be self-excited; that is, the motion itself produces the forcing condition or excitation. In this case if the system is clamped, the forcing condition will vanish.

A system is dynamically unstable if the motion diverges, that is, when the displacement or amplitude increases continuously with time. If the motion converges or remains steady, it is dynamically stable. When the self-excitation increases the energy of the system, the motion will diverge and thus will be unstable. The condition of dynamic instability will eventually cause the system to be seriously disrupted or destroyed. Because of this, it is desirable to determine the limiting condition for stability.

Actually, a case of self-excitation was investigated in Section 2-9 in which the nonrestoring force was a function of displacement. The conditions and motion will now be analyzed for a self-exciting force which is a function of the velocity. Examples of this type of self-excited motion include airplane wing flutter, airplane nose-gear shimmy, automobile wheel shimmy, aerodynamically induced motion of bridges, and electric oscillations of electronic equipment.

Consider a single-degree-of-freedom damped spring-mass system to be excited by the force $P = P_0 \dot{x}$ acting on the mass. The differential equation of motion will be

$$m\ddot{x} + c\dot{x} + kx = P_0\dot{x} \tag{4-128}$$

which can be rearranged as

$$\ddot{x} + \left(\frac{c - P_0}{m}\right)\dot{x} + \frac{k}{m}x = 0 \tag{4-129}$$

Taking the solution as $x = Ce^{st}$ and substituting gives the auxiliary equation

$$s^2 + \left(\frac{c - P_0}{m}\right)s + \frac{k}{m} = 0 \tag{4-130}$$

This will have the roots

$$s_{1,2} = \left(\frac{P_0 - c}{2m}\right) \pm \sqrt{\left(\frac{P_0 - c}{2m}\right)^2 - \frac{k}{m}} \tag{4-131}$$

Consider now the case for which $P_0 > c$. This represents a condition of negative damping, as can be noted from the coefficient of the velocity term in Eq. 4-129. Two possibilities prevail here, as follows:

1. For $(k/m) < [(P_0 - c)/2m]^2$, both roots for s are real and positive, and both parts of the solution are exponentials that increase with time. The solution would be written as

$$x = C_1 e^{\{[(P_0 - c)/2m] + \sqrt{[(P_0 - c)/2m]^2 - (k/m)}\}t}$$
$$+ C_2 e^{\{[(P_0 - c)/2m] - \sqrt{[(P_0 - c)/2m]^2 - (k/m)}\}t} \tag{4-132}$$

This represents a diverging nonoscillatory motion, which is unstable.

2. For $(k/m) > [(P_0 - c)/2m]^2$, the roots can be written as

$$s_{1,2} = \frac{P_0 - c}{2m} \pm i\sqrt{\frac{k}{m} - \left(\frac{P_0 - c}{2m}\right)^2} \tag{4-133}$$

and the solution becomes

$$x = Xe^{[(P_0 - c)/2m]t}\sin\left[\sqrt{\frac{k}{m} - \left(\frac{P_0 - c}{2m}\right)^2}\,t + \phi\right] \tag{4-134}$$

Since the exponent is positive, this represents a diverging oscillatory motion, that is, a vibration having an amplitude which increases exponentially with time. Thus the system is dynamically unstable.

If $P_0 < c$, the solution will be of the same form as that for a damped free system, and the motion will be either an aperiodic movement that diminishes with time or a decaying oscillation, both of which are stable. If $P_0 = c$, then the coefficient of $\dot{x}$ vanishes in the differential equation and the solution is that for a free undamped system, which is also stable. Thus the condition for dynamic stability is that

$$P_0 \leq c \tag{4-135}$$

It is simple to verify and interpret this condition physically.

For systems of several degrees of freedom, the dynamic stability cannot be established readily on a physical basis, but must be determined mathematically. This more general approach to the problem will be presented here for the single-degree-of-freedom case, in order to introduce the method in a simple manner. Consider the auxiliary equation in the general form

$$s^2 + a_1 s + a_0 = 0 \tag{4-136}$$

having been obtained for the assumed solution $x = Ce^{st}$. Equation 4-136 will have two roots. For a diverging aperiodic motion, the roots would be real and positive. If a_0 and a_1 are both positive, Eq. 4-136 can have no positive real roots, and this type of dynamic instability cannot occur. On the other hand, a divergent oscillation will occur if the roots of the auxiliary equation are complex with positive real parts. To investigate this, assume the roots to be

$$s_1 = p + iq \qquad s_2 = p - iq \tag{4-137}$$

where p and q are real. The auxiliary equation can be written as

$$(s - s_1) \cdot (s - s_2) = 0$$

whence

$$s^2 - (s_1 + s_2)s + s_1 s_2 = 0 \tag{4-138}$$

Comparison with Eq. 4-136 shows that

$$a_1 = -(s_1 + s_2) \qquad a_0 = s_1 s_2$$

and substituting in Eq. 4-137 gives

$$a_1 = -2p \qquad a_0 = p^2 + q^2 \tag{4-139}$$

For the oscillation to be convergent, p must be negative. This requires only that a_1 be positive. Thus, summarizing the foregoing, if both the coefficients a_0 and a_1 are positive, the motion of the system will be dynamically stable.

PROBLEMS

4-1. An undamped system consists of a body weighing 14.475 lb supported by an elastic member having a modulus of 15 lb/in. It is acted on by a harmonic force having a maximum value of 24 lb. (a) Determine the equilibrium displacement of the spring due to the body weight. (b) Calculate the static displacement of the spring due to the maximum impressed force. (c) Obtain the amplitude of the forced motion for the following forced frequencies f_f: 60, 120, 130, and 300 Hz.

4-2. The mass of an undamped mass-spring system is subjected to a harmonic force having a maximum value of 45 lb and a period of 0.25 sec. The body weighs 2.5 lb and the spring constant is 15 lb/in. Determine the amplitude of the forced motion.

4-3. An undamped system is composed of mass $m = 1.1$ kg and spring $k = 4400$ N/m. It is acted on by a harmonic force having maximum value of 440 N and a frequency of 180 cycles/min. Determine the amplitude of the forced motion.

4-4. A weight of 8 lb is suspended from a spring having a constant of 25 lb/in. The system is undamped but the mass is subjected to a harmonic force with a frequency of 7 Hz, resulting in a forced-motion amplitude of 1.59 in. Determine the maximum value of the impressed force.

4-5. The body of an undamped system is driven by a harmonic force having an amplitude of 36 N and a frequency of 450 cycles/min. The body has a mass of 8 kg and exhibits a forced-displacement amplitude of 4.06 mm. Obtain the value of the spring modulus.

4-6. An undamped system consists of an 8.75-kg mass suspended from a spring having a constant of 3500 N/m. A harmonic force acting on the base and having a maximum value of 187 N causes the system to vibrate with a forced amplitude of 7.6 cm. Determine the frequency of the impressed force.

4-7. A weight W is suspended from a spring having a constant of 25 lb/in. The system is undamped, but the mass is driven by a harmonic force having a frequency of 2 Hz and a maximum value of 10 lb, causing a forced-motion amplitude of 0.5 in. Determine the value of weight W.

4-8. The mass of an undamped system is acted on by force $P = P_0 \cos \omega_f t$, having a maximum value of 10 lb. The spring constant is 3 lb/in. The forced frequency can be any integral multiple of the natural frequency of the system other than 1, thereby avoiding resonance. Only a single forced frequency can exist at a time. Determine the maximum amplitude of the forced motion that could occur.

4-9. Express the relation for the complete motion of an undamped system, excited by a force $P = P_0 \sin \omega_f t$, for the initial conditions of $x = 0$, $\dot{x} = 0$, at $t = 0$. ($r \neq 1$.)

4-10. Write the relation that defines the complete motion of an undamped system, subjected to the harmonic force $P = P_0 \sin \omega_f t$, for the initial conditions of $x = 0$, $\dot{x} = \dot{x}_0$, at $t = 0$. ($r \neq 1$.)

4-11. Express the equation for the complete motion of an undamped system, subjected to the forcing condition $P = P_0 \sin \omega_f t$, for the initial conditions of $x = x_0$, $\dot{x} = 0$, at $t = 0$. ($r \neq 1$.)

4-12. Write the complete solution for the motion of an undamped system, driven by the force $P = P_0 \sin \omega_f t$, for the initial conditions of $x = x_0$, $\dot{x} = \dot{x}_0$, at $t = 0$. ($r \neq 1$.)

4-13. An undamped system consists of a weight of 19.3 lb and a spring having a modulus of 10 lb/in. The system mass is driven at resonance by harmonic force $P = P_0 \sin \omega_f t$ having a maximum value of 4 lb. Determine the amplitude of the forced motion at the end of (a) $\frac{1}{2}$ cycle, (b) $2\frac{1}{2}$ cycles, (c) $4\frac{1}{2}$ cycles, and (d) $6\frac{1}{2}$ cycles.

4-14. An undamped system is composed of a 4.375-kg mass and a spring having a modulus of 3500 N/m. The mass is driven at resonance by a harmonic force having a maximum value of 14 N. Determine the amplitude of the forced motion at the end of (a) 1 cycle, (b) 5 cycles, and (c) 10 cycles.

4-15. An undamped system is harmonically forced near resonance, resulting in a beating condition. The natural frequency of the system is 1765 cycles/min, and the forced frequency is 1752 cycles/min. Determine the beat period of the motion.

4-16. For the arrangement shown, x represents the absolute displacement of the mass m, and y is the absolute displacement of the end A of the spring k. Point A is moved according to the relation $y = Y \sin \omega_f t$. (a) Construct the free-body diagram for m. (b) Write the differential equation of motion for m. (c) Obtain the solution for the steady-state motion of m. (d) Determine the relation for the impressed force at A. (e) Obtain the expression for the force transmitted to the support at B.

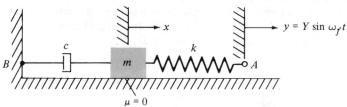

Problem 4-16

4-17. For the arrangement shown, x represents the absolute displacement of the mass m, and y is the absolute displacement of the end A of the dashpot c. The motion of point A is defined by $y = Y \sin \omega_f t$. (a) Construct the free-body diagram for m. (b) Write the differential equation of motion for m. (c) Obtain the solution for the steady-state motion of m. (d) Determine the relation for the impressed force at A. (e) Obtain the expression for the force transmitted to the support at B.

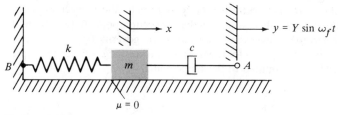

Problem 4-17

4-18. The spring k and the dashpot c of the accompanying diagram are fastened together at A; x represents the absolute displacement of m, and y is the absolute displacement of the point A. The motion of A is defined by $y = Y \sin \omega_f t$. (a) Construct the free-body diagram for m. (b) Write the differential equation of motion for m. (c) Obtain the solution for the steady-state motion of m. (d) Determine the relation for the impressed force at A.

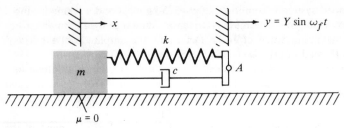

$$\mu = 0$$

Problem 4-18

4-19. For the arrangement shown, x represents the absolute displacement of the mass m, and y is the absolute displacement of the point B. The impressed force P moves m in accordance with the relation $x = X \sin \omega_f t$. (a) Construct the free-body diagram for m. (b) Write the differential equation for the dynamic condition of m. (c) Construct the free-body diagram for the connection point B. (d) Write the differential equation for the connection point B. (e) Obtain the solution for part (d), representing the relation that governs the motion of point B. (f) Determine the relation for the impressed force P. (g) Obtain the expression for the force transmitted to the support at D.

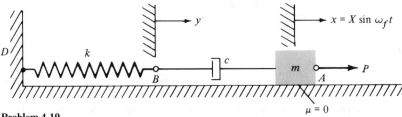

$$\mu = 0$$

Problem 4-19

4-20. For the arrangement shown, x represents the absolute displacement of the mass m, and y is the absolute displacement of the point B. The impressed force P moves m in accordance with the relation $x = X \sin \omega_f t$. (a) Construct the free-body diagram for m. (b) Write the differential equation for the dynamic condition of m. (c) Construct the free-body diagram for the connection point B. (d) Write the differential equation for the connection point B. (e) Obtain the solution for part (d), representing the relation that governs the motion of point B. (f) Determine the relation for the impressed force P. (g) Obtain the expression for the force transmitted to the support at D.

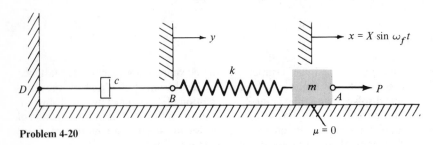

Problem 4-20

$$\mu = 0$$

4-21. A damped system is composed of a mass of 8.75 kg, a spring having a modulus of 1750 N/m, and a dashpot having a damping constant of 37.13 N · s/m. The mass is acted on by a harmonic force $P = P_0 \sin \omega_f t$ having a maximum value of 220 N and a frequency of 4.50 Hz. Using the symbols of X' and ϕ for the arbitrary constants of the transient, write the complete solution representing the motion of the mass.

4-22. A body having a mass of 100 kg moves along a straight line according to the relation

$$x = 0.12 \sin (3t - \beta) + \frac{0.05 \sin (5.96t + \pi/6)}{e^{0.6t}}$$

where x is the displacement along the line in meters and t is time in seconds. (a) Obtain the physical constants and parameters of the system, including appropriate frequencies, and so on. (b) Obtain the initial displacement and velocity.

4-23. For a damped system excited by the harmonic force $P = P_0 \sin \omega_f t$, plot the magnification-factor curve for $\zeta = 0.3$. Do this from $r = 0$ to $r = 4$, carefully determining the peak value.

4-24. For a damped system driven by the harmonic force $P = P_0 \sin \omega_f t$, plot the phase-angle curve for $\zeta = 0.3$. Do this from $r = 0$ to $r = 4$.

4-25. Develop a relation for the ratio of the maximum amplitude to the resonant amplitude for steady-state motion for the case of a harmonically forced damped system.

4-26. For a damped system excited by the harmonic force $P = P_0 \sin \omega_f t$, plot the curve of the steady-state amplitude against the spring constant k for $c = 0.707 m\omega_f$. Carry this out from $k = 0$ to $k = 4m\omega_f^2$.

4-27. A damped system has a spring modulus of 24 lb/in. and a damping constant of 0.88 lb sec/in. It is subjected to a harmonic force having a force amplitude of 15 lb. When excited at resonance, the steady-state amplitude is measured as 2.8409 in. Determine (a) the damping factor, (b) the natural undamped frequency of the system, and (c) the damped natural frequency.

4-28. A damped system is subjected to a harmonic force for which the frequency can be adjusted. It is determined experimentally that at twice the resonant frequency, the steady amplitude is one-tenth of that which occurs at resonance. Determine the damping factor for the system.

4-29. A damped torsional system is composed of a shaft having a torsional spring constant $k_T = 60\,000$ lb in./rad, a disk with a mass moment of inertia $I = 24$ lb in. sec², and a torsional damping device having a torsional damping constant $c_T = 840$ lb in. sec/rad. A harmonic torque with a maximum value of 2700 lb in., acting on the disk, produces a sustained angular oscillation of 3.368-degree amplitude. (a) Determine the frequency of the impressed torque. (b) Obtain the maximum torque transmitted to the support.

4-30. A torsional system consists of a shaft having a torsional spring constant $k_T = 12\,800$ N · m/rad, a disk with a mass moment of inertia $I = 8$ kg · m², and a torsional damping device with a torsional damping constant $c_T = 192$ N · m · s/rad. A harmonic torque with a maximum value of 480 N · m produces a steady angular oscillation of 1.5-degree amplitude. (a) Determine the frequency of the impressed torque. (b) Obtain the maximum torque transmitted to the support.

4-31. A machine having a mass of 70 kg is mounted as shown on springs having a total stiffness of 33 880 N/m. The damping factor is $\zeta = 0.20$. A harmonic force $P = 450 \sin 13.2t$ (where P is in newtons and t is in seconds) acts on the mass. For the sustained or steady-state vibration, determine (a) the amplitude of the motion of the machine, (b) its phase with respect to the exciting force, (c) the transmissibility, (d) the maximum dynamic force transmitted to the foundation, and (e) the maximum velocity of the motion.

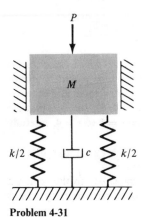

Problem 4-31

4-32. A machine having a total weight of 96.5 lb is mounted as shown on a spring having a modulus of 900 lb/in. and is connected to a dashpot with a damping *factor* of 0.25. The machine contains a rotating unbalance $(w_0 e)$ of 5 lb in. If the speed of rotation is 401.1 rpm, obtain (a) the amplitude of the steady-state motion, (b) the maximum dynamic force transmitted to the foundation, and (c) the angular position of the arm when the structure goes upward through its neutral position.

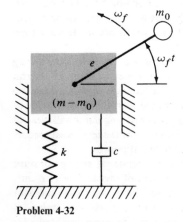

Problem 4-32

4-33. A machine having a total weight of 128.67 lb contains a part that rotates at 859.44 rpm. The machine is supported equally on four identical springs and is connected to a dashpot for which the damping constant is 10 lb sec/in. In operation the machine has a steady amplitude of 0.116 in. and transmits a maximum dynamic force of 174 lb to the supporting base. (a) Determine the modulus for the springs. (b) Calculate the unbalanced moment ($w_0 e$) for the rotating part.

4-34. A machine with a total mass of 50 kg contains a shaft mechanism that rotates at 1800 rpm. The machine is supported by springs for which the equivalent modulus is 20 000 N/m but the damping constant is unknown. Due to the rotating unbalance, the steady-state amplitude is 1.32 cm and the maximum dynamic force carried by the base is 278 N. Determine (a) the damping constant and (b) the unbalanced moment $m_0 e$.

4-35. A typical vibration exciter is composed of two eccentric masses that rotate oppositely, as represented in the top of the accompanying figure. Thus an oscillation is produced in the vertical direction only, since the horizontal effect cancels. This device is used to determine the vibrational characteristics of the structure to which it is attached. The unbalance ($ew_0/2$) of each exciter wheel is 3 lb in. The total arrangement has a weight $mg = 200$ lb. The exciter speed (of eccentric rotation) was adjusted until a stroboscope showed the structure to be moving upward through its equilibrium position at the instant the eccentric weights were at their top position. The exciter speed then was 840 rpm, and the steady amplitude was 0.75 in. Determine (a) the natural frequency of the structure and the damping factor for the structure. If the speed were changed to 1260 rpm, (b) obtain the steady-state amplitude of the structure and (c) the angular position of the eccentrics as the structure moves upward through its equilibrium position.

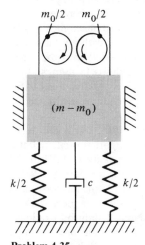

Problem 4-35

4-36. A small compressor weighs 69 lb and runs at 875 rpm. It is to be supported by four springs (equally) and no damping is provided. Design the springs, thus determining k, so that only 15% of the shaking force is transmitted to the supporting foundation or structure.

4-37. Solve Prob. 4-36 considering that damping is also to be included such that the damping factor will be 0.4 for the system. Equal damping at each spring will be provided by a plug or block of energy-absorbing material. Determine the required damping constant as well as the spring constant.

4-38. A refrigerator compressor unit weighs 179 lb and operates at 590 rpm. The maximum dynamic force imposed on the compressor is 20 lb. The unit is to be supported equally by three springs and is to be undamped. (a) Determine the required spring constant if 10% of the dynamic force is to be transmitted to the supporting base. (b) Determine the clearance that must be provided for the unit.

4-39. Solve Prob. 4-38, considering that *each* spring contains a viscous damping absorber having a damping constant of 0.8333 lb sec/in.

4-40. A small machine is known to contain a rotating unbalance on its main shaft. The machine weighs 65 lb and when placed on elastic vibration isolators (which serve as a viscous damping member as well as the elastic support), the static equilibrium displacement is 2.166 in. Also, when the machine is displaced further and released, the subsequent vibration diminishes from an amplitude of 2.350 in. to 0.135 in. in exactly 3 cycles. Operating the machine at resonance produces a sustained amplitude of 0.0193 in. (a) Determine the damping constant for the system. (b) Considering that the shaft diameter is 3 in. so that the eccentric moment arm is 1.5 in., calculate the unbalanced weight. (c) Determine the steady amplitude at an operation speed of twice the resonant speed.

4-41. A damped system with a rotating unbalance is composed of a 96.5-lb body and a spring having a constant of 80 lb/in. but the damping constant is unknown. When operated at resonance, the sustained amplitude is 2.143 in. If the speed of operation is greatly increased, the steady amplitude eventually approaches the value of 0.3572 in. Determine the damping constant for the system.

4-42. The support for a damped system (see Fig. 4-26) is oscillated harmonically with a frequency of 75 cycles/min, causing the mass to vibrate with a steady-state amplitude of 0.6 in. and a maximum dynamic force of 48.75 lb to be transmitted to the support. For the system, the spring modulus is 13 lb/in. and the damping constant is 0.454 lb sec/in. Determine (a) the natural frequency of the system, (b) the mass of the system, and (c) the amplitude of oscillation of the support.

4-43. The support for a damped system is oscillated harmonically. The body weighs 20 lb and the spring constant is 14 lb/in. When the support is oscillated at a rate of 300 cycles/min, the sustained amplitude of the body is 0.793 in. (a) Determine the maximum dynamic force carried by the support. (b) If the support is oscillated at resonance with an amplitude of 1.121 in., the body has a steady amplitude of 3.069 in. Obtain the damping constant for the system.

4-44. A mass-spring system, arranged horizontally, is damped only by Coulomb friction (refer to Fig. 4-27). The system is driven by a harmonic force having a frequency of 54.356 cycles/min and a maximum value of 30 lb, causing the body to oscillate with a steady amplitude of 3.09 in. The spring constant is 12 lb/in. and the body weighs 28.95 lb. Determine the average coefficient of sliding friction for the surfaces involved.

4-45. A machine is designed to produce an abrasive action on a horizontal surface in order to wear and smooth the surface. It consists of a sliding member weighing 20 lb fixed to an elastic member having a stiffness coefficient of 8 lb/in. The average coefficient of sliding friction of the body against the surface is 0.3. A harmonic force having a maximum value of 15 lb actuates the body at a frequency of 0.9, the resonant value. Determine the total excursion or stroke of the member.

4-46. A certain mass-spring system exhibits hysteresis damping only. The spring modulus is 310 lb/in. When excited harmonically at resonance, the steady amplitude is 1.540 in. for an energy input of 32 in. lb. When the resonant energy input is increased to 87 in. lb, the sustained amplitude changes to 2.464 in. Determine the hysteresis coefficient β and the exponent γ.

4-47. A small built-up structure shows solid damping characteristics with $\gamma = 2$. A 1000-lb load on the structure causes a static displacement of 2.500 in., and a supported machine that is subjected to a harmonic force having a constant amplitude of 75 lb produces a resonant amplitude of 6.818 in. (a) Determine the hysteresis damping coefficient β and (b) the energy dissipated per cycle by hysteresis at resonance. (c) Calculate the steady amplitude at twice the resonant frequency and (d) at one-half the resonant frequency.

chapter
5

General Forcing Condition
and Response

5-1. INTRODUCTION

In the preceding chapter, the response of the system to a harmonic forcing function was considered. Although this type of forcing condition is fairly common, forcing functions of a more general nature also occur. These may be periodic or nonperiodic, as indicated in the examples of Fig. 5-1. It is often necessary to determine the response of mass-elastic systems to these general forcing conditions. The periodic case will be considered first.

5-2. FOURIER SERIES AND HARMONIC ANALYSIS

Any periodic function can be written in a series form having harmonic components. Thus a forcing function $P = P(z)$, such as that shown in Fig. 5-1(a), can be expressed as the trigonometric series:

$$P = a_0 + a_1 \cos z + a_2 \cos 2z + \cdots + a_n \cos nz$$
$$+ \cdots + b_1 \sin z + b_2 \sin 2z + \cdots + b_n \sin nz + \cdots \tag{5-1}$$

where n is an integer. This assumes that the various parts of the sine and cosine curves combine so as to properly define the P curve. Equation 5-1 is known as a Fourier series, having been named after the famous French mathematician who showed its correctness in 1822. In order to have the series fit a particular periodic function, it is only necessary to evaluate the coefficients $a_0, a_1, a_2, \ldots, a_n, \ldots; b_1, b_2, \ldots, b_n, \ldots$.

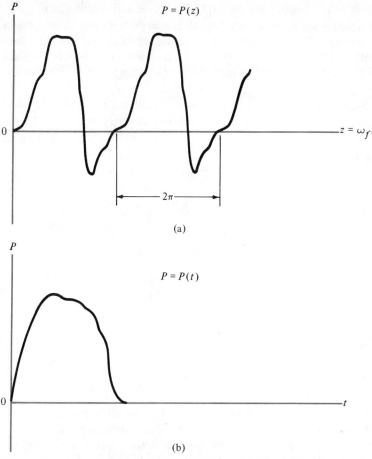

$P = P(z)$

$z = \omega_f t$

2π

(a)

$P = P(t)$

(b)

Figure 5-1. (a) Periodic forcing function. (b) Nonperiodic forcing function.

The coefficient a_0 of the series can be evaluated by multiplying both sides of Eq. 5-1 by dz and integrating over one cycle. Thus

$$\int_0^{2\pi} P\, dz = \int_0^{2\pi} (a_0 + a_1 \cos z + a_2 \cos 2z$$

$$+ \cdots + a_n \cos nz + \cdots + b_1 \sin z + b_2 \sin 2z$$

$$+ \cdots + b_n \sin nz + \cdots)\, dz$$

$$= a_0 z \Big|_0^{2\pi} = 2\pi a_0 \qquad (5\text{-}2)$$

Then solving for a_0 gives

$$a_0 = \frac{1}{2\pi} \int_0^{2\pi} P\, dz \qquad (5\text{-}3)$$

In the integration represented by Eq. 5-2, all sine and cosine terms on the right-hand side vanish, since they are integrated over a cycle or an integral multiple thereof. The significance of a_0 in the series is that it represents a vertical offsetting term of the curve; that is, it offsets the axis of the remaining parts (the sine and cosine terms) by the amount a_0. Stated differently, a_0 is the average ordinate of the curve for a cycle.

The coefficients $a_1, a_2, \ldots, a_n, \ldots$, can be evaluated by multiplying both sides of Eq. 5-1 by $\cos nz \, dz$ (where n denotes the integers 1, 2, 3, ...) and integrating over a cycle. Thus

$$\int_0^{2\pi} P \cos nz \, dz = \int_0^{2\pi} (a_0 \cos nz + a_1 \cos z \cos nz$$

$$+ a_2 \cos 2z \cos nz + \cdots + a_n \cos^2 nz$$

$$+ \cdots + b_1 \sin z \cos nz + b_2 \sin 2z \cos nz$$

$$+ \cdots + b_n \sin nz \cos nz + \cdots) \, dz$$

$$= \int_0^{2\pi} a_n \cos^2 nz \, dz = a_n \left(\frac{z}{2} + \frac{\sin nz \cos nz}{2n} \right) \Big|_0^{2\pi}$$

$$= \pi a_n \tag{5-4}$$

from which

$$a_n = \frac{1}{\pi} \int_0^{2\pi} P \cos nz \, dz \tag{5-5}$$

Note that in the integration represented by Eq. 5-4 all terms on the right-hand side vanish except the $\cos^2 nz$ part. This is because the parts that vanish are cyclic curves of symmetric form, with equal positive and negative areas under the curve during the integration cycle.

The coefficients $b_1, b_2, \ldots, b_n, \ldots$ can be evaluated by multiplying both sides of Eq. 5-1 by $\sin nz \, dz$ (where $n = 1, 2, 3, \ldots$) and integrating over a cycle. Thus

$$\int_0^{2\pi} P \sin nz \, dz = \int_0^{2\pi} (a_0 \sin nz + a_1 \cos z \sin nz$$

$$+ a_2 \cos 2z \sin nz + \cdots + a_n \cos nz \sin nz$$

$$+ \cdots + b_1 \sin z \sin nz + b_2 \sin 2z \sin nz$$

$$+ \cdots + b_n \sin^2 nz + \cdots) \, dz$$

$$= \int_0^{2\pi} b_n \sin^2 nz \, dz = b_n \left(\frac{z}{2} - \frac{\sin nz \cos nz}{2n} \right) \Big|_0^{2\pi}$$

$$= \pi b_n \tag{5-6}$$

and

$$b_n = \frac{1}{\pi} \int_0^{2\pi} P \sin nz \, dz \tag{5-7}$$

Here, all terms on the right side except $\sin^2 nz$ vanish in the integration because of the positive-negative symmetry of the other parts, as explained for the preceding case.

If the function $P = P(z)$ is of a form such that the integrals of Eqs. 5-3, 5-5, and 5-7 can be performed without difficulty, the coefficients of the series can be readily determined.

5-3. JUSTIFICATION OF PRECEDING INTEGRALS

In the foregoing analysis it was observed that most of the terms on the right side of Eqs. 5-2, 5-4, and 5-6 vanished in the integration over a cycle. In order to show this explicitly, let j and k both represent the integers 1, 2, 3, ... Then, subject to the restrictions noted below on j and k, the following can be written

$$\int_0^{2\pi} \cos kz \, dz = 0 \qquad k \neq 0$$

$$\int_0^{2\pi} \sin kz \, dz = 0 \qquad k \neq 0$$

$$\int_0^{2\pi} \cos jz \cos kz \, dz = \frac{1}{2} \left[\frac{\sin (j-k)z}{j-k} + \frac{\sin (j+k)z}{j+k} \right]_0^{2\pi} = 0 \qquad j \neq k$$

$$\int_0^{2\pi} \cos^2 kz \, dz = \left[\frac{z}{2} + \frac{\sin kz \cos kz}{2k} \right]_0^{2\pi} = \pi \qquad k \neq 0$$

$$\int_0^{2\pi} \sin jz \cos kz \, dz = -\frac{1}{2} \left[\frac{\cos (j-k)z}{j-k} + \frac{\cos (j+k)z}{j+k} \right]_0^{2\pi} = 0 \qquad j \neq k$$

$$\int_0^{2\pi} \sin kz \cos kz \, dz = \frac{\sin^2 kz}{2k} \bigg|_0^{2\pi} = 0 \qquad k \neq 0$$

$$\int_0^{2\pi} \sin jz \sin kz \, dz = \frac{1}{2} \left[\frac{\sin (j-k)z}{j-k} - \frac{\sin (j+k)z}{j+k} \right]_0^{2\pi} = 0 \qquad j \neq k$$

$$\int_0^{2\pi} \sin^2 kz \, dz = \left[\frac{z}{2} - \frac{\sin kz \cos kz}{2k} \right]_0^{2\pi} = \pi \qquad k \neq 0$$

Application of the foregoing relations to Eqs. 5-2, 5-4, and 5-6 verifies the integrals of the right-hand side for these equations so that the coefficients a_0, a_n, and b_n are then given by Eqs. 5-3, 5-5, and 5-7.

5-4. RESPONSE TO GENERAL PERIODIC FORCING CONDITION

For a damped vibrational system subjected to a periodic forcing condition of a general type, the differential equation of motion would be

$$m\ddot{x} + c\dot{x} + kx = P$$

$$= a_0 + a_1 \cos \omega_f t + a_2 \cos 2\omega_f t + \cdots$$

$$+ a_n \cos n\omega_f t + \cdots + b_1 \sin \omega_f t + b_2 \sin 2\omega_f t$$

$$+ \cdots + b_n \sin n\omega_f t + \cdots \tag{5-8}$$

The complete differential equation is linear, including the series on the right-hand side. Consequently, the principle of superposition applies, and the particular solution is the sum of the particular solutions obtained by taking each part on the right-hand side separately. The general solution will consist of the complementary function and the sum of the particular solutions. The complementary function will represent a transient that will eventually die out, and the sum of the particular solutions will be the steady state which will remain.

The particular solution for the a_0 term can be verified as

$$x_b = \frac{a_0}{k} \tag{5-9}$$

The solution for the $a_n \cos n\omega_f t$ part would be

$$x_b = \frac{a_n/k}{\sqrt{(1 - r_n^2)^2 + (2\zeta r_n)^2}} \cos \left(n\omega_f t - \psi_n\right) \tag{5-10}$$

where

$$\tan \psi_n = \frac{2\zeta r_n}{1 - r_n^2} \quad \text{and} \quad r_n = nr = n\left(\frac{\omega_f}{\omega}\right) \tag{5-11}$$

The solution for the $b_n \sin n\omega_f t$ part would be

$$x_b = \frac{b_n/k}{\sqrt{(1 - r_n^2)^2 + (2\zeta r_n)^2}} \sin \left(n\omega_f t - \psi_n\right) \tag{5-12}$$

where ψ_n and r_n are given by Eq. 5-11.

The steady-state solution for Eq. 5-8 is then the sum of Eqs. 5-9, 5-10, and 5-12, or

$$x_b = \frac{a_0}{k} + \sum_{n=1}^{\infty} \frac{a_n/k}{\sqrt{(1 - r_n^2)^2 + (2\zeta r_n)^2}} \cos \left(n\omega_f t - \psi_n\right)$$

$$+ \sum_{n=1}^{\infty} \frac{b_n/k}{\sqrt{(1 - r_n^2)^2 + (2\zeta r_n)^2}} \sin \left(n\omega_f t - \psi_n\right) \tag{5-13}$$

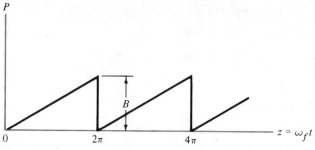

Figure 5-2

The form of the response curve defined by Eq. 5-13 will differ from that of the forcing curve for two reasons: First, the amplitude coefficient of the sine and cosine terms in Eq. 5-13 depends on r_n and hence on n. Second, the phase shift is defined by ψ_n, which also depends on r_n and thus on n.

As n becomes larger, the amplitude coefficient of the sine and cosine parts becomes smaller, so that such parts tend to vanish. Because of this, generally only the first few (perhaps only three or four) terms are sufficient to define the response fairly accurately.

The transient part of the solution can also be included. However, the evaluation of the arbitrary constants is awkward, since it involves the initial value of all parts of the complete solution and of its time derivative, and results in rather complicated coefficients for the transient portion of the complete solution. A different approach, with less complications, can be used to express the complete solution. This will be explained in a later section.

In order to illustrate the foregoing, consider the case defined by the curve of Fig. 5-2, representing the forcing condition P. For such a plot, ω_f is first defined by dividing the cycle length of 2π by the time for one cycle. That is,

$$\omega_f = \frac{2\pi}{\tau} \tag{5-14}$$

The plot can then be made with $\omega_f t$ as the independent variable. The forcing function can be defined within the first cycle as

$$P = \frac{B}{2\pi} z \qquad 0 \le z \le 2\pi \tag{5-15}$$

where, for convenience, $z = \omega_f t$. The coefficients of the Fourier series can now be obtained, from Eqs. 5-3, 5-5, and 5-7, as follows:

$$a_0 = \frac{1}{2\pi} \int_0^{2\pi} P \, dz$$

$$= \frac{1}{2\pi} \int_0^{2\pi} \frac{B}{2\pi} z \, dz = \frac{B}{(2\pi)^2} \cdot \frac{z^2}{2} \bigg|_0^{2\pi} = \frac{B}{2} \tag{5-16}$$

$$a_n = \frac{1}{\pi} \int_0^{2\pi} P \cos nz \, dz$$

$$= \frac{1}{\pi} \int_0^{2\pi} \frac{B}{2\pi} z \cos nz \, dz = \frac{B}{2\pi^2} \left[\frac{\cos nz}{n^2} + \frac{z \sin nz}{n} \right]_0^{2\pi} = 0 \qquad (5\text{-}17)$$

$$b_n = \frac{1}{\pi} \int_0^{2\pi} P \sin nz \, dz$$

$$= \frac{1}{\pi} \int_0^{2\pi} \frac{B}{2\pi} z \sin nz \, dz = \frac{B}{2\pi^2} \left[\frac{\sin nz}{n^2} - \frac{z \cos nz}{n} \right]_0^{2\pi} = -\frac{B}{n\pi} \qquad (5\text{-}18)$$

(where $n = 1, 2, 3, \ldots$). The series then becomes

$$P = \frac{B}{2} - \frac{B}{\pi} \sin \omega_f t - \frac{B}{2\pi} \sin 2\omega_f t - \cdots$$

$$= \frac{B}{\pi} \left[\frac{\pi}{2} - \left(\sin \omega_f t + \frac{1}{2} \sin 2\omega_f t + \cdots \right) \right] \qquad (5\text{-}19)$$

The first three terms of the series are plotted in Fig. 5-3(a), where for convenience, B has been taken as π. Observe how well this approaches the sawtooth shape of Fig. 5-2, even for the small number of terms included. Also note how the a_0 term offsets the axis of the remaining parts upward by $B/2$.

The steady response to the forcing function defined by Eq. 5-19 would be

$$x = \frac{B}{\pi} \left\{ \frac{\pi}{2k} - \left[\frac{1/k}{\sqrt{(1 - r_1^2)^2 + (2\zeta r_1)^2}} \sin (\omega_f t - \psi_1) \right. \right.$$

$$\left. \left. + \frac{1/k}{2\sqrt{(1 - r_2^2)^2 + (2\zeta r_2)^2}} \sin (2\omega_f t - \psi_2) + \cdots \right] \right\} \qquad (5\text{-}20)$$

where

$$\tan \psi_1 = \frac{2\zeta r_1}{1 - r_1^2}, \quad \tan \psi_2 = \frac{2\zeta r_2}{1 - r_2^2}, \quad \ldots$$

and

$$r_1 = \frac{\omega_f}{\omega}, \quad r_2 = 2\frac{\omega_f}{\omega}, \quad \ldots \qquad (5\text{-}21)$$

This can be written as

$$x = \frac{B}{\pi k} \left\{ \frac{\pi}{2} - \left[\frac{\sin (\omega_f t - \psi_1)}{\sqrt{(1 - r^2)^2 + (2\zeta r)^2}} \right. \right.$$

$$\left. \left. + \frac{\sin (2\omega_f t - \psi_2)}{2\sqrt{(1 - 4r^2)^2 + (4\zeta r)^2}} + \cdots \right] \right\} \qquad (5\text{-}22)$$

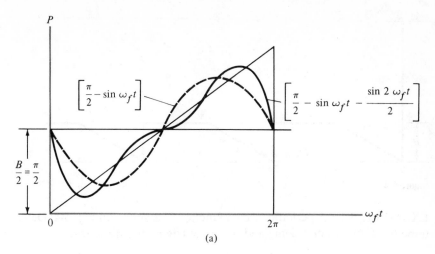

(a)

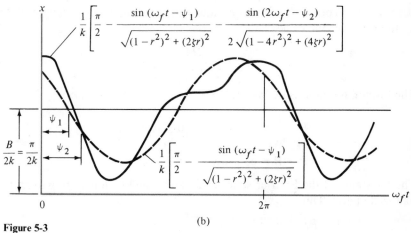

(b)

Figure 5-3

where

$$\tan \psi_1 = \frac{2\zeta r}{1 - r^2}, \quad \tan \psi_2 = \frac{4\zeta r}{1 - 4r^2}, \quad \cdots \qquad (5\text{-}23)$$

and

$$r = \frac{\omega_f}{\omega}$$

The response curve represented by Eq. 5-22 is plotted in Fig. 5-3(b), with terms only through the second sine part included. This curve differs in form from the forcing curve for two reasons: first, because the amplitude coefficient changes for each part and, second, due to the difference in phase shift of each part.

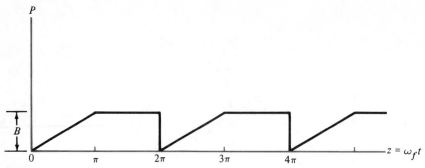

Figure 5-4

EXAMPLE 5-1   A periodic forcing function P is shown in Fig. 5-4. Determine the Fourier coefficients and write the trigonometric series for P.

SOLUTION   The value of ω_f is obtained by dividing the cycle length of 2π by the time for one cycle. Thus

$$\omega_f = \frac{2\pi}{\tau}$$

The forcing function is defined as

$$P = \frac{B}{\pi} z \qquad\qquad 0 \le \omega_f t \le \pi$$

$$P = B = \text{constant} \qquad \pi \le \omega_f t \le 2\pi$$

The constants of the Fourier series can be obtained by integrating over the half-cycle parts. Thus

$$a_0 = \frac{1}{2\pi} \left\{ \int_0^\pi \frac{Bz}{\pi} dz + \int_\pi^{2\pi} B \, dz \right\}$$

$$= \frac{B}{2\pi} \left\{ \frac{z^2}{2\pi} \Big|_0^\pi + z \Big|_\pi^{2\pi} \right\} = \frac{3}{4} B$$

$$a_n = \frac{1}{\pi} \left\{ \int_0^\pi \frac{Bz}{\pi} \cos nz \, dz + \int_\pi^{2\pi} B \cos nz \, dz \right\}$$

$$= \frac{B}{\pi} \left\{ \frac{1}{\pi} \left[\frac{\cos nz}{n^2} + \frac{z \sin nz}{n} \right]_0^\pi + \left[\frac{\sin nz}{n} \right]_\pi^{2\pi} \right\}$$

$$= \frac{B}{\pi} \left\{ \frac{1}{\pi} \left[\frac{(-1)}{n^2} - \frac{(1)}{n^2} \right] \right\} = \frac{-2B}{n^2\pi^2} \qquad \text{For } n \text{ odd}$$

$$= \frac{B}{\pi} \left\{ \frac{1}{\pi} \left[\frac{(1)}{n^2} - \frac{(1)}{n^2} \right] \right\} = 0 \qquad \text{For } n \text{ even}$$

$$b_n = \frac{1}{\pi}\left\{\int_0^\pi \frac{Bz}{\pi}\sin nz\, dz + \int_\pi^{2\pi} B\sin nz\, dz\right\}$$

$$= \frac{B}{\pi}\left\{\frac{1}{\pi}\left[\frac{\sin nz}{n^2} - \frac{z\cos nz}{n}\right]_0^\pi - \left[\frac{\cos nz}{n}\right]_\pi^{2\pi}\right\}$$

$$= \frac{B}{\pi}\left\{\frac{1}{\pi}\left[\frac{-\pi(-1)}{n}\right] - \left[\frac{(1)}{n} - \frac{(-1)}{n}\right]\right\} = -\frac{B}{n\pi} \qquad \text{For } n \text{ odd}$$

$$= \frac{B}{\pi}\left\{\frac{1}{\pi}\left[\frac{-\pi(1)}{n}\right] - \left[\frac{(1)}{n} - \frac{(1)}{n}\right]\right\} = -\frac{B}{n\pi} \qquad \text{For } n \text{ even}$$

The trigonometric series for P is then

$$P = B\left\{\frac{3}{4} - \frac{2}{\pi^2}\left[\cos z + \frac{\cos 3z}{9} + \frac{\cos 5z}{25} + \cdots\right]\right.$$

$$\left. - \frac{1}{\pi}\left[\sin z + \frac{\sin 2z}{2} + \frac{\sin 3z}{3} + \cdots\right]\right\}$$

5-5. CYCLIC FORCING CONDITION OF IRREGULAR FORM

In certain instances, the forcing condition may be obtained experimentally and cannot be expressed analytically; that is, the curve is of irregular form which can be represented graphically but not as an explicit function $P(t)$. An example of this is shown in Fig. 5-5, exhibiting a curve of irregular form but one that is repeated. It is possible to obtain the Fourier coefficients by arithmetical summation and then write the trigonometric series.

The curve can be approximated properly by dividing the cycle length 2π into a sufficient number of intervals. If the number of intervals is designated by v, then each segment has a width $q = 2\pi/v$. The abscissa of the jth interval is jq, and its ordinate may be designated as P_j. The Fourier coefficients are then expressed as follows:

$$a_0 = \frac{1}{2\pi}\sum_{j=1}^{v} P_j q = \frac{1}{v}\sum_{j=1}^{v} P_j \tag{5-24}$$

$$a_n = \frac{1}{\pi}\sum_{j=1}^{v} P_j[\cos(n\cdot jq)]q = \frac{2}{v}\sum_{j=1}^{v} P_j\cos(njq) \tag{5-25}$$

$$b_n = \frac{1}{\pi}\sum_{j=1}^{v} P_j[\sin(n\cdot jq)]q = \frac{2}{v}\sum_{j=1}^{v} P_j\sin(njq) \tag{5-26}$$

The terms v and q will be numerical values determined by dividing the curve into segments. Then to obtain a certain coefficient such as b_5, the ordinate P_j and the term $\sin 5jq$ will depend on the particular segment j (for example,

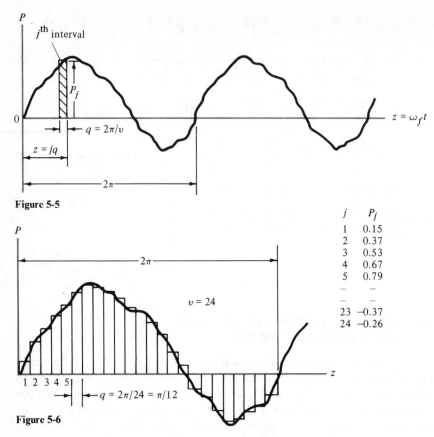

Figure 5-5

j	P_j
1	0.15
2	0.37
3	0.53
4	0.67
5	0.79
–	–
23	–0.37
24	–0.26

Figure 5-6

the third) in mind, and thus will yield numerical values. Summing the value of $P_j \sin 5jq$ for all the segments will yield the numerical value of coefficient b_5. In this manner all b_n coefficients can be obtained. Coefficients a_0 and a_n may be similarly evaluated. The Fourier series can then be written as Eq. 5-1.

EXAMPLE 5-2 An experimentally obtained periodic curve is shown in Fig. 5-6. The curve has been divided into 24 parts and the ordinates of the segments are listed. From these data, determine the value of the Fourier coefficient b_3 in the trigonometric series.

SOLUTION

$$b_3 = \frac{2}{24} \left[0.15 \sin \left(3 \times 1 \times \frac{\pi}{12} \right) + 0.37 \sin \left(3 \times 2 \times \frac{\pi}{12} \right) \right.$$
$$\left. + 0.53 \sin \left(3 \times 3 \times \frac{\pi}{12} \right) + \cdots - 0.26 \sin \left(3 \times 24 \times \frac{\pi}{12} \right) \right]$$
$$= -0.017$$

Thus the third sine term in the series will be $-0.017 \sin 3z$.

5-6. **RESPONSE TO GENERAL DISTURBING FORCE**

In the preceding sections the steady-state response to a general periodic forcing condition was considered. The method outlined was convenient for such determinations, but it would be cumbersome if the complete solution were to be expressed by including the transient. A different approach, outlined in the sections that follow, is advantageous for determining the complete motion of a system.

Also of importance is the response of a mass-elastic system to a nonperiodic force, representing the most general type of forcing condition. In general, the Fourier analysis is not suitable for this problem. The method presented below is generally convenient for this condition.

5-7. **RESPONSE TO UNIT IMPULSE**

Consider the undamped single-degree-of-freedom system shown in Fig. 5-7(a). Assume that this is subjected to an instantaneous force of large magnitude but short duration, and that the linear impulse, represented by the area within the force-time curve shown in Fig. 5-7(b), is unity. Since the force acts for instantaneous time only, the equation of motion is

$$m\ddot{x} = -kx$$

$$\ddot{x} + \frac{k}{m}x = 0 \tag{5-27}$$

If the mass m is at rest (so that $x = 0 = \dot{x}$ for $t < 0$), and the unit impulse is imposed at $t = 0$, the initial conditions are

$$\left.\begin{array}{c} x = 0 \\[2mm] m\dot{x} = 1 \\[2mm] \dot{x} = \dfrac{1}{m} \end{array}\right\} \quad \text{at } t = 0 \tag{5-28}$$

or

The condition $m\dot{x} = 1$ is based on equating the linear momentum change $m\dot{x}$ to the linear impulse of unity which is imposed at $t = 0$.

The solution to the differential equation is

$$x = A \sin \omega t + B \cos \omega t \tag{5-29}$$

Determining the arbitrary constants from the initial conditions results in

$$A = \frac{1}{m\omega} = \frac{1}{\sqrt{mk}} \quad \text{and} \quad B = 0$$

so that

$$x = \frac{1}{\sqrt{mk}} \sin \omega t = h(t) \tag{5-30}$$

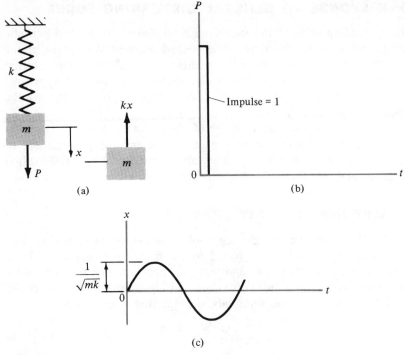

Figure 5-7

The term $h(t)$ is called either the impulsive response or the *response to a unit impulse,* and it represents the response, or motion change, due to a unit impulse being imposed on the system. This is shown in Fig. 5-7(c).

5-8. DUHAMEL'S INTEGRAL

Consider next a force of a general nature, defined by the curve of Fig. 5-8(a), to be applied to the system. Assume that this force is composed of a series of discrete segments, as shown. Time is now defined as follows:

τ = time at which the impulse represented by the general
　　segment (shown shaded) is imposed†
t = time at which the response is measured
$t - \tau$ = elapsed time from the moment the impulse is imposed
　　　　until the response is measured

For the general segment of width $d\tau$ and height P (a variable defined by the curve), the impulse is $P\,d\tau$. For this impulse, from Eq. 5-30, the response will

---

† The use of τ, in Sections 5-8 through 5-9, for *time* (a variable) is not to be confused with the same symbol representing *period*, as employed elsewhere in the text.

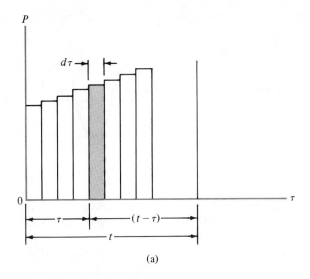

(a)

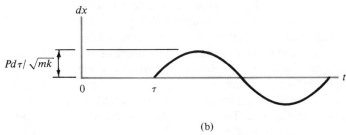

(b)

Figure 5-8

be

$$dx = (P\,d\tau)\frac{1}{\sqrt{mk}}\sin \omega(t-\tau)$$

$$= \frac{1}{\sqrt{mk}}P \sin \omega(t-\tau)\,d\tau \qquad (5\text{-}31)$$

It should be noted that the elapsed time is used here. The response is dx, representing the change in x due to the incremental impulse, as there would already be motion due to the impulse of previous segments. In order to obtain the response to all the segments, Eq. 5-31 would be summed or integrated over the entire interval, so that

$$\int dx_{\scriptscriptstyle\bullet} = \int \frac{1}{\sqrt{mk}}P \sin \omega(t-\tau)\,d\tau$$

$$x = \frac{1}{\sqrt{mk}}\int_{0}^{t} P \sin \omega(t-\tau)\,d\tau \qquad (5\text{-}32)$$

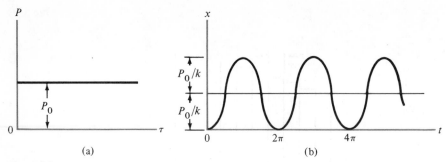

Figure 5-9

This is known as Duhamel's integral, or the convolution integral. An alternative method of deriving this relation employs the concept of indicial admittance, representing the motion due to imposing a unit-step force.[†]

Careful note should be made that τ is the independent variable; accordingly, the integration is performed with respect to τ, with t being treated as a constant in this process. In applying Eq. 5-32 to determine the response of the system, it is presumed that the function $P = P(\tau)$ is explicitly defined and that the integration represented can be performed conveniently. The response determined in this manner will be the complete motion, but only for the initial conditions of zero displacement and zero velocity.

If the system has initial motion when the forcing condition is imposed, the complete motion will be a continuation of the initial motion plus the response as defined by Eq. 5-32. Thus if the initial displacement and velocity are x_0 and $\dot{x}_0$, the complete motion is given by

$$x = \frac{\dot{x}_0}{\omega}\sin \omega t + x_0 \cos \omega t + \frac{1}{\sqrt{mk}}\int_0^t P \sin \omega(t - \tau)\, d\tau \qquad (5\text{-}33)$$

EXAMPLE 5-3 Determine the response of an undamped system to the force represented in Fig. 5-9(a).

SOLUTION The forcing condition is defined by $P = P_0 = $ constant. Then, from Eq. 5-32,

$$x = \frac{1}{\sqrt{mk}}\int_0^t P_0 \sin \omega(t - \tau)\, d\tau$$

$$= \frac{P_0}{\sqrt{mk}}\left[\frac{\cos \omega(t - \tau)}{\omega}\right]_0^t = \frac{P_0}{k}(1 - \cos \omega t)$$

This motion will plot as shown in Fig. 5-9(b).

[†] See T. von Karman and M. A. Biot, *Mathematical Methods in Engineering* (New York: McGraw-Hill, 1940), pp. 397–398, 403–404.

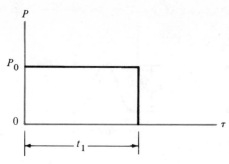

Figure 5-10

EXAMPLE 5-4   Obtain the response to the constant force P_0 applied for the time t_1, as shown in Fig. 5-10.

SOLUTION   The force is defined by

$$P = P_0 \qquad 0 \le \tau \le t_1$$

$$P = 0 \qquad \tau \ge t_1$$

In order to determine the response at some time after t_1 (that is, for $t \ge t_1$), Duhamel's integral can be set up as

$$x = \frac{1}{\sqrt{mk}} \int_0^{t_1} P_0 \sin \omega(t - \tau)\, d\tau$$

$$+ \frac{1}{\sqrt{mk}} \int_{t_1}^{t} [0] \sin \omega(t - \tau)\, d\tau \qquad \text{where } t \ge t_1$$

$$= \frac{P_0}{\sqrt{mk}} \left[\frac{\cos \omega(t - \tau)}{\omega} \right]_0^{t_1}$$

$$= \frac{P_0}{k} [\cos \omega(t - t_1) - \cos \omega t]$$

The motion depends on the value of t_1. Thus, for example,

a.

$$x = -2\frac{P_0}{k} \cos \omega t \qquad \text{for } t_1 = \frac{\pi}{\omega}$$

b.

$$x = 0 \qquad \text{for } t_1 = \frac{2\pi}{\omega}$$

c.

$$x = \frac{P_0}{k} (+\sin \omega t - \cos \omega t)$$

$$= \sqrt{2}\frac{P_0}{k} \left[\sin \left(\omega t - \frac{\pi}{4} \right) \right] \qquad \left. \right\} \quad \text{for } t_1 = \frac{\pi}{2\omega}$$

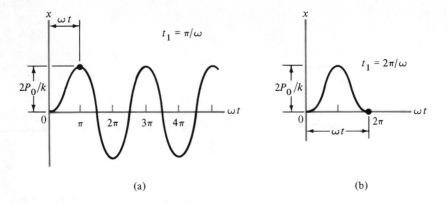

(a)

(b)

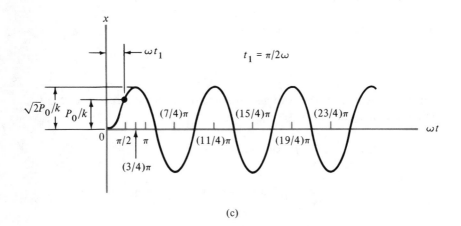

(c)

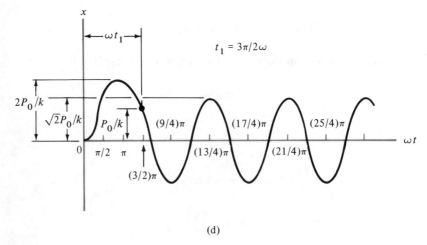

(d)

Figure 5-11

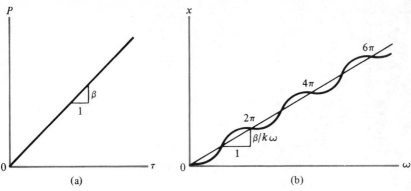

Figure 5-12

d.
$$x = \frac{P_0}{k}(-\sin \omega t - \cos \omega t)$$

$$= \sqrt{2}\,\frac{P_0}{k}\left[-\sin\left(\omega t + \frac{\pi}{4}\right)\right]$$

$$\text{for } t_1 = \frac{3}{2}\frac{\pi}{\omega}$$

In each case the motion from $t = 0$ to $t = t_1$ is obtained from the solution for Example 5-3. The motion for these examples is shown in Fig. 5-11.

EXAMPLE 5-5 Determine the response of an undamped system to the forcing function $P = \beta\tau$, shown in Fig. 5-12(a).

SOLUTION Duhamel's integral is then

$$x = \frac{1}{\sqrt{mk}}\int_0^t \beta\tau \sin \omega(t - \tau)\,d\tau$$

$$= \frac{\beta}{\sqrt{mk}}\left[\frac{\sin \omega(t - \tau)}{\omega^2} + \frac{\tau \cos \omega(t - \tau)}{\omega}\right]_0^t$$

$$= \frac{\beta}{\sqrt{mk}}\left(\frac{t}{\omega} - \frac{\sin \omega t}{\omega^2}\right)$$

$$= \frac{\beta}{k\omega}(\omega t - \sin \omega t)$$

Figure 5-12(b) shows the motion represented by this expression.

5-9. RESPONSE OF DAMPED SYSTEM TO GENERAL FORCING CONDITION

The approach presented in Section 5-8 may be applied to a viscously damped system. Consider that the system of Fig. 5-13(a) is subjected to a unit impulse at time zero. The equation of motion is

$$m\ddot{x} + c\dot{x} + kx = 0 \qquad (5\text{-}34)$$

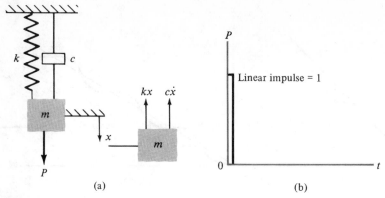

(a)

(b)

Figure 5-13

If m is at rest previous to time zero and the unit impulse is imposed on the mass at $t = 0$, the initial conditions are

$$\left. \begin{array}{c} x = 0 \\ m\dot{x} = 1 \\ \dot{x} = \dfrac{1}{m} \end{array} \right\} \quad \text{at } t = 0 \qquad (5\text{-}35)$$

or

For subcritical damping, the solution to Eq. 5-34 is

$$x = e^{-\zeta\omega t}(A \sin \omega_d t + B \cos \omega_d t) \qquad (5\text{-}36)$$

From the initial conditions, the arbitrary constants are found to be

$$A = \frac{1}{m\omega_d} \quad \text{and} \quad B = 0$$

The response is then

$$x = \frac{1}{m\omega_d} e^{-\zeta\omega t} \sin \omega_d t \qquad (5\text{-}37)$$

Referring to Fig. 5-14, for the forcing function $P = P(\tau)$ the response at time t to the impulse $P \, d\tau$ of the general segment imposed at time τ would be

$$dx = \frac{1}{m\omega_d} e^{-\zeta\omega(t-\tau)} \sin \omega_d(t - \tau)P \, d\tau \qquad (5\text{-}38)$$

and the complete response would be

$$x = \frac{1}{m\omega_d} \int_0^t P e^{-\zeta\omega(t-\tau)} \sin \omega_d(t - \tau) \, d\tau \qquad (5\text{-}39)$$

EXAMPLE 5-6 Determine the response of a viscously damped system to the force $P = P_0 = $ constant, represented in Fig. 5-15.

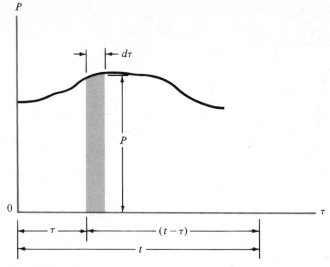

Figure 5-14

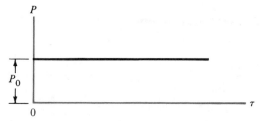

Figure 5-15

SOLUTION   From Eq. 5-39, the motion would be expressed as

$$x = \frac{P_0}{m\omega_d} \int_0^t e^{-\zeta\omega(t-\tau)} \sin \omega_d(t - \tau) \, d\tau$$

$$= \frac{P_0}{m\omega_d} \left\{ \frac{e^{-\zeta\omega(t-\tau)}[\zeta\omega \sin \omega_d(t - \tau) + \omega_d \cos \omega_d(t - \tau)]}{(\zeta\omega)^2 + \omega_d^2} \right\}\bigg|_0^t$$

$$= \frac{P_0}{m\omega_d} \left[\frac{\omega_d - e^{-\zeta\omega t}(\zeta\omega \sin \omega_d t + \omega_d \cos \omega_d t)}{(\zeta\omega)^2 + \omega_d^2} \right]$$

$$= \frac{P_0}{k\sqrt{1 - \zeta^2}} [\sqrt{1 - \zeta^2} - e^{-\zeta\omega t}(\zeta \sin \omega_d t + \sqrt{1 - \zeta^2} \cos \omega_d t)]$$

$$= \frac{P_0}{k\sqrt{1 - \zeta^2}} [\sqrt{1 - \zeta^2} - e^{-\zeta\omega t} \cos (\omega_d t - \beta)]$$

where $\tan \beta = \zeta/\sqrt{1 - \zeta^2}$. This response will plot as shown in Fig. 5-16. Note that increased damping reduces the magnitude of the movement and increases the period of the motion.

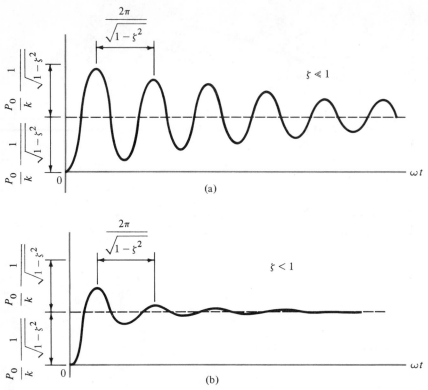

Figure 5-16. (a) For small damping, ($\zeta \ll 1$). (b) For medium damping, ($\zeta < 1$).

5-10. IRREGULAR FORCING CONDITION AND RESPONSE; NUMERICAL ANALYSIS

It is possible that the forcing condition may be determined experimentally, resulting in a curve of irregular form that cannot be expressed as an explicit function. In this case the curve can be broken into segments and the response determined by a summation corresponding to the convolution integral. However, this is generally unsatisfactory as it requires such a summation, involving numerous trigonometric terms, for the evaluation of each point on the response curve.

A preferable method is the numerical integration of the differential equation of motion. The displacement and derivative terms are evaluated progressively in time increments from the differential equation and finite-difference relations, starting from a time when the motion conditions are known and proceeding for the desired time span. Thus numerical values of the integral are obtained at regular intervals along the time axis. Although

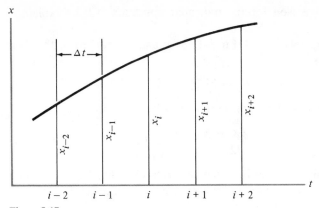

Figure 5-17

the solution obtained is approximate, as the time interval is reduced the determinations approach those for the exact solution.†

Finite differences are algebraic approximations for the derivatives of a function. These may be obtained by the Taylor series. Referring to Fig. 5-17, in which the curve $x = x(t)$ is divided into time increments Δt and points are designated by the index i, as indicated, the Taylor expansions for x_{i+1}, x_{i-1}, and x_{i-2} in terms of the central point i are written as follows:

$$x_{i+1} = x_i + \dot{x}_i(\Delta t) + \ddot{x}_i \frac{(\Delta t)^2}{2!} + \dddot{x}_i \frac{(\Delta t)^3}{3!} + \ddddot{x}_i \frac{(\Delta t)^4}{4!} + \cdots \qquad (5\text{-}40)$$

$$x_{i-1} = x_i - \dot{x}_i(\Delta t) + \ddot{x}_i \frac{(\Delta t)^2}{2!} - \dddot{x}_i \frac{(\Delta t)^3}{3!} + \ddddot{x}_i \frac{(\Delta t)^4}{4!} - \cdots \qquad (5\text{-}41)$$

$$x_{i-2} = x_i - \dot{x}_i(2\Delta t) + \ddot{x}_i \frac{(2\Delta t)^2}{2!} - \dddot{x}_i \frac{(2\Delta t)^3}{3!} + \ddddot{x}_i \frac{(2\Delta t)^4}{4!} - \cdots \qquad (5\text{-}42)$$

Also, the following identity is included:

$$x_i \equiv x_i \qquad (5\text{-}43)$$

Adding Eqs. 5-40 and 5-41 results in

$$\ddot{x}_i(\Delta t)^2 = x_{i+1} - 2x_i + x_{i-1} - \ddddot{x}_i \frac{(\Delta t)^4}{12} - \cdots \qquad (5\text{-}44)$$

Ignoring higher-order terms and rearranging gives

$$\ddot{x}_i = \frac{x_{i+1} - 2x_i + x_{i-1}}{(\Delta t)^2} \qquad i \geq 2 \qquad (5\text{-}45)$$

† For further explanation of finite differences and related numerical methods see M. G. Salvadori and M. L. Baron, *Numerical Methods in Engineering* (Englewood Cliffs, N.J.: Prentice-Hall, 1952), chaps. 2 and 3.

(The meaning and reason for the indicated limitation will be explained later.)

Multiplying Eq. 5-41 by −4, Eq. 5-42 by 1, and Eq. 5-43 by 3, and then adding these, yields

$$3x_i - 4x_{i-1} + x_{i-2} = 2\dot{x}_i(\Delta t) - \tfrac{2}{3}\dddot{x}_i(\Delta t)^3 + \cdots \qquad (5\text{-}46)$$

Discarding the higher-order terms here results in

$$\dot{x}_i = \frac{3x_i - 4x_{i-1} + x_{i-2}}{(2\Delta t)} \qquad i \geq 3 \qquad (5\text{-}47)$$

(The indicated limitation will be explained subsequently.) Equations 5-45 and 5-47 express the derivatives in terms of finite differences of the function, or ordinates of the $x = x(t)$ curve.

For purposes of the numerical solution procedure, Eq. 5-47 can be used to calculate the velocity $\dot{x}_i$ at point i.

For a damped mass-elastic system subjected to the forcing condition $P(t)$, the differential equation of motion may be written in the form

$$\ddot{x}_i = \frac{P(t_i)}{m} - \frac{c}{m}\dot{x}_i - \frac{k}{m}x_i \qquad (5\text{-}48)$$

The acceleration at point i can then be calculated.

The displacement at the next point $i + 1$ can now be determined from Eq. 5-45 rearranged as

$$x_{i+1} = \ddot{x}_i(\Delta t)^2 + 2x_i - x_{i-1} \qquad i \geq 2 \qquad (5\text{-}49)$$

Thus by successive use of Eqs. 5-47, 5-48, and 5-49, calculations can proceed from point i to the next point $i + 1$. However, unless there is some form of initial motion or force, the calculations cannot be started and an independent condition must be established.

Because of the prospect of setting up a digital computer program for solving the problem, the subscript 1 will be used for the initial point as subzero is not available. Accordingly, the reason for the indicated limitation of $i \geq 2$ for Eqs. 5-45 and 5-49 and $i \geq 3$ for Eq. 5-47 is that the finite difference relations cannot involve values of the function preceding the initial point on the curve since these values are unavailable.

Initial Conditions

Consider the case for which the initial displacement and velocity are zero. For a small time increment, the acceleration can be assumed to increase uniformly during the first time interval. Then for time t during the first interval the acceleration may be expressed as

$$\ddot{x} = \ddot{x}_1 + \eta t \qquad (5\text{-}50)$$

where $\ddot{x}_1$ is the initial acceleration and η is an unknown constant. Integrating to obtain the velocity $\dot{x}$ at time t within the interval results in

$$\dot{x} = \int_0^t \ddot{x}\, dt = \int_0^t (\ddot{x}_1 + \eta t)\, dt = \ddot{x}_1 t + \eta \frac{t^2}{2} \tag{5-51}$$

The displacement x during the interval is then

$$x = \int_0^t \dot{x}\, dt = \int_0^t \left(\ddot{x}_1 t + \frac{\eta t^2}{2} \right) dt = \frac{\ddot{x}_1 t^2}{2} + \eta \frac{t^3}{6} \tag{5-52}$$

Eliminating η from Eqs. 5-51 and 5-52 yields

$$x = \frac{\ddot{x}_1 t^2}{2} + \left[(\dot{x} - \ddot{x}_1 t) \frac{2}{t^2} \right] \frac{t^3}{6}$$

$$= \tfrac{1}{6}(\ddot{x}_1 t + 2\dot{x})t \tag{5-53}$$

Similarly, eliminating η from Eqs. 5-50 and 5-52 results in

$$x = \frac{\ddot{x}_1 t^2}{2} + \left(\frac{\ddot{x} - \ddot{x}_1}{t} \right) \frac{t^3}{6}$$

$$= \tfrac{1}{6}(2\ddot{x}_1 + \ddot{x})t^2 \tag{5-54}$$

At the end of the first interval, $t = \Delta t$ and $x = x_2, \dot{x} = \dot{x}_2, \ddot{x} = \ddot{x}_2$. Substituting these into Eqs. 5-53 and 5-54 results in the following:

$$x_2 = \tfrac{1}{6}(\ddot{x}_1 \Delta t + 2\dot{x}_2)\, \Delta t \tag{5-55}$$

and

$$x_2 = \tfrac{1}{6}(2\ddot{x}_1 + \ddot{x}_2)\,(\Delta t)^2 \tag{5-56}$$

Two situations will now be considered. First, if the initial acceleration $\ddot{x}_1$ is not zero, the acceleration is usually taken to be constant during the first interval, so that $\ddot{x}_2 = \ddot{x}_1$ and Eq. 5-56 becomes

$$x_2 = \ddot{x}_1 \frac{(\Delta t)^2}{2} \qquad \ddot{x}_1 \neq 0 \tag{5-57}$$

Also, since the acceleration is taken as constant for the interval, then $\eta = 0$, and Eq. 5-51 yields

$$\dot{x}_2 = \ddot{x}_1 \Delta t \tag{5-58}$$

The initial acceleration $\ddot{x}_1$ can be obtained from the differential equation (Eq. 5-48) since $\dot{x}_1 = 0 = x_1$. Thus

$$\ddot{x}_1 = \frac{P(t_1)}{m} \tag{5-59}$$

Then x_2 and $\dot{x}_2$ are calculated from Eqs. 5-57 and 5-58, respectively.

Next, $\ddot{x}_2$ is determined from the differential equation (Eq. 5-48). Then x_3 and $\dot{x}_3$ are calculated from finite difference relations (Eqs. 5-49 and 5-47) and $\ddot{x}_3$ is determined from the differential equation (Eq. 5-48), and so on.

Second, if $P(t_1) = 0$, as well as $\dot{x}_1 = 0 = x_1$, then from Eq. 5-48 the initial acceleration $\ddot{x}_1 = 0$. In this case the initial motion relations Eq. 5-55 and 5-56 become

$$x_2 = \frac{\dot{x}_2}{3} \Delta t \qquad (5\text{-}60)$$

and

$$x_2 = \frac{\ddot{x}_2}{6} (\Delta t)^2 \qquad (5\text{-}61)$$

The differential equation (Eq. 5-48) for point 2 would be

$$\ddot{x}_2 = \frac{P(t_2)}{m} - \frac{c}{m}\dot{x}_2 - \frac{k}{m}x_2 \qquad (5\text{-}62)$$

The values of x_2, $\dot{x}_2$, and $\ddot{x}_2$ can be determined by simultaneous solution of these three relations (Eqs. 5-60, 5-61, and 5-62). Then the numerical solution procedure can be continued, using finite-difference relations (Eqs. 5-49 and 5-47) to calculate x_3 and $\dot{x}_3$, and the differential equation (Eq. 5-48) to determine $\ddot{x}_3$, and so on.

The numerical processes described are continued for the desired time span. The time increment selected depends on the accuracy required. In general, one should select $\Delta t \leq 0.05\tau$, where $\tau = 2\pi/\sqrt{k/m}$ is the period for the free undamped system.

EXAMPLE 5-7 For a certain damped mass-elastic system, the differential equation of motion is $5\ddot{x} + 30\dot{x} + 2000x = P(t)$, where the forcing function is specified by the curve of Fig. 5-18. The initial conditions are $\dot{x}_1 = 0 = x_1$. Solve the differential equation numerically, thereby evaluating the displacement x and plotting a curve showing x against time t.

SOLUTION The natural period of the undamped motion would be

$$\omega = \sqrt{\frac{k}{m}} = \sqrt{\frac{2000}{5}} = 20 \text{ rad/sec}$$

$$\tau = \frac{2\pi}{\omega} = \frac{2\pi}{20} = 0.3142 \text{ sec}$$

The computations can be conveniently carried out on a small electronic calculator. Accordingly, the interval is chosen as $\Delta t = 0.02$ sec, which is approximately 0.05τ.

Rearranging the differential equation defines the acceleration as

$$\ddot{x} = 0.2P(t) - 6\dot{x} - 400x$$

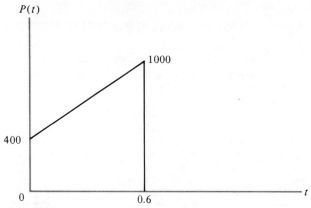

Figure 5-18

For the initial conditions of $\dot{x}_1 = 0 = x_1$, and $P(t_1) = 400$ from the curve,

$$\ddot{x}_1 = 0.2P(t_1) - 6\dot{x}_1 - 400x_1$$
$$= 0.2(400) - 6(0) - 400(0) = 80$$

Then from Eqs. 5-57 and 5-58,

$$x_2 = \tfrac{1}{2}\ddot{x}_1 \,(\Delta t)^2 = \tfrac{1}{2}(80)(0.02)^2 = 0.016$$
$$\dot{x}_2 = \ddot{x}_1 \,\Delta t = 80(0.02) = 1.6$$

and from the differential equation,

$$\ddot{x}_2 = 0.2P(t_2) - 6\dot{x}_2 - 400x_2$$
$$= 0.2(420) - 6(1.6) - 400(0.016) = 68$$

Next, x_3 and $\dot{x}_3$ are calculated from the finite-difference relations (Eqs. 5-49 and 5-47).

$$x_3 = \ddot{x}_2 \,(\Delta t)^2 + 2x_2 - x_1$$
$$= 68(0.02)^2 + 2(0.016) - 0 = 0.0592$$
$$\dot{x}_3 = \frac{3x_3 - 4x_2 + x_1}{2\,\Delta t}$$
$$= \frac{3(0.0592) - 4(0.016) + 0}{2(0.02)} = 2.84$$

Then from the differential equation

$$\ddot{x}_3 = 0.2P(t_3) - 6\dot{x}_3 - 400x_3$$
$$= 0.2(440) - 6(2.84) - 400(0.0592) = 47.28$$

Table 5-1

i	t	P	x	$\dot{x}$	$\ddot{x}$
1	0	400	0	0	80
2	0.02	420	0.016^a	1.6^b	68
3	0.04	440	0.0592	2.84	47.28
4	0.06	460	0.121 312	3.5784	22.0048
5	0.08	480	0.192 226	3.765 74	−3.484 83
6	0.10	500	0.261 746	3.441 15	−25.3453
7	0.12	520	0.321 128	2.715 64	−40.7450
8	0.14	540	0.364 212	1.746 75	−48.1651
9	0.16	560	0.388 030	0.709 240	−47.4673
10	0.18	580	0.392 860	−0.233 126	−39.7454
—	—	—	—	—	—

^{a} From Eq. 5-57.
^{b} From Eq. 5-58.

Continuing in this manner,

$$x_4 = \ddot{x}_3 \, (\Delta t)^2 + 2x_3 - x_2$$
$$= 47.28(0.02)^2 + 2(0.0592) - 0.016 = 0.121\,312$$

$$\dot{x}_4 = \frac{3x_4 - 4x_3 + x_2}{2\,\Delta t}$$

$$= \frac{3(0.121\,312) - 4(0.0592) + 0.016}{2(0.02)} = 3.5784$$

$$\ddot{x}_4 = 0.2P(t_4) - 6\dot{x}_4 - 400x_4$$
$$= 0.2(460) - 6(3.5784) - 400(0.121\,312) = 22.0048$$

The procedure is then repeated continuously until the desired time has been covered. The calculations can be conveniently carried out in tabular form, as shown in Table 5-1. Since the typical calculator is capable of sequential algebraic entry procedure, it is unnecessary to record intermediate calculated values in the table. The order of these calculations is self-explanatory and should be studied with this in mind. To save space, the calculations for the first several steps only are shown.

From the calculated values, the displacement-time curve is plotted and shown in Fig. 5-19.

EXAMPLE 5-8 The differential equation for a mass-spring system is $4\ddot{x} + 2500x = P(t)$, where $P(t)$ is given by the irregular curve of Fig. 5-20, which was determined experimentally. Carry out a numerical solution of the differential equation for this case, taking $x_1 = 0 = \dot{x}_1$.

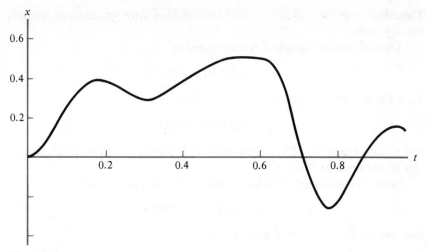

Figure 5-19

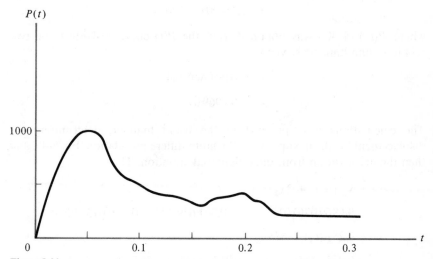

Figure 5-20

SOLUTION   The natural frequency is

$$\omega = \sqrt{\frac{k}{m}} = \sqrt{\frac{2500}{4}} = 25 \text{ rad/sec}$$

and the period is

$$\tau = \frac{2\pi}{\omega} = \frac{2\pi}{25} = 0.2513 \text{ sec}$$

The calculation interval of $\Delta t = 0.01$ sec is chosen, corresponding to approximately 0.04τ.

The differential equation is rearranged as

$$\ddot{x} = 0.25P(t) - 625x$$

Then for the initial point 1,

$$\ddot{x}_1 = 0.25P(t_1) - 625x_1$$

and it is observed that $P(t_1) = 0$ so that $\ddot{x}_1$ is also zero. However, $\ddot{x}_2$ and x_2 can be obtained as follows.

Since the initial acceleration is zero, Eq. 5-61 applies and

$$x_2 = \tfrac{1}{6}\ddot{x}_2 \, (\Delta t)^2 = \tfrac{1}{6}\ddot{x}_2(0.01)^2$$

also writing the differential equation here

$$\ddot{x}_2 = 0.25P(t_2) - 625x_2$$
$$= 0.25(400) - 625x_2$$

where $P(t_2)$ of 400 was obtained from the $P(t)$ curve. Solving these two relations simultaneously yields

$$x_2 = 0.001\,649\,484$$
$$\ddot{x}_2 = 98.969\,07$$

The calculations now proceed in the usual manner, determining the displacement for the next point by the finite-difference relation, Eq. 5-49, and then the acceleration from the differential equation. Thus

$$x_3 = \ddot{x}_2 \, (\Delta t)^2 + 2x_2 - x_1$$
$$= 98.969\,07(0.01)^2 + 2 \times 0.001\,649\,484 - 0 = 0.013\,195\,88$$
$$\ddot{x}_3 = 0.25P(t_3) - 625x_3$$
$$= 0.25 \times 660 - 625 \times 0.013\,195\,88 = 156.7526$$

Next

$$x_4 = 156.7526 \times (0.01)^2 + 2 \times 0.013\,195\,88 - 0.001\,649\,484 = 0.040\,417\,84$$
$$\ddot{x}_4 = 0.25 \times 840 - 625 \times 0.040\,417\,84 = 184.7388$$

and so on. The procedure can be carried out conveniently in tabular form as in Table 5-2, and may also be readily programmed on a digital computer.

The results of these calculations are shown plotted as the displacement-time curve of Fig. 5-21.

Table 5-2

i	t	P	x	$\ddot{x}$
1	0	0	0	0
2	0.01	400	0.001 649 484[a]	98.969 07[a]
3	0.02	660	0.013 195 88	156.7526
4	0.03	840	0.040 417 84	184.7388
5	0.04	960	0.086 113 70	186.1789
6	0.05	1000	0.150 427 4	155.9828

[a] From Eqs. 5-61 and 5-62.

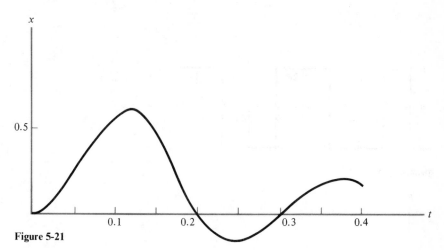

Figure 5-21

PROBLEMS

5-1 through 5-12. Each curve shown represents a periodic forcing function. For any assigned problem, determine the Fourier coefficients and write the trigonometric series for the forcing function.

For Probs. 5-13 through 5-24, express the steady response of a viscously damped system to the forcing function defined by the corresponding curve.

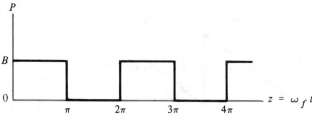

Problem 5-1

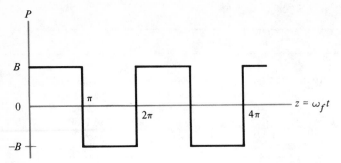

Problem 5-2

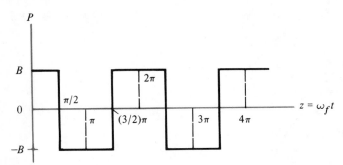

Problem 5-3

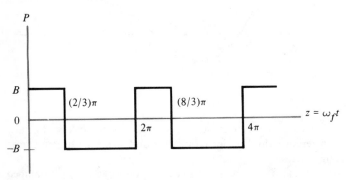

Problem 5-4

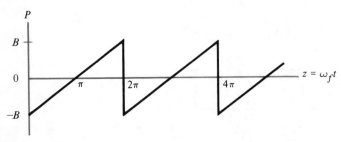

Problem 5-5

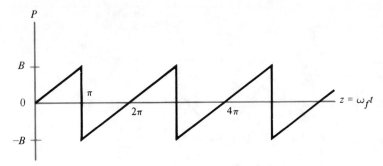

Problem 5-6

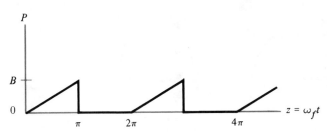

Problem 5-7

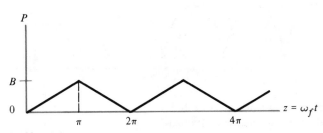

Problem 5-8

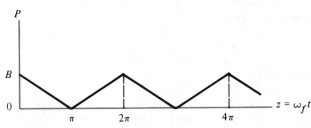

Problem 5-9

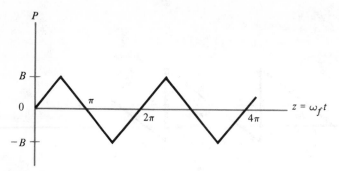

Problem 5-10

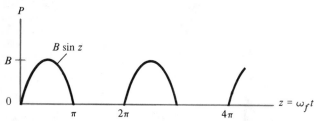

Problem 5-11

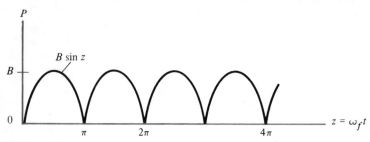

Problem 5-12

5-13. Carry out for the forcing function of Prob. 5-1.
5-14. Carry out for the forcing function of Prob. 5-2.
5-15. Carry out for the forcing function of Prob. 5-3.
5-16. Carry out for the forcing function of Prob. 5-4.
5-17. Carry out for the forcing function of Prob. 5-5.
5-18. Carry out for the forcing function of Prob. 5-6.
5-19. Carry out for the forcing function of Prob. 5-7.
5-20. Carry out for the forcing function of Prob. 5-8.
5-21. Carry out for the forcing function of Prob. 5-9.
5-22. Carry out for the forcing function of Prob. 5-10.
5-23. Carry out for the forcing function of Prob. 5-11.
5-24. Carry out for the forcing function of Prob. 5-12.
5-25. Using the arithmetical-summation method outlined in Section 5-5, determine the Fourier coefficient b_3 of the trigonometric series for the curve of Prob. 5-2.

Assume that $B = 1$, and divide one cycle of the curve into 24 parts. Compare this with the exact value of $4/3\pi$.

5-26. Using the arithmetical-summation method, determine the Fourier coefficient b_2 of the series for the curve of Fig. 5-2. Assume that $B = 1$, and divide one cycle of the curve into 24 segments. Compare this with the exact value obtained in Section 5-4.

5-27. Using the arithmetical-summation method, determine the Fourier coefficient a_3 of the series for the curve of Prob. 5-9. Assume that $B = 1$, and divide one cycle of the curve into 24 segments. Compare this with the exact value of $4/(3\pi)^2$.

5-28. Determine the response of an undamped system to the forcing function represented by the curve: (a) for the interval $0 < \omega t < 2\pi$, and (b) for $\omega t > 2\pi$. (c) Plot the response x against ωt from $\omega t = 0$ to $\omega t > 2\pi$.

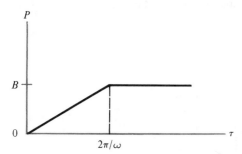

Problem 5-28

5-29. Determine the response of an undamped system to the forcing condition shown, as follows: (a) for the interval $0 < \omega t < 2\pi$, and (b) for the interval $2\pi < \omega t < 4\pi$. (c) Plot the response x against ωt from $\omega t = 0$ to $\omega t = 4\pi$.

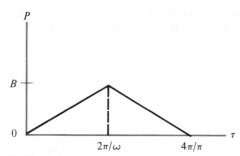

Problem 5-29

5-30. For the preceding problem, (a) determine the response for $\omega t > 4\pi$. (b) Plot this response x against ωt.

5-31. Determine the response of an undamped system to the forcing condition represented by the curve: (a) for the interval $0 < \omega t < 2\pi$ and (b) for $\omega t > 2\pi$. (c) Plot the response x against ωt from $\omega t = 0$ to $\omega t > 2\pi$.

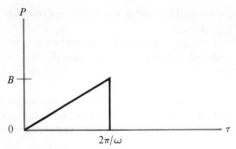

Problem 5-31

5-32. Determine the response of an undamped system to the forcing function represented by the curve shown: (a) for the interval $0 < \omega t < 2\pi$, and (b) for the interval $2\pi < \omega t < 4\pi$. (c) Plot the response x against ωt from $\omega t = 0$ to $\omega t = 4\pi$.

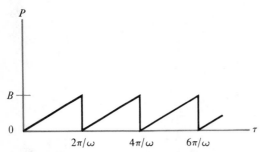

Problem 5-32

5-33. For the preceding problem, (a) determine the response for $4\pi < \omega t < 6\pi$. (b) Plot this response x against ωt.

5-34. The differential equation of motion for a certain undamped mass-elastic system is $4\ddot{x} + 14\,400x = P(t)$, where the forcing function $P(t)$ is defined by the curve

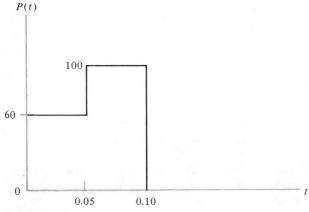

Problem 5-34

shown. The initial conditions are $\dot{x}_1 = 0 = x_1$. Solve the differential equation numerically, thereby evaluating the displacement x and plotting a curve showing x against t, for the range $t = 0$ to $t = 0.15$. (Suggestion: Take $\Delta t \approx 0.1\tau$).

5-35. Solve Prob. 5-34 if the system is also damped so that the differential equation is $4\ddot{x} + 72\dot{x} + 14\,400x = P(t)$.

5-36. The differential equation of motion for a certain undamped mass-elastic system is $3\ddot{x} + 12\,675x = P(t)$, where the forcing function $P(t)$ is defined by the curve shown. The initial conditions are $\dot{x}_1 = 0 = x_1$. Solve the differential equation numerically, evaluating the displacement x and plotting a curve showing x against t, for the range of $t = 0$ to $t = 0.13$. (Suggestion: Take $\Delta t \approx 0.1\tau$.)

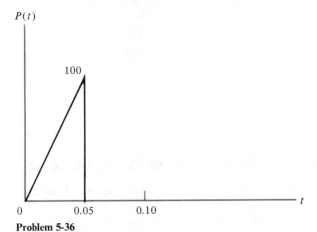

Problem 5-36

5-37. Solve Prob. 5-36 if the system is also damped so that the differential equation is $3\ddot{x} + 65\dot{x} + 12\,675x = P(t)$.

chapter
6

Vibration-Measuring Instruments

6-1. MOTION MEASUREMENT AND BASIC RELATIONS

It is often necessary to investigate the vibrations of existing machines and structures, experimental models, or equipment which is under development. Instruments for this purpose have long been available, but they have been considerably refined in recent years. A basic understanding of instrumentation should be attained by anyone working in the field of vibrations. The essential components of a vibration-measuring instrument are shown in Fig. 6-1(a). It is contemplated that the instrument base is placed on, or fastened to, a machine or structure that has its absolute motion defined by $y = Y \sin \omega_f t$. The instrument parts are restrained to move vertically only. The small mass of the base can be disregarded, since it becomes part of the mass of the machine to which it is fastened. The elements of the instrument are m, k, and c, as indicated. The absolute displacements of the mass m and the base are x and y, respectively. Then $(x - y) = z$ is the displacement of m relative to the base. The output or indication of the instrument is a function of this relative motion, as will be seen later. From the free-body diagram of m, shown in Fig. 6-1(b), the differential equation of motion would be

$$m\ddot{x} = -k(x - y) - c(\dot{x} - \dot{y}) \tag{6-1}$$

Substituting $(x - y) = z$ and its derivatives and rearranging gives

$$m(\ddot{z} + \ddot{y}) = -kz - c\dot{z}$$

$$m\ddot{z} + c\dot{z} + kz = -m\ddot{y}$$

$$= m\omega_f^2 Y \sin \omega_f t \tag{6-2}$$

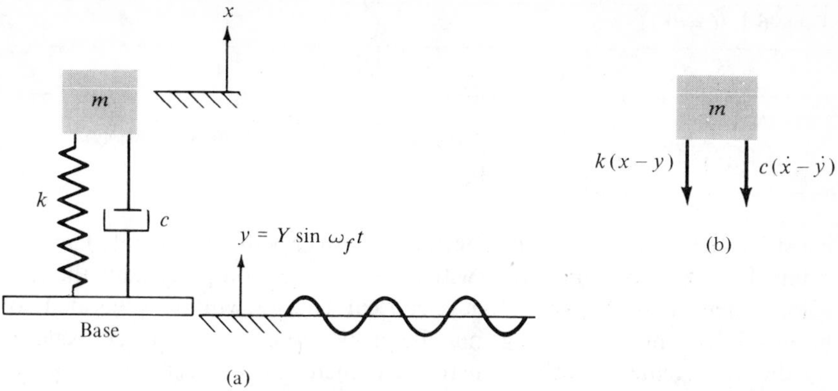

Figure 6-1

The right-hand side becomes the forcing condition specified by the motion of the base, and the left-hand side defines the relative motion of the instrument. The steady-state solution for this is

$$z = \frac{(m\omega_f^2 Y/k) \sin (\omega_f t - \psi)}{\sqrt{(1 - r^2)^2 + (2\zeta r)^2}}$$

$$= \frac{r^2}{\sqrt{(1 - r^2)^2 + (2\zeta r)^2}} Y \sin (\omega_f t - \psi) \qquad (6\text{-}3)$$

where

$$\tan \psi = \frac{2\zeta r}{1 - r^2}, \quad r = \frac{\omega_f}{\omega}, \quad \omega = \sqrt{\frac{k}{m}} \qquad (6\text{-}4)$$

The system arrangement and analysis here are essentially the same as previously presented in Section 4-15. However, the present concern is not with the absolute or relative motion of m, as such, but with the basis this forms for vibration-measuring devices.

Equation 6-3 represents the basic relation for the vibration instrument. The symbol y and its derivatives have to do with the motion of the machine or equipment to which the base is fixed. For such motion, it may be desired to measure the displacement, velocity, or acceleration, and from this the instrument is designated, respectively, as a *vibrometer*, a *velocity meter* or *velometer*, or an *accelerometer*. These are also sometimes referred to as vibration pickups.

On the other hand, z and its derivatives are concerned with the motion of the instrument mass m relative to the base. It is presumed that this relative motion can be observed or recorded in some manner. The observation, or output, of the instrument may be the relative displacement or the relative velocity. Such output may actuate a mechanical or electrical system insofar as the actual relative motion is concerned. For example, a pointer or stylus can be fastened to m and if a drum is fixed to the base and rotated at a

Table 6-1 $(\zeta = 0.7)$

r	1	2	3	4	5	∞
$\dfrac{r^2}{\sqrt{(1 - r^2)^2 + (2\zeta r)^2}}$	0.7143	0.9747	0.9961	0.9993	1.0000	1

constant rate, the relative-displacement z curve will be generated on the drum. This is a mechanical gage with a displacement output. Again, the gage arrangement may be such that electrical energy will be generated, or modified, for which the voltage may be proportional to z or to $\dot{z}$, depending on the arrangement. Such an instrument then has an electrical output or signal. Several different types of instruments are discussed below. Only those instruments that are practical and generally available are included.

6-2. VIBROMETER

From the basic relation (Eq. 6-3), if

$$\frac{r^2}{\sqrt{(1 - r^2)^2 + (2\zeta r)^2}} \approx 1 \tag{6-5}$$

then

$$z \approx Y \sin (\omega_f t - \psi) \tag{6-6}$$

Comparing this to $y = Y \sin \omega_f t$ shows that z observes directly the motion y, except for the phase lag ψ. The corresponding time t' by which the instrument displacement z lags behind the base displacement y is given by

$$t' = \frac{\psi}{\omega_f} \tag{6-7}$$

For a single harmonic motion of a machine (and the instrument base), this lag is of no consequence, and z will give a proper record of the y motion.

In order to meet the requirement represented by Eq. 6-5, r should be large, as can be seen from Fig. 6-2. (This family of curves is of the same form as that shown in Fig. 4-20, and previously discussed.) This means that ω should be small; hence the natural frequency of the instrument is low. Damping is not needed in order to satisfy Eq. 6-5, although it does improve the range of application somewhat. For $\zeta = 0.7$, the frequency ratio r should be on the order of 3 or greater, as indicated by Table 6-1. The range is thus quite large. Since $\omega = \sqrt{k/m}$ is a constant for the instrument, then ω_f and r are proportional, and the range of application of the gage can be specified in terms of ω_f. It would appear that there is no upper limit to the frequency range for which the gage might be applicable. However, high-frequency

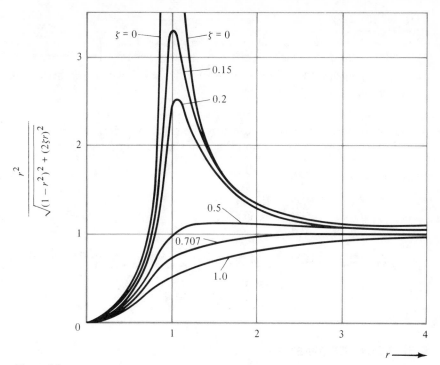

Figure 6-2

vibrations are usually of small amplitude and hence eventually not within the sensitivity of the gage. If the high-frequency oscillation is not of low amplitude, then the energy level is too great for the gage. For these reasons, it is necessary to specify an upper limit of frequency for which the instrument may be properly used.

Since $\omega = \sqrt{k/m}$, the small natural frequency required for the gage results in a large mass and a spring having a low modulus, causing the instrument to be too bulky and cumbersome for many applications.

The factor $r/\sqrt{(1 - r^2)^2 + (2\zeta r)^2}$ can be used to calculate the error in the reading of the vibrometer or in determining the correct value of the displacement.

For r greater than 3 or 4, the phase angle ψ would be 180 degrees for $\zeta = 0$, and it approximates this value for $\zeta = 0.7$ as r becomes larger. This means that for the vibrometer, the motion shown by the instrument is approximately opposite to that of the machine to which it is fixed. This generally is of no significance except where complex waves are measured, as will be discussed later.

EXAMPLE 6-1   A certain vibrometer has a damping factor of $\zeta = 0.60$ and a natural frequency of 5.00 Hz. When placed on a machine vibrating at

Table 6-2 ($\zeta = 0.7$)

r	0	0.1	0.2	0.3	0.4	0.5	0.6	0.7	0.8	0.9	1.0
$\dfrac{1}{\sqrt{(1 - r^2)^2 + (2\zeta r)^2}}$	1	1.0002	1	0.9978	0.9905	0.9747	0.9469	0.9052	0.8500	0.7848	0.7143

450 cycles/min, the amplitude shown by the instrument is 0.825 cm. Determine the correct value for the amplitude of the machine oscillation.

SOLUTION

$$r = \frac{450/60}{5.00} = 1.5$$

$$\frac{r^2}{\sqrt{(1 - r^2)^2 + (2\zeta r)^2}} = \frac{(1.5)^2}{\sqrt{[1 - (1.5)^2]^2 + (2 \times 0.6 \times 1.5)^2}} = 1.027$$

$$Y = \frac{0.825}{1.027} = 0.803 \text{ cm}$$

6-3. ACCELEROMETER

The basic relation (Eq. 6-3) can be rearranged as

$$-z\omega^2 = \frac{1}{\sqrt{(1 - r^2)^2 + (2\zeta r)^2}} [-Y\omega_f^2 \sin (\omega_f t - \psi)] \qquad (6\text{-}8)$$

and if

$$\frac{1}{\sqrt{(1 - r^2)^2 + (2\zeta r)^2}} \approx 1 \qquad (6\text{-}9)$$

then

$$-z\omega^2 \approx -Y\omega_f^2 \sin (\omega_f t - \psi) \qquad (6\text{-}10)$$

Comparison of this with $\ddot{y} = -Y\omega_f^2 \sin \omega_f t$ reveals that except for the phase lag ψ, the term $(-z\omega^2)$ measures the acceleration $\ddot{y}$ of the base. Furthermore, since $\omega^2 = k/m$ is an instrument constant, $(-z\omega^2)$ is simply proportional to and opposite the gage displacement z. By calibrating the gage to read $(-z\omega^2)$, the acceleration $\ddot{y}$ is read directly. The time t' by which the instrument lags the motion is expressed by

$$t' = \frac{\psi}{\omega_f} \qquad (6\text{-}11)$$

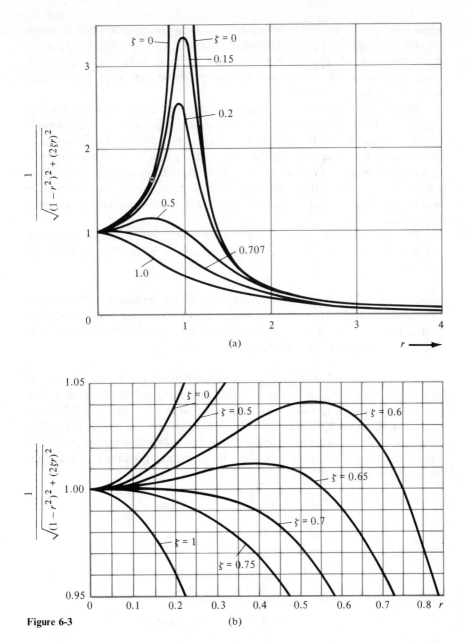

Figure 6-3

This is not of consequence where a single harmonic curve is measured. It will be considered later in connection with the observation of a complex wave.

The requirement of Eq. 6-9 can be fulfilled by $\zeta = 0.7$ for $0 \leq r \leq 0.6$, as can be seen from Fig. 6-3(a) (which is identical to Fig. 4-15), and from Fig. 6-3(b) and Table 6-2 (for $\zeta = 0.7$). Thus the range is somewhat limited. Here again it would be customary to specify the range of application in terms of ω_f. The damping, which is required in this case, will also effectively reduce the transients of the instrument, and it will reduce the resonant magnification of any components having frequencies near the natural frequency of the instrument.

Since r is small here, it is required that ω be large. Thus the natural frequency of the instrument is high. This results in poor mechanical sensitivity of the gage, as the large value of ω^2 is accompanied by small z displacements, since the gage observations are $(-z\omega^2)$. However, if the output is electrical, then amplification may result in good terminal sensitivity. Also, since ω must be large, the instrument spring is short (for large k) and mass m is small, so that the gage is small in size, which is an advantage in many applications.

The factor $1/\sqrt{(1-r^2)^2 + (2\zeta r)^2}$ can be used to determine the error in the instrument reading or in calculating the correct value of the acceleration measured.

EXAMPLE 6-2   Design an accelerometer that will measure vibrations in the range of 0 to 50 Hz with a maximum error of 1%. Select the value of the suspended mass so that it will be small ($w \approx 1$ oz) and determine the required spring modulus and damping constant.

SOLUTION $\zeta = 0.7$ is chosen as the desired damping factor. The corresponding value of $1/\sqrt{(1-r^2)^2 + (2\zeta r)^2}$ and curve will be above 1, with a maximum of 1.0002 at $r = 0.1414$, within the range of $r = 0$ to $r = 0.2$. (Refer to Eqs. 4-68 and 4-69 of Section 4-10.) This is an error of less than 1%. The curve has a value of unity at $r = 0.2$ and lies below this value for $r > 0.2$. Hence the error of 1% occurs beyond this point, and the following may then be set down:

$$\frac{1}{\sqrt{(1-r^2)^2 + (2\zeta r)^2}} = 0.99$$

Rearranging and expanding gives

$$r^4 - 0.04r^2 - 0.020\,304\,0 = 0$$

Obtaining the r^2 roots and then r results in

$$r^2 = 0.163\,888\,95, \qquad r = 0.404\,83$$

(The negative root for r^2 is discarded.) Then

$$\omega_f = 2\pi f_f = 2\pi \times 50 = 100\pi \text{ rad/sec}$$

$$\omega = \frac{\omega_f}{r} = \frac{100\pi}{0.404\,83} = 776.0269 \text{ rad/sec}$$

Selecting $w = 0.05$ lb gives

$$k = m\omega^2 = \frac{0.05}{386}(776.0269)^2 = 78.0075 \text{ lb/in.}$$

$$c = c_c\zeta = 2m\omega\zeta = 2 \times \frac{0.05}{386} \times 776.0269 \times 0.7 = 0.140\,730 \text{ lb sec/in.}$$

EXAMPLE 6-3   A certain piece of equipment is vibrating according to the relation

$$y = 0.25 \sin 3\pi t + 0.10 \sin 6\pi t$$

y being in centimeters and t in seconds. An accelerometer is fixed to the vibrating equipment. If the accelerometer has a natural frequency of 391 cycles/min and a damping factor of $\zeta = 0.65$, determine the record which the accelerometer will show, assuming a zero error.

SOLUTION   The equipment vibration is expressed by the time derivative

$$\ddot{y} = -0.25(3\pi)^2 \sin 3\pi t - 0.10(6\pi)^2 \sin 6\pi t$$

$$= -(22.2 \sin 3\pi t + 35.5 \sin 6\pi t)$$

For the instrument, $\omega = (391/60) \times 2\pi = 13.03\pi$ rad/sec

$$r_1 = \frac{3\pi}{13.03\pi} = 0.230, \qquad r_2 = \frac{6\pi}{13.03\pi} = 0.460$$

Both of these frequency ratios are within the proper range for the accelerometer.

$$\psi_1 = \tan^{-1}\frac{2\zeta r_1}{1 - r_1^2} = \tan^{-1}\frac{2 \times 0.65 \times 0.230}{1 - (0.230)^2} = \tan^{-1} 0.316 = 17.5°$$

$$\psi_2 = \tan^{-1}\frac{2\zeta r_2}{1 - r_2^2} = \tan^{-1}\frac{2 \times 0.65 \times 0.460}{1 - (0.460)^2} = \tan^{-1} 0.758 = 37.2°$$

The acceleration record would be

$$\ddot{y} = -[22.2 \sin (3\pi t - 17.5°) + 35.5 \sin (6\pi t - 37.2°)]$$

6-4. VELOMETER

Taking the time derivative of Eq. 6-3 results in

$$\dot{z} = \frac{r^2}{\sqrt{(1 - r^2)^2 + (2\zeta r)^2}} Y\omega_f \cos (\omega_f t - \psi) \tag{6-12}$$

and if

$$\frac{r^2}{\sqrt{(1 - r^2)^2 + (2\zeta r)^2}} \approx 1 \tag{6-13}$$

then

$$\dot{z} \approx Y\omega_f \cos (\omega_f t - \psi) \tag{6-14}$$

Since $\dot{y} = Y\omega_f \cos \omega_f t$, then except for the phase difference ψ, Eq. 6-14 shows that $\dot{z}$ measures $\dot{y}$ directly, provided Eq. 6-13 holds. This is the same requirement as Eq. 6-5 for the vibrometer. Accordingly, r should be large (3 or greater), as noted in Section 6-2. The remainder of the discussion there, regarding phase lag, and so on, also applies to the present case. Note that $\dot{y}$ is the machine velocity and is measured by the instrument relative velocity $\dot{z}$. Hence the instrument is a velocity meter with a relative velocity sensitivity or output.

EXAMPLE 6-4   A velocity meter has a natural frequency of 16.0 Hz and a damping factor of $\zeta = 0.625$. Determine the lowest frequency of structural vibration for which the velocity may be measured with an error of 2%.

SOLUTION   The curve for $\zeta = 0.625$ would be similar to, but above the curve for $\zeta = 0.707$ of Fig. 6-2. Assume that there are two places where this $\zeta = 0.625$ curve has the value of 1.02, one point on the way up and one point after passing the peak. Then set

$$\frac{r^2}{\sqrt{(1 - r^2)^2 + (2 \times 0.625r)^2}} = 1.02$$

Squaring and rearranging gives

$$0.038\,183r^4 - 0.4375r^2 + 1 = 0$$

This has the roots

$$r^2 = 3.190,\ 8.078$$

The larger value must be selected in determining the minimum frequency for the specified error. Then

$$r = \sqrt{8.078} = 2.842$$

$$f = 2.842 \times 16 = 45.47 \text{ Hz}$$

Immediately below this frequency, the error would be greater than 2%.

6-5. PHASE-SHIFT ERROR

It was observed that vibration-measuring instruments exhibit phase lag. That is, the instrument response or record lags behind the motion which it measures. The time for this lag is defined by dividing the phase angle by the circular frequency ω_f. The lag is unimportant if a single harmonic wave is being measured. However, it is often necessary to measure a motion which has a complex wave form. Since a complex wave contains harmonic components of different frequencies, and the lag depends on frequency, the components of the output wave of the instrument may lag by different amounts. This would result in the instrument curve or record being at variance with the motion being measured. This is called phase-shift error or phase distortion. Fortunately, instruments have characteristics which cause the phase distortion to be negligible in most cases, as will be shown.

Consider the measurement of a symmetric complex wave. There will be no offsetting term, and hence the wave is defined by the series

$$y = a_1 \cos \omega_f t + a_2 \cos 2\omega_f t + \cdots$$
$$+ b_1 \sin \omega_f t + b_2 \sin 2\omega_f t + \cdots \tag{6-15}$$

Measurement of this displacement by using a vibrometer will be discussed first. The gage will respond to each component of the series, according to Eq. 6-6. Thus the instrument output would be

$$z \approx a_1 \cos (\omega_f t - \psi_1) + a_2 \cos (2\omega_f t - \psi_2) + \cdots$$
$$+ b_1 \sin (\omega_f t - \psi_1) + b_2 \sin (2\omega_f t - \psi_2) + \cdots \tag{6-16}$$

where

$$\tan \psi_1 = \frac{2\zeta(\omega_f/\omega)}{1 - (\omega_f/\omega)^2}$$

$$\tan \psi_2 = \frac{2\zeta(2\omega_f/\omega)}{1 - (2\omega_f/\omega)^2}$$

$$\vdots \qquad \vdots$$

$$\tan \psi_n = \frac{2\zeta(n\omega_f/\omega)}{1 - (n\omega_f/\omega)^2} \tag{6-17}$$

For proper application of this instrument, ω_f/ω must be large. Referring to Fig. 6-4, which is identical to Fig. 4-16, this results in

$$\psi_1 \approx \pi, \qquad \psi_2 \approx \pi, \qquad \ldots, \qquad \psi_n \approx \pi \tag{6-18}$$

Equation 6-16 then becomes

$$z \approx -(a_1 \cos \omega_f t + a_2 \cos 2\omega_f t + \cdots$$
$$+ b_1 \sin \omega_f t + b_2 \sin 2\omega_f t + \cdots) \tag{6-19}$$
$$\approx -y \tag{6-20}$$

Thus all parts—and hence the entire output curve z of the instrument—are simply opposite to the y motion of the machine or structure. It should be

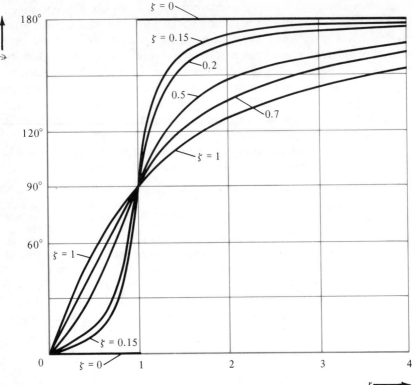

Figure 6-4

noted that damping is undesirable here insofar as phase distortion is concerned. The reversal of direction is generally unimportant and can be corrected easily.

Similar analysis of the velocity-sensitive velometer will show that

$$\dot{z} \approx -\dot{y} \tag{6-21}$$

for the case of a complex wave, so that the instrument velocity output $\dot{z}$ is equal and opposite to the machine velocity $\dot{y}$ being observed.

The problem of phase distortion for an accelerometer is of a somewhat different nature. For this instrument, the phase lag ψ does not approach 180 degrees but varies with the frequency, as can be seen from Fig. 6-4. Within the specified frequency range of $0 \leq r \leq 0.6$ for $\zeta = 0.7$, the value of ψ varies from zero to around 50 degrees. Accordingly, the phase lag depends on the frequency of the component being observed. The complex acceleration curve to be measured can be expressed, from the second time derivative of Eq. 6-15, as

$$\ddot{y} = -a_1 \omega_f^2 \cos \omega_f t - a_2 (2\omega_f)^2 \cos 2\omega_f t - \cdots$$
$$- b_1 \omega_f^2 \sin \omega_f t - b_2 (2\omega_f)^2 \sin 2\omega_f t - \cdots \tag{6-22}$$

The instrument will respond to each part of this series. Accordingly, from Eq. 6-10, the gage output would be

$$-z\omega^2 = -a_1\omega_f^2 \cos{(\omega_f t - \psi_1)} - a_2(2\omega_f)^2 \cos{(2\omega_f t - \psi_2)} - \cdots$$
$$- b_1\omega_f^2 \sin{(\omega_f t - \psi_1)} - b_2(2\omega_f)^2 \sin{(2\omega_f t - \psi_2)} - \cdots \quad (6\text{-}23)$$

The phase lag ψ, being a function of frequency as defined by Eq. 6-17, will be different for each component of the output series. That is, $\psi_1, \psi_2, \psi_3, \ldots$, will all differ. Fortunately, however, the time lag is approximately the same for the main components of the complex wave, as will be shown. The time lag for a component is defined by

$$t' = \frac{\psi}{\omega_f} \quad (6\text{-}24)$$

Note that in Fig. 6-4, for $\zeta = 0.7$ the curve for ψ against r is almost straight from the origin to 90 degrees at $r = 1$. Hence ψ can be expressed by the linear relation

$$\psi \approx \eta r = \eta \frac{\omega_f}{\omega} = \gamma \omega_f \quad (6\text{-}25)$$

where η and γ are constants. Substituting into Eq. 6-24 gives

$$t' = \frac{\gamma \omega_f}{\omega_f} = \gamma \quad (6\text{-}26)$$

Thus for the various components the time offset of the instrument is independent of the frequency, provided the frequency lies within the range for which $0 \leq r \leq 1$. That is, each component of the gage output will be offset by the same amount of time γ with respect to the corresponding acceleration component being measured. Hence the record will be a true representation of the acceleration of the machine or structure. This is evident from substitution of $\psi = \gamma \omega_f$ into Eq. 6-23 and by letting $(t - \gamma) = t''$, so that

$$-z\omega^2 = -a_1\omega_f^2 \cos{\omega_f t''} - a_2(2\omega_f)^2 \cos{2\omega_f t''} - \cdots$$
$$- b_1\omega_f^2 \sin{\omega_f t''} - b_2(2\omega_f)^2 \sin{2\omega_f t''} - \cdots \quad (6\text{-}27)$$

provided the highest frequency $n\omega_f$ involved is less than ω, so that the frequency of all components falls within the range for which $0 \leq r \leq 1$. The time t'' merely has a different reference origin, being offset from t by the amount γ. In a complex wave, generally only the first few components are important, since the amplitudes of the higher modes are small and contribute very little to the form of the curve, and also may be beyond the sensitivity of the instrument. Hence no appreciable distortion would result if the frequencies of the higher-order components lie above the specified range.

6-6. GAGE OUTPUT AND COMMON TYPES OF INSTRUMENTS

For the vibration-measuring devices discussed in the preceding sections, the apparent output of the instrument was the relative mechanical motion of the instrument mass. This can be read or recorded directly, or it may be converted into an electrical signal. Such a conversion device is generally called a *transducer*. The reading given may be amplitude or double amplitude, or it may be the wave form. The reading may require instantaneous observation or it may give a permanent record.

Using the mechanical output directly to measure the vibration is simple in principle. As shown in Fig. 6-5(a) and (b), the relative motion of the instrument mass can be noted by the movement of an attached pointer along a scale fixed to the base, or by the indicator of a dial gage connected between the mass and the base. Such a device would require the attention of an observer in order to obtain the reading, and only amplitude would be read. On the other hand, a stylus attached to the mass pointer will generate the motion curve on recording paper affixed to the surface of a drum which rotates at a constant velocity on an axle attached to the base, as shown in Fig. 6-5(c). These instruments have a relative displacement output, but whether they measure displacement or acceleration depends on the characteristics (that is, natural frequency, damping, and so on) of the gage, as previously explained. Mechanical gages such as this tend to be large and cumbersome, and their use is confined to applications in which size is not a factor, such as in seismic instruments for measuring earthquakes.

Converting the mechanical output to an electrical signal results in considerable reduction in instrument size, and an improvement in accuracy and sensitivity as well as convenience. It also permits taking measurements which would not be possible with other systems, and it enables the differentiation and integration of the measurements to be made readily and conveniently. The electrical signal generated is proportional to the relative displacement or velocity of the instrument mass, depending on the arrangement. In one type of gage a permanent magnet is fixed to the instrument base and a wire coil is fastened to the mass, as shown in Fig. 6-6. As the coil moves, magnetic-flux lines are cut, thereby generating a voltage in the coil circuit which is proportional to the velocity of the mass relative to the base. The instrument thus has a relative velocity output.

A principle known as the piezoelectric effect is also used to generate an electrical signal in vibration pickups. When certain crystals, such as quartz and barium titanate, are strained, they become polarized so that a difference in electrical potential can be measured. The voltage is proportional to the force or pressure on the element or to the strain of the element. Thus for the arrangement of Fig. 6-7(a) where the force on the element is kz, the output voltage depends on z. The crystal may serve both as the spring and

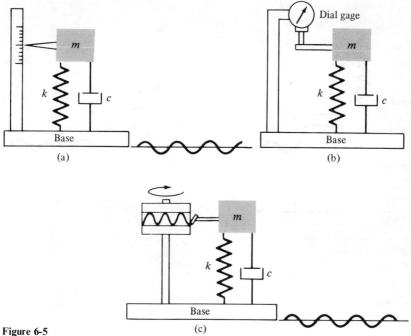

Figure 6-5

(a)

(b)

(c)

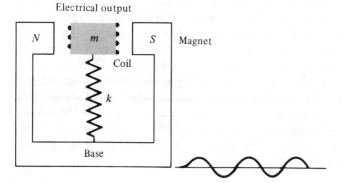

Figure 6-6

the sensor, as shown in Fig. 6-7(b). The electrical potential developed is then proportional to the strain of the crystal element. The output voltage from a piezoelectric device may be sufficient to produce the vibration record directly, or an amplifier may be used to strengthen the signal. These gages are enclosed and filled with a fluid of the proper viscosity to provide the desired damping for the instrument mass.

Electrical-resistance strain gages are widely used in vibration measurements. The gage consists of a grid of fine electric wire or thin metal foil, on

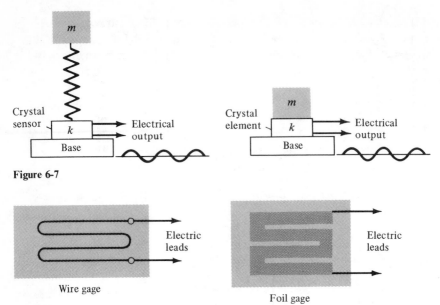

Figure 6-7

Wire gage

Foil gage

Figure 6-8

paper backing, as shown in Fig. 6-8, which can be firmly cemented to a machine or structural part. The elastic strain of the metal gage is equal to the strain of the element to which it is cemented. The accompanying change in the electrical resistance of the gage wire or foil, due to change in length and section, can be calibrated for the circuit so that the strain is read directly and easily. Vibration strains or displacements can be readily obtained in this manner. The gages are extremely small and light, and permit measurements, including those for internal and confined locations, which could not be made by other methods. This type of strain gage can also be used as the sensing element of a vibration instrument having a displacement output. A carrier circuit is then used, and the voltage change from the instrument modifies the carrier voltage to produce the vibration record.

6-7. ELECTRICAL SIGNAL AND MEASUREMENT SYSTEM

Depending on the characteristics of the instrument, the electrical signal from a vibration instrument may represent the displacement, velocity, or acceleration of the member on which it is placed. However, by suitable electric circuitry, any of these may be differentiated or integrated.† In general, the process of integration is more feasible than that of differentiation. Thus a

---

† See R. Burton, *Vibration and Impact* (Reading, Mass.: Addison-Wesley, 1958), pp. 155–156. C. M. Harris and 'C. E. Crede, eds., *Shock and Vibration Handbook*, 2nd ed. (New York: McGraw-Hill, 1976), p. 18–22.

velocity meter can be set up to give a displacement record, an accelerometer will yield a velocity record, and so on. Two successive integrations of an accelerometer output will produce a displacement record. Sometimes the derivative of the acceleration, known as the jerk, is obtained because it is significant in shock and comfort determinations.

The electrical signal from vibration instruments can be used to produce permanent records of the wave form or to actuate an oscilloscope so that the vibration or shock curve can be observed or monitored. Permanent records of shock can be obtained by high-speed photographs of oscilloscope response, or by recording on magnetic tape. The amount of electric circuitry and electronic equipment used in vibration and shock measurements is often quite extensive and costly.

6-8. OTHER TYPES OF INSTRUMENTS

Instruments for measuring torsional vibrations are also available. These are similar to those for measuring rectilinear motion, except that the elements are rotational. Thus the instrument mass is replaced by the inertia of a disk, and the elastic element becomes a torsional spring or shaft. Torsional gages have the same characteristics, limitations, and so on, as were discussed for rectilinear gages.

Although frequency can be determined from the wave-form record and often is obtained in this manner, instruments are available which measure only the frequency. A frequency meter, or tachometer, usually consists of a series of metal reeds or strips, representing cantilever beams of different natural frequencies, mounted firmly on the instrument base. When the meter is placed on a vibrating structure, the reed having the same frequency as that of the structure responds resonantly. This is readily observed, and the corresponding frequency value can be read.

6-9. VIBRATION RECORDS AND ANALYSIS

The output from vibration-measuring devices can be conveniently recorded in continuous form. Experimental investigation of the vibration of machines and structures is often quite extensive. For example, in the flight testing of an airplane of new design, multitudinous records are taken of the vibration at many locations in the structure. These are generally in the form of multiple trace records, such as those shown in Fig. 6-9. The number and length of such recorded wave forms may be very large. These curves are analyzed later in great detail to determine the vibration characteristics.

From the analysis of a recorded wave form, determinations are made of the frequency and maximum amplitude of the curve; the wave is broken down into its harmonic components, and the frequency and amplitude of these parts are obtained, the decay rate and damping are established wherever applicable, and so on. This may be done for velocity and acceleration

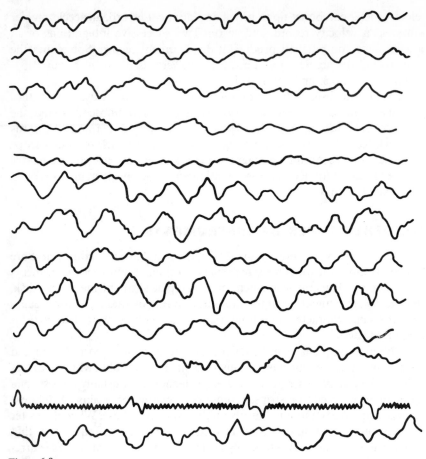

Figure 6-9

traces as well as for displacement records. In such analyses, various methods† are employed, including numerical procedures, superposition, the envelope method, and so forth. Mechanical and other aids are also used to facilitate these determinations.

PROBLEMS

6-1. Determine the percent error of a vibrometer for the frequency ratios of 3, 5, and 10 for each of the following damping factors: (a) $\zeta = 0$, (b) $\zeta = 0.6$, and (c) $\zeta = 0.7$.

6-2. For a vibrometer, determine the greatest percent error that occurs for the range of frequency ratio from $r = 3$ to $r = \infty$. Also obtain the corresponding value of r at which this occurs. Do this for each of the following damping factors: (a) $\zeta = 0$, (b) $\zeta = 0.6$, and (c) $\zeta = 0.7$.

† See R. G. Manley, *Waveform Analysis* (New York: John Wiley, 1945).

6-3. Determine the percent error of an accelerometer for the frequency ratios of 0, 0.2, 0.4, and 0.6 for each of the following damping ratios: (a) $\zeta = 0$, (b) $\zeta = 0.66$, and (c) $\zeta = 0.71$.

6-4. For an accelerometer, determine the greatest percent error in measurement that occurs for the range of the frequency ratio from $r = 0$ to $r = 0.6$. Also obtain the value of r at which this occurs. Carry this out for (a) $\zeta = 0$, (b) $\zeta = 0.66$, and (c) $\zeta = 0.71$.

6-5. An accelerometer is composed of weight $W = 0.042$ lb suspended by a spring for which $k = 5$ lb/in., and the fluid damping factor for the instrument is $\zeta = 0.65$. When used on a machine vibrating at a frequency of 13.647 Hz, the instrument reading for the maximum acceleration is 50 in./sec². Determine the correct value of the maximum acceleration of the machine.

6-6. A vibrometer is composed of a spring for which $k = 28.8$ lb/in., a suspended weight $W = 19.3$ lb, and a viscous damper for which $c = 0.96$ lb sec/in. The amplitude indicated by the instrument is 0.73 in. when attached to a structure oscillating with a frequency of 14 Hz. Determine the correct value for the amplitude of displacement of the structure.

6-7. An accelerometer is composed of weight $W = 0.0965$ lb suspended by a spring for which $k = 0.355$ lb/in., and the fluid damping factor for the instrument is $\zeta = 0.65$. When used on a machine vibrating at a frequency of 2.4 Hz, the instrument reading for the maximum acceleration is 20 in./sec². Determine the correct value for the maximum acceleration of the machine.

6-8. A vibrometer is composed of a spring for which $k = 1.44$ lb/in., a suspended weight of $W = 3.86$ lb, and a viscous damper for which $c = 0.096$ lb sec/in. The amplitude indicated by the instrument is 0.5 in. when attached to a structure oscillating with a frequency of 7 Hz. Determine the correct value for the amplitude of displacement of the structure.

6-9. A vibrometer is to be designed, based on a maximum error of 3% when used for frequencies of 1200 cycles/min and above. The instrument mass is to weigh 1 lb, and the damping factor is to be 0.6. Determine the required spring modulus and damping constant for the instrument.

6-10. An accelerometer is composed of a suspended weight of 0.193 lb and a spring having a modulus of 5 lb/in., but the damping is not known. When tested on a vibration having a frequency of 382 cycles/min, the acceleration measurement is found to be 1% greater than the correct value. Determine the damping constant for the instrument.

6-11. It is desired to design an accelerometer that will have an optimum *range* of frequency ratio for a maximum accelerometer error of 1%. (a) Calculate the required damping factor. (b) Determine this optimum range of frequency ratio.

6-12. An accelerometer is to be designed that will have an optimum *range* of frequency ratio for a maximum accelerometer error of 2%. (a) Calculate the required damping factor. (b) Determine this optimum range of frequency ratio.

6-13. Design an accelerometer based on a maximum error of 1% when used for vibration measurements having frequencies in the range from 0 to 1800 cycles/min. The instrument mass is to have a weight of 0.05 lb. Determine the required spring modulus and damping constant for the instrument. (Suggestion: Refer to Prob. 6-11.)

6-14. Design an accelerometer based on a maximum error of 2% when used for vibration measurements having frequencies in the range from 0 to 6000 cycles/min. The spring modulus for the gage is to be 100 lb/in. Determine the required damping constant and suspended weight for the instrument. (Suggestion: Refer to Prob. 6-12.)

6-15. Design an accelerometer based on a maximum error of 4% for measurements having frequencies in the range from 0 to 4200 cycles/min. The damping constant is to be 45 N · s/m. Determine the required spring constant and the suspended mass.

6-16. An accelerometer has an undamped natural frequency of 100 Hz and the suspended mass is known to weigh 0.04 lb. When attached to a vibrating test device that has an acceleration amplitude of 0.280 in./sec² and a frequency of 55 Hz, the acceleration amplitude is measured to be 0.294 in./sec². Determine for the accelerometer (a) the damping constant and (b) the spring constant.

6-17. An accelerometer has a spring modulus of 25 lb/in., damping constant of 0.073 26 lb sec/in., and the suspended mass weighs 0.05 lb. Determine the frequency range (a) for 1% instrument error and (b) for 2% gage error.

6-18. An accelerometer has an undamped frequency of 70 Hz and the damping constant is 9.81 N · s/m. It is attached to an oscillating machine which has an acceleration amplitude of 6.50 m/s² and a frequency of 42 Hz, and the instrument records an acceleration amplitude of 7.02 m/s². For the accelerometer, determine (a) the suspended mass and (b) the spring constant.

6-19. A machine is vibrating in accordance with the relation $x = 0.90 \sin 8\pi t + 0.45 \sin 16\pi t + 0.30 \sin 24\pi t$, where x is in inches and t is in seconds. Determine the record of this vibration that would be made by a vibrometer having a natural frequency of 1 Hz and a damping factor of 0.2. Discuss the accuracy of this record, noting the reason for any discrepancies.

6-20. A structure is vibrating in accordance with the expression $x = 1.0 \sin 20\pi t + 0.5 \sin 60\pi t + 0.3 \sin 100\pi t$, where the units for x and t are centimeters and seconds, respectively. The motion is measured by a vibrometer having a natural frequency of 5 Hz and a damping factor of 0.6. Determine the record that the instrument would give. Discuss the accuracy of this record, noting the reason for any discrepancies.

6-21. A body is vibrating in accordance with the equation $x = 0.01 \sin 7.5\pi t + 0.003 \sin 22.5\pi t$, where x is in meters, and t is in seconds. An accelerometer having a natural frequency of 25 Hz and a damping factor of 0.65 is attached to the body. Determine the record made by the instrument, considering that it is calibrated to read acceleration directly in meters per second squared.

6-22. For the accelerometer of Prob. 6-21, determine the phase shift and the time shift of each part of the complex wave. Compare these times, and discuss the distortion of the acceleration record.

6-23. A body is vibrating according to the relation $x = 0.050 \sin 20\pi t + 0.020 \sin 40\pi t + 0.008 \sin 60\pi t$, where the units of x and t are inches and seconds, respectively. An accelerometer having a natural frequency of 50 Hz and a damping factor of 0.7 is attached to the body. Determine the record made by the instrument, assuming that it is calibrated to show acceleration directly, in inches per second squared.

6-24. For the accelerometer of Prob. 6-23, determine the phase shift and time shift of each part of the acceleration record. Compare these times, and discuss the distortion of the acceleration record.

chapter
7

Two-Degree-of-Freedom
Systems

7-1. INTRODUCTION

So far, only systems having a single degree of freedom have been considered. For the free undamped case, it was found that such a system could oscillate only harmonically, and with a single natural frequency.

Systems having two degrees of freedom will now be considered. This requires two independent coordinates to define the configuration of the system. For the free undamped case, it will be found that there are two ways in which the system can vibrate harmonically. These are designated as *principal modes*. The frequency of motion for a principal mode is called the natural frequency.

In addition, it will be discovered that there are many ways in which the system can vibrate nonharmonically. The motion is then of complex wave form. Such periodic motion is composed of the principal-mode components.

7-2. PRINCIPAL-MODE VIBRATIONS

Consider the two-degree-of-freedom system, represented by the model of Fig. 7-1(a), and the corresponding free-body diagrams shown in Fig. 7-1(b). In the free-body diagrams, x_2 has been assumed to be greater than x_1 so that the spring k_2 is in tension, as shown, the tensile force being $k_2(x_2 - x_1)$. On the other hand if x_1 had been taken as greater than x_2, the k_2 spring would be in compression, with the spring force expressed as $k_2(x_1 - x_2)$. Since both the force and the expression for the spring force are reversed, either method is proper and will give correct results when the force summation is written.

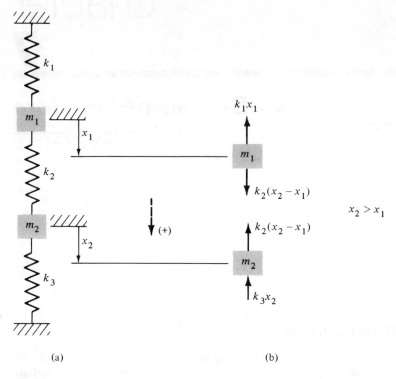

(a)   (b)

Figure 7-1

Proceeding with the analysis, the differential equations of motion are written as

$$m_1 \ddot{x}_1 = -k_1 x_1 + k_2 (x_2 - x_1)$$
$$m_2 \ddot{x}_2 = -k_2 (x_2 - x_1) - k_3 x_2 \qquad (7\text{-}1)$$

Rearranging gives

$$\ddot{x}_1 + \left(\frac{k_1 + k_2}{m_1} \right) x_1 - \frac{k_2}{m_1} x_2 = 0$$

$$\ddot{x}_2 - \frac{k_2}{m_2} x_1 + \left(\frac{k_2 + k_3}{m_2} \right) x_2 = 0 \qquad (7\text{-}2)$$

These are simultaneous linear homogeneous equations with constant coefficients. There are several ways of obtaining the solution. In the method that follows, the solution will be obtained by physical reasoning and contemplation of the form of the differential equations. It is anticipated that harmonic motion of the system is probable. Also, if the solution returns to the same form in the second derivative, it will tend to satisfy the differential equations. These ideas lead to expressing the solution form as

$$x_1 = A_1 \sin (\omega t + \phi)$$
$$x_2 = A_2 \sin (\omega t + \phi) \tag{7-3}$$

where A_1 and A_2 represent amplitudes, ϕ is the phase, and ω is the circular frequency, as yet undefined. Although other forms of the solution could be used, it is deemed that Eq. 7-3 is the most convenient and meaningful. The apparent restrictiveness of the expressions assumed, with regard to the common frequency and same phase for the two parts, will be discussed later.

Substituting the assumed solutions into the differential equations and rearranging results in

$$\left[\left(\omega^2 - \frac{k_1 + k_2}{m_1} \right) A_1 + \frac{k_2}{m_1} A_2 \right] \sin (\omega t + \phi) = 0$$

$$\left[\frac{k_2}{m_2} A_1 + \left(\omega^2 - \frac{k_2 + k_3}{m_2} \right) A_2 \right] \sin (\omega t + \phi) = 0 \tag{7-4}$$

Since the solutions must hold for all values of t, the sine term cannot vanish and hence may be cancelled. Thus for the solution to be general requires that

$$\left(\omega^2 - \frac{k_1 + k_2}{m_1} \right) A_1 + \frac{k_2}{m_1} A_2 = 0$$

$$\frac{k_2}{m_2} A_1 + \left(\omega^2 - \frac{k_2 + k_3}{m_2} \right) A_2 = 0 \tag{7-5}$$

These are known as the *amplitude equations* since they will define the amplitudes A_1 and A_2, provided the frequency ω has been determined. Leaving out the trivial condition of $A_1 = 0 = A_2$, Eq. 7-5 can be satisfied only if the determinant of the amplitudes vanishes.† Thus

$$\begin{vmatrix} \left(\omega^2 - \dfrac{k_1 + k_2}{m_1} \right) & \left(\dfrac{k_2}{m_1} \right) \\ \left(\dfrac{k_2}{m_2} \right) & \left(\omega^2 - \dfrac{k_2 + k_3}{m_2} \right) \end{vmatrix} = 0 \tag{7-6}$$

Expanding this gives

$$\left(\omega^2 - \frac{k_1 + k_2}{m_1} \right) \left(\omega^2 - \frac{k_2 + k_3}{m_2} \right) - \left(\frac{k_2}{m_1} \right) \left(\frac{k_2}{m_2} \right) = 0$$

$$\omega^4 - \left(\frac{k_1 + k_2}{m_1} + \frac{k_2 + k_3}{m_2} \right) \omega^2 + \left(\frac{k_1 k_2 + k_1 k_3 + k_2 k_3}{m_1 m_2} \right) = 0 \tag{7-7}$$

† For a set of n simultaneous linear equations in n unknowns with all the constant terms zero, the determinant of the coefficients of the unknowns must vanish. See C. R. Wylie, Jr., *Advanced Engineering Mathematics*, 3rd ed. (New York: McGraw-Hill, 1966), Corollary 1 of Theorem 7, p. 454.

For the general problem, this is known as the *characteristic equation*, having roots of the characteristic number ω^2. For the vibration case, this is usually referred to as the *frequency equation*.

For two degrees of freedom, there are two differential equations and hence two amplitude equations. Also, the determinant is second order, and the frequency equation is of second degree in ω^2. Furthermore, keeping in mind that the k's and m's are positive physical constants, a study of the sign changes for the frequency equation reveals that for ω^2 two positive real roots are possible but that no negative real roots could exist.

In order to promote a clear understanding of the solution, the problem will be simplified by taking equal spring constants and equal masses for the system. Thus, set $k_1 = k_2 = k_3 = k$ and $m_1 = m_2 = m$. The frequency equation then becomes

$$\omega^4 - 4\frac{k}{m}\omega^2 + 3\left(\frac{k}{m}\right)^2 = 0 \tag{7-8}$$

This may be solved by factoring, giving the roots

$$\omega_1^2 = \frac{k}{m}, \qquad \omega_2^2 = 3\frac{k}{m} \tag{7-9}$$

whence

$$\omega_1 = \sqrt{\frac{k}{m}} \quad \text{and} \quad \omega_2 = \sqrt{\frac{3k}{m}} \tag{7-10}$$

These are the natural circular frequencies of the harmonic motion for the system.

If the lower natural frequency $\omega_1 = \sqrt{k/m}$ is substituted into *either* amplitude equation, the result is

$$A_2 = A_1 \quad \text{or} \quad \frac{A_2}{A_1} = 1 \tag{7-11}$$

This defines the *fundamental* or *first mode* of harmonic vibration, representing the pattern of the motion. This is evidenced by substituting Eq. 7-11 into the solutions, resulting in

$$x_1 = A_1 \sin(\omega_1 t + \phi_1)$$
$$x_2 = A_1 \sin(\omega_1 t + \phi_1) \tag{7-12}$$

with

$$\omega_1 = \sqrt{\frac{k}{m}}$$

and where ϕ_1 represents the phase for the first mode. The values of A_1 and ϕ_1 depend on the initial conditions. The motions of the two equal masses are

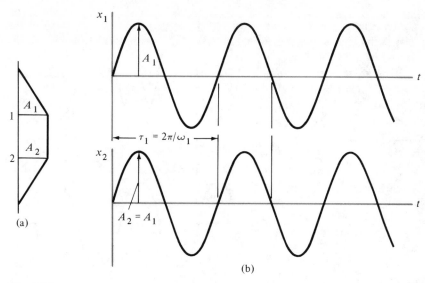

Figure 7-2

identical with respect to time, amplitude, and so on. The mode pattern is represented in Fig. 7-2(a) by plotting the amplitudes horizontally, and the two motions are shown in Fig. 7-2(b), where ϕ_1 has been taken as zero, for convenience. Note that the two bodies move up and down together.

If $\omega_2 = \sqrt{3k/m}$ is substituted into *either* amplitude equation, it is found that

$$A_2 = -A_1 \qquad \text{or} \qquad \frac{A_2}{A_1} = -1 \tag{7-13}$$

This defines the second vibration mode. The solution then becomes

$$x_1 = A_1 \sin (\omega_2 t + \phi_2)$$
$$x_2 = -A_1 \sin (\omega_2 t + \phi_2) \tag{7-14}$$

where

$$\omega_2 = \sqrt{\frac{3k}{m}}$$

Here again, the values of A_1 and ϕ_2 depend on the initial conditions. The second mode is represented in Fig. 7-3(a), and the motions are shown in part (b), with $\phi_2 = 0$ for convenience. In this case the two masses move oppositely to each other. Note that there is a fixed point or *node* located at the midpoint of the central spring.

The solution for the case in which the spring constants are unequal and the masses are different can be readily obtained, although it is somewhat

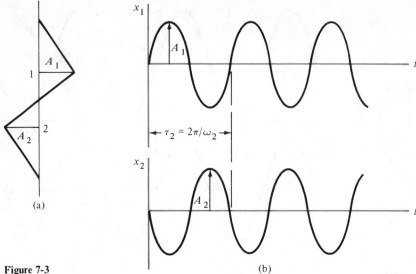

Figure 7-3

cumbersome. The solution process can be simplified by setting

$$2\alpha = \frac{k_1 + k_2}{m_1} \qquad \gamma = \frac{k_2}{m_1}$$

$$2\beta = \frac{k_2 + k_3}{m_2} \qquad \varepsilon = \frac{k_2}{m_2} \tag{7-15}$$

The amplitude equations (Eq. 7-5) then become

$$(\omega^2 - 2\alpha)A_1 + \gamma A_2 = 0$$

$$\varepsilon A_1 + (\omega^2 - 2\beta)A_2 = 0 \tag{7-16}$$

Setting the determinant to zero and expanding results in the frequency equation

$$\omega^4 - 2(\alpha + \beta)\omega^2 + (4\alpha\beta - \gamma\varepsilon) = 0 \tag{7-17}$$

which yields the natural frequencies

$$\omega_1 = \sqrt{(\alpha + \beta) - \sqrt{(\alpha - \beta)^2 + \gamma\varepsilon}}$$

$$\omega_2 = \sqrt{(\alpha + \beta) + \sqrt{(\alpha - \beta)^2 + \gamma\varepsilon}} \tag{7-18}$$

Substituting the expression for ω_1 into the first amplitude equation defines the first mode as

$$\frac{A_2}{A_1} = \frac{(\alpha - \beta) + \sqrt{(\alpha - \beta)^2 + \gamma\varepsilon}}{\gamma}$$

$$= \eta_1 \qquad \text{where } \eta_1 > 0 \tag{7-19}$$

Note that α, β, γ, and ε are all positive. Then the radical term is larger than $(\alpha - \beta)$ and is positive; the numerator is positive, and the amplitude ratio η_1 is also positive, as indicated. In general, η_1 will have a value different from unity, except for certain special cases. The first-mode motion is defined by

$$x_1 = A_1' \sin (\omega_1 t + \phi_1)$$

$$x_2 = \eta_1 A_1' \sin (\omega_1 t + \phi_1) \tag{7-20}$$

Thus in the first mode the bodies move together in a general manner but with different amplitudes.

Similarly, substituting the ω_2 relation of Eq. 7-18 into the first amplitude equation gives the second-mode amplitude ratio as

$$\frac{A_2}{A_1} = \frac{(\alpha - \beta) - \sqrt{(\alpha - \beta)^2 + \gamma\varepsilon}}{\gamma}$$

$$= -\eta_2 \qquad \text{where } \eta_2 > 0 \tag{7-21}$$

Since the radical term is necessarily positive and greater than $(\alpha - \beta)$, then the numerator is negative and the amplitude ratio is also negative, as noted. Again, except for special cases, η_2 is not unity. Motion for the second mode is given by

$$x_1 = A_1'' \sin (\omega_2 t + \phi_2)$$

$$x_2 = -\eta_2 A_1'' \sin (\omega_2 t + \phi_2) \tag{7-22}$$

Accordingly, in the second mode the masses move oppositely and with different amplitudes.

In the first part of this section it was mentioned that other solution forms could be assumed. These would include $x_1 = C_1 e^{st}$, $x_2 = C_2 e^{st}$; $x_1 = C_1 e^{ist}$, $x_2 = C_2 e^{ist}$; $x_1 = \bar{X}_1 e^{i\omega t}$, $x_2 = \bar{X}_2 e^{i\omega t}$; and so on. All these will satisfy the differential equations in much the same manner as the solution form used. The same amplitude equations and frequency equation will result. Furthermore, they all represent harmonic functions and define the same principal modes, frequencies, and so on.

7-3. GENERALITY OF THE ASSUMED SOLUTION

It was noted that the solution (refer to Eq. 7-3) assumed in Section 7-2 imposed certain restrictions; namely, that the harmonic motion of the mass elements would occur with the same frequency and with no phase difference. In order to justify this, the following more general solution form may first be assumed:

$$x_1 = A_1 \sin \omega' t$$

$$x_2 = A_2 \sin (\omega'' t + \beta) \tag{7-23}$$

where ω' is taken to be different from ω''. There is no loss in generality in assuming no phase for x_1 and only a phase difference β between the two motions.

Consider the differential equations of motion to be

$$\ddot{x}_1 + b_{11}x_1 + b_{12}x_2 = 0$$
$$\ddot{x}_2 + b_{21}x_1 + b_{22}x_2 = 0 \qquad (7\text{-}24)$$

where the b terms are coefficients defined by the physical constants of the system. Substituting Eq. 7-23 into Eq. 7-24 gives

$$[b_{11} - (\omega')^2]A_1 \sin \omega't + b_{12}A_2 \sin (\omega''t + \beta) = 0$$
$$b_{21}A_1 \sin \omega't + [b_{22} - (\omega'')^2]A_2 \sin (\omega''t + \beta) = 0 \qquad (7\text{-}25)$$

For the solution to be general, these relations must hold for all values of time. Setting $t = 0$ in the first equation results in

$$b_{12}A_2 \sin \beta = 0$$

Since, by definition b_{12} and A_2 cannot be zero, then

$$\sin \beta = 0$$

whence

$$\beta = 0 \qquad (7\text{-}26)$$

(The value $\beta = \pi$ can be ruled out, because it constitutes only a sign change.) Thus there can be no phase difference in the harmonic motions of the two parts. The first expression may then be written as

$$[b_{11} - (\omega')^2]A_1 \sin \omega't + b_{12}A_2 \sin \omega''t = 0 \qquad (7\text{-}27)$$

This may be rearranged as

$$\frac{\sin \omega''t}{\sin \omega't} = \frac{[(\omega')^2 - b_{11}]A_1}{b_{12}A_2}$$

$$= \text{constant} \qquad (7\text{-}28)$$

Thus the left-hand side must be constant for all values of t. This is possible only if $\omega'' = \omega'$, and consequently the harmonic motion of the two parts takes place at the same frequency.

It may then be stated that the apparent restrictions as to frequency and phase which were assumed in Eq. 7-3 were proper.

7-4. GENERAL SOLUTION AND RELATION TO PRINCIPAL-MODE SOLUTION

In the preceding analysis the principal-mode solutions were obtained. These were then considered separately. This was appropriate, since the motion of either mode could exist alone. This, however, requires starting the system with the proper initial conditions of the particular mode. It is implied that

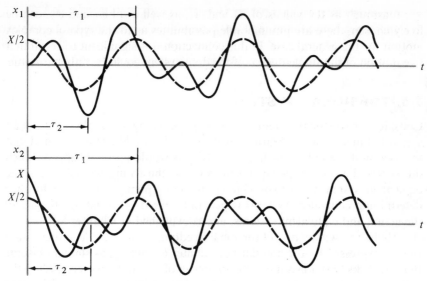

Figure 7-4

any other initial conditions would produce not a principal-mode motion but some other form of movement. The question then arises as to the form of motion produced and how this is accounted for in the solutions obtained.

Each principal mode obtained satisfies the differential equations separately. Since the differential equations are linear, the principle of superposition applies. Hence the sum of the principal modes satisfies the differential equations. These sums will represent the general solution. For the foregoing case, the general solution is then

$$x_1 = A'_1 \sin (\omega_1 t + \phi_1) + A''_1 \sin (\omega_2 t + \phi_2)$$
$$x_2 = \eta_1 A'_1 \sin (\omega_1 t + \phi_1) - \eta_2 A''_1 \sin (\omega_2 t + \phi_2) \qquad (7\text{-}29)$$

The solution contains four arbitrary constants $(A'_1, A''_1, \phi_1,$ and $\phi_2)$, and thus meets the requirement for the general solution for two simultaneous second-order differential equations. These four constants can be evaluated from the available initial conditions of velocity and displacement for the two masses.

The type of motion represented by each relation of Eq. 7-29 is of complex wave form, due to the different frequencies of the parts. An example of the type of motion is shown in Fig. 7-4, being based on the initial conditions of $x_{1_0} = \dot{x}_{1_0} = 0$, $x_{2_0} = X$ and $\dot{x}_{2_0} = 0$. Also, for the sake of simplicity, the case of equal springs and equal masses was taken (hence $\eta_1 = 1 = \eta_2$). The solutions then become

$$x_1 = \frac{X}{2} \cos \omega_1 t - \frac{X}{2} \cos \omega_2 t$$

$$x_2 = \frac{X}{2} \cos \omega_1 t + \frac{X}{2} \cos \omega_2 t \qquad (7\text{-}30)$$

Inasmuch as the values of A_1' and A_1'', as well as of ϕ_1 and ϕ_2, can be freely chosen, there are innumerable possibilities as to the type of complex motion for the general case. In this connection it is significant to recall that free motion of the single-degree-of-freedom system could be only harmonic.

7-5. TORSIONAL SYSTEM

Consider the torsional system, Fig. 7-5(a), consisting of two disks on a shaft supported in frictionless bearings at the ends. It should be kept in mind that this is a model, and the machine it represents would probably appear much different and more complicated. For example, the arrangement might be the equivalent system for a two-cylinder engine or pump. (The procedure for determining equivalent systems is generally based on equating energies of the actual and equivalent systems, and is explained in Chapter 8.)

There are two degrees of torsional freedom, so the system configuration can be expressed by the angular coordinates θ_1 and θ_2, as shown. Assume that the disks have mass moments of inertia of I_1 and I_2, with respect to the rotation axis, as indicated. The torsional stiffness for the portion of the shaft between the disks is expressed by K_T, representing the torsional spring constant. With these things in mind, the free-body diagrams of Fig. 7-5(b) have been constructed, θ_1 having been assumed to be greater than θ_2. The differential equations of motion are

$$I_1 \ddot{\theta}_1 = -K_T(\theta_1 - \theta_2)$$
$$I_2 \ddot{\theta}_2 = K_T(\theta_1 - \theta_2) \tag{7-31}$$

Rearranging gives

$$\ddot{\theta}_1 + \frac{K_T}{I_1}\theta_1 - \frac{K_T}{I_1}\theta_2 = 0$$

$$\ddot{\theta}_2 - \frac{K_T}{I_2}\theta_1 + \frac{K_T}{I_2}\theta_2 = 0 \tag{7-32}$$

The assumed harmonic solutions can be expressed as

$$\theta_1 = A_1 \sin(\omega t + \phi)$$
$$\theta_2 = A_2 \sin(\omega t + \phi)$$

Substituting these yields the amplitude equations

$$\left(\frac{K_T}{I_1} - \omega^2\right)A_1 - \frac{K_T}{I_1}A_2 = 0$$

$$-\frac{K_T}{I_2}A_1 + \left(\frac{K_T}{I_2} - \omega^2\right)A_2 = 0 \tag{7-33}$$

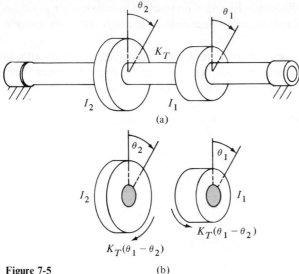

Figure 7-5 (b)

Setting the determinant to zero and expanding leads to the frequency equation

$$\omega^4 - \left(\frac{K_T}{I_1} + \frac{K_T}{I_2}\right)\omega^2 = 0$$

or

$$\omega^2\left[\omega^2 - \frac{K_T}{I_1 I_2}(I_1 + I_2)\right] = 0 \qquad (7\text{-}34)$$

The roots of this are

$$\omega = 0, \qquad \sqrt{\frac{K_T}{I_1 I_2}(I_1 + I_2)} \qquad (7\text{-}35)$$

The frequency value $\omega = 0$ is of interest here. This represents a nonoscillatory condition, which may be defined by substituting $\omega = 0$ into either amplitude equation, resulting in $A_1 = A_2$. Substituting this, in turn, into the assumed solutions yields $\theta_1 = \theta_2$, without defining the solutions. Finally, substituting $\theta_1 = \theta_2$ into the differential equations results in $\ddot{\theta}_1 = 0$ and $\ddot{\theta}_2 = 0$. Solving these by integration, and making use of the condition that $\theta_1 = \theta_2$, gives

$$\theta_1 = Bt + C$$

$$\theta_2 = Bt + C \qquad (7\text{-}36)$$

which will satisfy the differential equations, Eq. 7-32. The two disks rotate together with the same angular velocity (equal to B) and displacement. This is the case of rotation of the system, with no distortion of the shaft K_T. The condition is one of constant-velocity equilibrium, and if $B = 0$, then it is one of static equilibrium.

When a vibration mode vanishes in the foregoing manner, it is called a degenerate mode. It is due here essentially to the absence of an end shaft connection. An arrangement of this nature is said to constitute a degenerate system or a semidefinite system.

For this example, the remaining mode is defined by the other root for ω; namely,

$$\omega = \sqrt{\frac{K_T}{I_1 I_2}(I_1 + I_2)} \qquad (7\text{-}37)$$

Substituting this into the amplitude equations yields

$$\frac{A_2}{A_1} = -\frac{I_1}{I_2} \qquad (7\text{-}38)$$

and the motion is specified by

$$\theta_1 = A_1 \sin\left(\omega t + \phi\right)$$

$$\theta_2 = -\frac{I_1}{I_2} A_1 \sin\left(\omega t + \phi\right) \qquad (7\text{-}39)$$

The mode is plotted in Fig. 7-6. The position of the node can be obtained by simple geometry of the figure.

EXAMPLE 7-1   Determine the principal modes and natural frequencies for the torsional system shown in Fig. 7-7(a).

SOLUTION   Assuming that $\theta_2 > \theta_1$, the free-body diagrams are constructed as indicated in Fig. 7-7(b). The differential equations are then

$$I\ddot{\theta}_1 = -2K_T\theta_1 + K_T(\theta_2 - \theta_1)$$

$$I\ddot{\theta}_2 = -K_T(\theta_2 - \theta_1)$$

Rearranging and substituting the harmonic solution $\theta_j = A_j \sin\left(\omega t + \phi\right)$ gives the amplitude equations

$$\left(3\frac{K_T}{I} - \omega^2\right) A_1 - \frac{K_T}{I} A_2 = 0$$

$$-\frac{K_T}{I} A_1 + \left(\frac{K_T}{I} - \omega^2\right) A_2 = 0$$

Expanding the determinant for this yields the frequency equation

$$\omega^4 - 4\frac{K_T}{I}\omega^2 + 2\left(\frac{K_T}{I}\right)^2 = 0$$

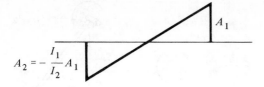

$$A_2 = -\frac{I_1}{I_2}A_1$$

Figure 7-6

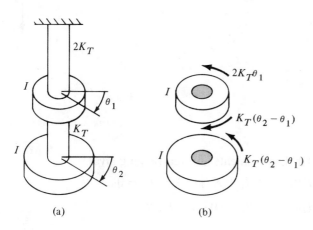

(a)   (b)

Figure 7-7

The roots are

$$\omega^2 = (2 - \sqrt{2})\frac{K_T}{I}, \qquad (2 + \sqrt{2})\frac{K_T}{I}$$

Substituting these into either amplitude equation gives the modes as follows:

$$\omega_1 = \sqrt{[2 - (2)^{1/2}]\frac{K_T}{I}} \qquad \frac{A_2}{A_1} = 1 + \sqrt{2} = 2.414$$

$$\omega_2 = \sqrt{[2 + (2)^{1/2}]\frac{K_T}{I}} \qquad \frac{A_2}{A_1} = 1 - \sqrt{2} = -0.414$$

EXAMPLE 7-2 The two pendulums shown in Fig. 7-8(a) are fastened to a shaft having a torsional stiffness K_T. The shaft rotates freely in end bearings. Obtain the differential equations that govern the motion of the system. Assume the masses to be small.

SOLUTION The free-body diagrams are constructed as represented in Fig. 7-8(b), assuming that $\theta_1 > \theta_2$. The differential equations are

$$m_1\,l_1^2\ddot{\theta}_1 = -W_1 \sin\theta_1\,l_1 - K_T(\theta_1 - \theta_2)$$

$$m_2\,l_2^2\ddot{\theta}_2 = -W_2 \sin\theta_2\,l_2 + K_T(\theta_1 - \theta_2)$$

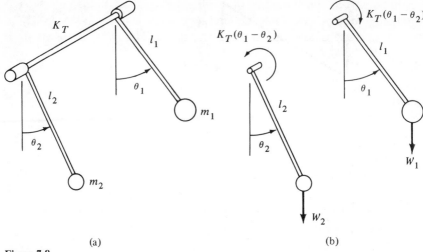

(a)   (b)

Figure 7-8

Assuming small angular displacements and rearranging gives

$$m_1 l_1^2 \ddot{\theta}_1 + (W_1 l_1 + K_T)\theta_1 - K_T\theta_2 = 0$$

$$m_2 l_2^2 \ddot{\theta}_2 - K_T\theta_1 + (W_2 l_2 + K_T)\theta_2 = 0$$

7-6. THE USE OF MATRIX NOTATION

Matrix notation enables the writing of sets of equations in a concise form. Its use involves different mathematical concepts and operations that are efficient for handling the relations for a large number of parameters such as those resulting from several degrees of freedom. One of the reasons for using matrix notation is that input data in this form can be readily utilized by the digital computer through the use of subscripted notation for a matrix array. Although it is inappropriate to employ matrix notation for only two degrees of freedom, it will be introduced here so the reader may become acquainted with it in its simplest form. It will then be used to a greater extent for systems of many degrees of freedom.

For the mass-spring system shown in Fig. 7-1 the differential equations of motion, Eq. 7-1, may be rearranged as follows:

$$m_1 \ddot{x}_1 + (k_1 + k_2)x_1 - k_2 x_2 = 0$$

$$m_2 \ddot{x}_2 - k_2 x_1 + (k_2 + k_3)x_2 = 0 \tag{7-40}$$

These can be written in matrix form as

$$\begin{bmatrix} m_1 & 0 \\ 0 & m_2 \end{bmatrix} \begin{Bmatrix} \ddot{x}_1 \\ \ddot{x}_2 \end{Bmatrix} + \begin{bmatrix} (k_1 + k_2) & -k_2 \\ -k_2 & (k_2 + k_3) \end{bmatrix} \begin{Bmatrix} x_1 \\ x_2 \end{Bmatrix} = \begin{Bmatrix} 0 \\ 0 \end{Bmatrix} \tag{7-41}$$

where it is understood that the coefficients and terms are grouped together in such a way as to form Eq. 7-1. The manner of such grouping can be readily observed and should be studied by the reader. Although a profound knowledge of matrices is not required here, some further understanding of matrix operations and definitions is helpful and may be obtained from Appendix A and related references.

In Eq. 7-41 the matrix

$$\begin{bmatrix} m_1 & 0 \\ 0 & m_2 \end{bmatrix}$$

represents the *mass matrix* and

$$\begin{bmatrix} (k_1 + k_2) & -k_2 \\ -k_2 & (k_2 + k_3) \end{bmatrix}$$

is the *stiffness matrix*. The arrays

$$\begin{Bmatrix} \ddot{x}_1 \\ \ddot{x}_2 \end{Bmatrix} \quad \text{and} \quad \begin{Bmatrix} x_1 \\ x_2 \end{Bmatrix}$$

are column matrices for the accelerations and displacements, respectively, and

$$\begin{Bmatrix} 0 \\ 0 \end{Bmatrix}$$

is a null column matrix.

Observe that in the stiffness matrix the off-diagonal elements of $-k_2$ are the same so that this is a symmetric matrix. The mass matrix is also symmetric and furthermore since the off-diagonal terms are zero, it is a diagonal matrix. General proof of these conditions is covered in Chapter 8.

The harmonic solutions,

$$x_1 = A_1 \sin (\omega t + \phi) \quad \text{and} \quad x_2 = A_2 \sin (\omega t + \phi)$$

may be expressed in matrix form as

$$\begin{Bmatrix} x_1 \\ x_2 \end{Bmatrix} = \begin{Bmatrix} A_1 \\ A_2 \end{Bmatrix} \sin (\omega t + \phi) \tag{7-42}$$

where $\sin (\omega t + \phi)$ is a scalar multiplier of the amplitude column matrix. The second time derivative of this matrix equation is

$$\begin{Bmatrix} \ddot{x}_1 \\ \ddot{x}_2 \end{Bmatrix} = \begin{Bmatrix} A_1 \\ A_2 \end{Bmatrix} (-\omega^2) \sin (\omega t + \phi)$$

Substituting these into Eq. 7-41 gives

$$-\omega^2 \sin (\omega t + \phi) \begin{bmatrix} m_1 & 0 \\ 0 & m_2 \end{bmatrix} \begin{Bmatrix} A_1 \\ A_2 \end{Bmatrix}$$

$$+ \sin (\omega t + \phi) \begin{bmatrix} (k_1 + k_2) & -k_2 \\ -k_2 & (k_2 + k_3) \end{bmatrix} \begin{Bmatrix} A_1 \\ A_2 \end{Bmatrix} = \begin{Bmatrix} 0 \\ 0 \end{Bmatrix} \tag{7-43}$$

Canceling the common scalar multiplier $\sin(\omega t + \phi)$ and factoring results in

$$\left(-\omega^2 \begin{bmatrix} m_1 & 0 \\ 0 & m_2 \end{bmatrix} + \begin{bmatrix} (k_1 + k_2) & -k_2 \\ -k_2 & (k_2 + k_3) \end{bmatrix}\right) \begin{Bmatrix} A_1 \\ A_2 \end{Bmatrix} = \begin{Bmatrix} 0 \\ 0 \end{Bmatrix} \tag{7-44}$$

The coefficient matrices may be combined to give the final matrix form of the amplitude equations as

$$\begin{bmatrix} \omega^2 m_1 - (k_1 + k_2) & k_2 \\ k_2 & \omega^2 m_2 - (k_2 + k_3) \end{bmatrix} \begin{Bmatrix} A_1 \\ A_2 \end{Bmatrix} = \begin{Bmatrix} 0 \\ 0 \end{Bmatrix} \tag{7-45}$$

This agrees with the algebraic form of the amplitude equations listed as Eq. 7-5, and can be pursued further in the manner of Section 7-2 to determine the principal-mode solutions by setting the determinant to zero, expanding to obtain the frequency equation, and so on.

7-7. DISPLACEMENT METHOD OF ANALYSIS FOR PRINCIPAL-MODE VIBRATIONS— AN ALTERNATIVE APPROACH

The method of Section 7-2 is a direct method of obtaining the differential equations. An inverse approach for setting up governing differential equations can be used. This involves the idea of the deformation or displacement of the elastic system. In order to illustrate this method, the mass-and-spring vibration model of Section 7-2 will be used. This is shown again in Fig. 7-9(a). Free-body diagrams of the masses, separated from the system, are shown in Fig. 7-9(b), where P_1 represents the force of the spring connection at the number 1 position acting on the mass m_1, and P_2 is the force of the spring connection at the number 2 position acting upon the mass m_2. From this the following may be written

$$m_1 \ddot{x}_1 = -P_1$$

$$m_2 \ddot{x}_2 = -P_2 \tag{7-46}$$

Figure 7-9(c) represents a deformation diagram of the spring system alone. Here P_1 is the force exerted by m_1 on the spring arrangement at the number 1 position, and P_2 is the force of the mass m_2 on the spring system at the number 2 location. A *flexibility coefficient* a_{ij} is now defined as the displacement at the i position due to a unit force applied at the j location. Displacement relations for the spring arrangement can then be written as

$$x_1 = a_{11} P_1 + a_{12} P_2$$

$$x_2 = a_{21} P_1 + a_{22} P_2 \tag{7-47}$$

Substituting Eq. 7-46 into Eq. 7-47 results in

$$x_1 = a_{11}(-m_1 \ddot{x}_1) + a_{12}(-m_2 \ddot{x}_2)$$

$$x_2 = a_{21}(-m_1 \ddot{x}_1) + a_{22}(-m_2 \ddot{x}_2) \tag{7-48}$$

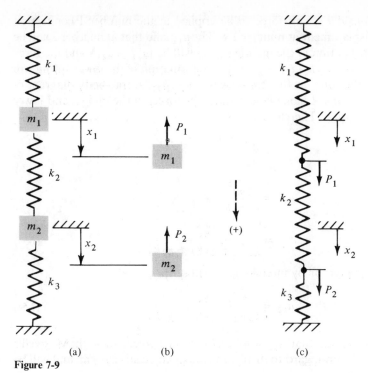

(a)   (b)   (c)

Figure 7-9

which may be rearranged as

$$a_{11}m_1\ddot{x}_1 + a_{12}m_2\ddot{x}_2 + x_1 = 0$$

$$a_{21}m_1\ddot{x}_1 + a_{22}m_2\ddot{x}_2 + x_2 = 0 \qquad (7\text{-}49)$$

These represent the differential equations that govern the motion of the system.

The flexibility coefficients will now be contemplated. Referring to Section 2-10, if a force is applied at the number 1 position, the equivalent spring constant for the spring system is

$$k'' = \frac{k_1 k_2 + k_1 k_3 + k_2 k_3}{k_2 + k_3} \qquad (7\text{-}50)$$

This is the force at number 1 location required to displace the number 1 point by a unit amount. The reciprocal of this will be the displacement at this position caused by a unit force at that point, or a_{11}. Thus

$$a_{11} = \frac{1}{k''} = \frac{k_2 + k_3}{k_1 k_2 + k_1 k_3 + k_2 k_3} \qquad (7\text{-}51)$$

Similarly, or by interchanging subscripts, the following can be written

$$a_{22} = \frac{k_1 + k_2}{k_1 k_2 + k_1 k_3 + k_2 k_3} \qquad (7\text{-}52)$$

Next, consider a unit force to be applied at the number 1 connection only. The displacement at number 1 will be a_{11}, and that at number 2 will be a_{21}. The deformation of the middle spring will be $(a_{11} - a_{21})$, and the force in this spring will be $k_2(a_{11} - a_{21})$. The deformation of the lower spring will be a_{21}, and the force in this spring will be $k_3 a_{21}$. A free-body diagram of connection number 2 alone would show the forces in the middle and lower springs to be equal. Therefore

$$k_3 a_{21} = k_2(a_{11} - a_{21})$$

whence

$$a_{21} = \frac{k_2}{k_2 + k_3} \cdot a_{11}$$

$$= \frac{k_2}{k_1 k_2 + k_1 k_3 + k_2 k_3} \tag{7-53}$$

In similar manner, or by interchanging subscripts,

$$a_{12} = \frac{k_2}{k_1 k_2 + k_1 k_3 + k_2 k_3} \tag{7-54}$$

Comparison reveals that $a_{12} = a_{21}$. This is in agreement with Maxwell's reciprocal law with regard to displacements of an elastic system, and will be proven in Chapter 8.

The expressions for the flexibility coefficients are now substituted into the differential equations (Eq. 7-49) and rearranged, resulting in

$$\left(\frac{k_2 + k_3}{k_1 k_2 + k_1 k_3 + k_2 k_3}\right) m_1 \ddot{x}_1 + \left(\frac{k_2}{k_1 k_2 + k_1 k_3 + k_2 k_3}\right) m_2 \ddot{x}_2 + x_1 = 0$$

$$\left(\frac{k_2}{k_1 k_2 + k_1 k_3 + k_2 k_3}\right) m_1 \ddot{x}_1 + \left(\frac{k_1 + k_2}{k_1 k_2 + k_1 k_3 + k_2 k_3}\right) m_2 \ddot{x}_2 + x_2 = 0 \tag{7-55}$$

For convenience of solution, the case of equal masses and equal springs will now be considered. Also, if the harmonic solutions

$$x_1 = A_1 \sin(\omega t + \phi)$$

$$x_2 = A_2 \sin(\omega t + \phi) \tag{7-56}$$

are substituted into the differential equations, the following amplitude equations are obtained:

$$\left(1 - \frac{2}{3}\frac{m}{k}\omega^2\right) A_1 - \frac{1}{3}\frac{m}{k}\omega^2 A_2 = 0$$

$$-\frac{1}{3}\frac{m}{k}\omega^2 A_1 + \left(1 - \frac{2}{3}\frac{m}{k}\omega^2\right) A_2 = 0 \tag{7-57}$$

These can be simplified by multiplying both equations by $k/m\omega^2$ and letting

$$\frac{k}{m\omega^2} = \lambda \tag{7-58}$$

where λ is an *inverse frequency factor*. The result is

$$(\lambda - \tfrac{2}{3})A_1 - \tfrac{1}{3}A_2 = 0$$
$$-\tfrac{1}{3}A_1 + (\lambda - \tfrac{2}{3})A_2 = 0 \tag{7-59}$$

Setting the determinant to zero and expanding yields the characteristic equation

$$\lambda^2 - \tfrac{4}{3}\lambda + \tfrac{1}{3} = 0 \tag{7-60}$$

having the roots

$$\lambda = 1, \quad \tfrac{1}{3} \tag{7-61}$$

Substituting the root $\lambda_1 = 1$ into either amplitude equation of Eq. 7-59, and using the definition of λ, gives

$$\frac{A_2}{A_1} = 1 \qquad \text{with } \omega_1 = \sqrt{\frac{k}{m}} \tag{7-62}$$

Similarly, the root $\lambda_2 = \tfrac{1}{3}$ results in

$$\frac{A_2}{A_1} = -1 \qquad \text{with } \omega_2 = \sqrt{\frac{3k}{m}} \tag{7-63}$$

These are the same as the principal-mode values given by the direct approach of Section 7-2. It is not proposed that the displacement method of analysis is superior to the direct approach; it has been presented simply as an alternative procedure, and for the sake of the concepts which are embodied in it.

7-8. COUPLING AND PRINCIPAL COORDINATES

The direct approach of Section 7-2 resulted in the differential equations

$$\ddot{x}_1 + \frac{k_1 + k_2}{m_1}x_1 - \frac{k_2}{m_1}x_2 = 0$$

$$\ddot{x}_2 - \frac{k_2}{m_2}x_1 + \frac{k_2 + k_3}{m_2}x_2 = 0 \tag{7-64}$$

Note that neither equation contains only a single variable. More specifically, the first equation contains x_2 in addition to x_1 and $\ddot{x}_1$, and the second equation has x_1 as well as x_2 and $\ddot{x}_2$. The equations are said to be *coupled*. The coupling is due to the presence of x_2 in the first equation and x_1 in the second equation. This is designated as *static* or *elastic coupling*, since it occurs in a displacement coordinate, the idea being that a static or elastic force is transmitted from one part or coordinate of the system to the other. The coupling is physically due to the presence of the spring k_2; if this spring were removed, then $k_2 = 0$, and the differential equations would become

$$\ddot{x}_1 + \frac{k_1}{m_1} x_1 = 0$$

$$\ddot{x}_2 + \frac{k_3}{m_2} x_2 = 0 \tag{7-65}$$

which are not coupled. These represent the differential equations of the two independent systems, and they could be solved separately.

On the other hand the inverse approach produced the differential equations

$$\left(\frac{k_2 + k_3}{k_1 k_2 + k_1 k_3 + k_2 k_3} \right) m_1 \ddot{x}_1 + \left(\frac{k_2}{k_1 k_2 + k_1 k_3 + k_2 k_3} \right) m_2 \ddot{x}_2 + x_1 = 0$$

$$\left(\frac{k_2}{k_1 k_2 + k_1 k_3 + k_2 k_3} \right) m_1 \ddot{x}_1 + \left(\frac{k_1 + k_2}{k_1 k_2 + k_1 k_3 + k_2 k_3} \right) m_2 \ddot{x}_2 + x_2 = 0$$

$$\tag{7-66}$$

Here the coupling occurs in the second-derivative terms, being due to the presence of $\ddot{x}_2$ in the first equation and $\ddot{x}_1$ in the second relation. This is said to be *dynamic* or *inertia coupling*, since it occurs in the acceleration or motion variable. This suggests the concept that the motion of one part is transmitted or linked to the movement of the other, or that mass times acceleration represents a dynamic force. Here again, removing the spring k_2 will eliminate the coupling, and the differential equations will then be the same as Eq. 7-65.

Although the coupling is due to a physical element or connection in the system, it appears that the type of coupling—static or dynamic—depends on the approach or method of deriving the differential equations. Furthermore, by simultaneous elimination Eq. 7-66 can be obtained from Eq. 7-64, and vice versa. In a similar manner either set can be combined so as to yield a set of equations which exhibit both static and dynamic coupling. For example, if the equations of Eq. 7-64 are successively added and subtracted, the following are obtained:

$$\ddot{x}_1 + \ddot{x}_2 + \left(\frac{k_1 + k_2}{m_1} - \frac{k_2}{m_2} \right) x_1 + \left(\frac{k_2 + k_3}{m_2} - \frac{k_2}{m_1} \right) x_2 = 0$$

$$\ddot{x}_1 - \ddot{x}_2 + \left(\frac{k_1 + k_2}{m_1} + \frac{k_2}{m_2} \right) x_1 - \left(\frac{k_2 + k_3}{m_2} + \frac{k_2}{m_1} \right) x_2 = 0 \tag{7-67}$$

The presence of both types of coupling is evident.

It is also possible to select a set of general coordinates such that the differential equations show no coupling. These are called *principal* or *normal coordinates*. To illustrate this consider Eq. 7-67, but with $k_1 = k_2 = k_3 = k$ and $m_1 = m_2 = m$, for the sake of simplification. Then

$$(\ddot{x}_1 + \ddot{x}_2) + \frac{k}{m}(x_1 + x_2) = 0$$

$$(\ddot{x}_1 - \ddot{x}_2) + \frac{3k}{m}(x_1 - x_2) = 0 \tag{7-68}$$

These still exhibit both types of coupling. However, setting

$$x_1 + x_2 = q_1$$

$$x_1 - x_2 = q_2 \tag{7-69}$$

results in

$$\ddot{q}_1 + \frac{k}{m}q_1 = 0$$

$$\ddot{q}_2 + \frac{3k}{m}q_2 = 0 \tag{7-70}$$

which show no coupling and which properly define the motion of the system. They may be solved separately, giving

$$q_1 = B_1 \sin(\omega_1 t + \phi_1) \quad \text{where } \omega_1 = \sqrt{\frac{k}{m}}$$

$$q_2 = B_2 \sin(\omega_2 t + \phi_2) \quad \text{where } \omega_2 = \sqrt{\frac{3k}{m}} \tag{7-71}$$

The motions in coordinates q_1 and q_2 are independent of each other.

It is always possible to obtain such a set of principal coordinates, but the general procedure for doing so is rather involved.[†] Principal coordinates are useful as a reference system for an arrangement of several degrees of freedom involving forcing conditions. (In this example, q_1 is proportional to the displacement of the mass center of the system and q_2 represents the displacement of the upper mass relative to the lower mass.)

The principal-coordinate solutions may be used to obtain the principal-mode solutions in terms of the ordinary coordinates. Thus after substituting

† See N. O. Myklestad, *Fundamentals of Vibration Analysis* (New York: McGraw-Hill, 1956), Art. 21, pp. 181–185.

Eq. 7-71 into Eq. 7-69 and eliminating x_2 and x_1 successively, the following are obtained:

$$x_1 = A'_1 \sin (\omega_1 t + \phi_1) + A''_2 \sin (\omega_2 t + \phi_2)$$
$$x_2 = A'_1 \sin (\omega_1 t + \phi_1) - A''_2 \sin (\omega_2 t + \phi_2) \qquad (7\text{-}72)$$

with

$$\omega_1 = \sqrt{\frac{k}{m}} \quad \text{and} \quad \omega_2 = \sqrt{\frac{3k}{m}}$$

These represent the general solutions for this case.

7-9. MIXED COORDINATES; COUPLED AND AND UNCOUPLED MOTION

Consider the system shown in Fig. 7-10, consisting of a large unsymmetrical body, supported on unequal springs and lying within the plane of the figure. A vehicle suspension may be represented by such a model. (The mass of the body is m, and its mass center is at G.) In defining the configuration of the system it is desirable to use a coordinate x to specify the linear position of some point on the body and coordinate θ to define the angular position of the body, both of these coordinates being measured from the neutral or equilibrium position. These coordinates are mixed; that is, one is linear and the other angular. Because of this, the coupling is of a different sort than was previously considered. Although it is presumed that the system may be coupled, the nature of this coupling is not immediately discernible. Also, the possibility that the system might be uncoupled—and what the condition would be for this—is not evident. These will be brought out by the analyses that follow.

Since the arrangement has the freedom of plane motion, the equations of motion are written accordingly, as listed in Eqs. 1-6 and 1-10. Their form depends on the coordinates selected and this influences the type of coupling which results. Three cases will be considered.

Static Coupling

Select coordinates x_G and θ, where x_G is the linear displacement of mass center G measured from its equilibrium position. The corresponding free-body diagram is shown in Fig. 7-11.

Let J_G be the moment of inertia of the mass with respect to the mass center axis at G. The differential equations of motion are then

$$m\ddot{x}_G = -k_1(x_G - b_1 \theta) - k_2(x_G + b_2 \theta)$$
$$J_G \ddot{\theta} = k_1(x_G - b_1 \theta)b_1 - k_2(x_G + b_2 \theta)b_2 \qquad (7\text{-}73)$$

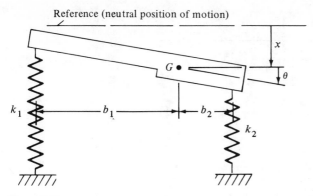

Figure 7-10

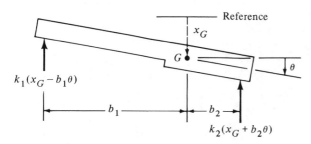

Figure 7-11

These may be rearranged as follows:

$$m\ddot{x}_G + (k_1 + k_2)x_G + (k_2 b_2 - k_1 b_1)\theta = 0$$
$$J_G\ddot{\theta} + (k_2 b_2 - k_1 b_1)x_G + (k_1 b_1^2 + k_2 b_2^2)\theta = 0 \qquad (7\text{-}74)$$

It is evident that these equations exhibit static coupling. Also, the coupling depends on the existence of $(k_2 b_2 - k_1 b_1)$, known as the "coupling coefficient," which is common to the two equations. That is, if

$$k_2 b_2 = k_1 b_1 \qquad (7\text{-}75)$$

the coupling vanishes. Thus when the moments about G of the spring constants are equal, there is no coupling. This condition could readily exist. For the uncoupled case the differential equations become

$$m\ddot{x}_G + (k_1 + k_2)x_G = 0$$
$$J_G\ddot{\theta} + (k_1 b_1^2 + k_2 b_2^2)\theta = 0 \qquad (7\text{-}76)$$

These are independent, and hence the x_G motion and θ motion could occur separately, or at the same time but independently.

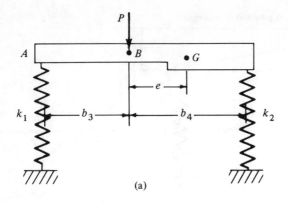

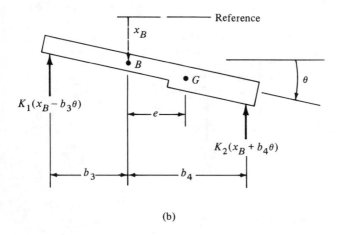

(b)

Figure 7-12

Dynamic Coupling

Choose coordinates x_B and θ, where B is a point on the bar where a static force applied normal to the bar would produce translation only. Referring to Fig. 7-12(a), this point would be where $k_1 b_3 = k_2 b_4$. The attendant free-body diagram is shown in Fig. 7-12(b). Let J_B be the moment of inertia of the mass with respect to an axis at B. The equations of motion are

$$m(\ddot{x}_B + e\ddot{\theta}) = -k_1(x_B - b_3\theta) - k_2(x_B + b_4\theta)$$

$$J_B\ddot{\theta} + m\ddot{x}_Be = k_1(x_B - b_3\theta)b_3 - k_2(x_B + b_4\theta)b_4 \qquad (7\text{-}77)$$

Rearranging and noting that $k_1 b_3 = k_2 b_4$ gives

$$m\ddot{x}_B + me\ddot{\theta} + (k_1 + k_2)x_B = 0$$

$$me\ddot{x}_B + J_B\ddot{\theta} + (k_1 b_3^2 + k_2 b_4^2)\theta = 0 \qquad (7\text{-}78)$$

These equations show dynamic coupling, due to the coordinates employed.

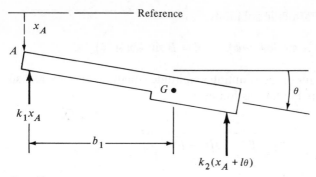

Figure 7-13

Static and Dynamic Coupling

Use coordinates x_A and θ, where A is a point on the bar at the left end. The corresponding free-body diagram is shown in Fig. 7-13. (The distance from G to A is again b_1.)

Letting J_A be the moment of inertia of the mass of the body with respect to A, the governing differential equations would be

$$m(\ddot{x}_A + b_1\ddot{\theta}) = -k_1 x_A - k_2(x_A + l\theta)$$
$$J_A\ddot{\theta} + m\ddot{x}_A b_1 = -k_2(x_A + l\theta) \cdot l \tag{7-79}$$

which may be rearranged as follows:

$$m\ddot{x}_A + mb_1\ddot{\theta} + (k_1 + k_2)x_A + k_2 l\theta = 0$$
$$mb_1\ddot{x}_A + J_A\ddot{\theta} + k_2 l x_A + k_2 l^2\theta = 0 \tag{7-80}$$

These equations exhibit both static and dynamic coupling.

Summarizing the foregoing, and also recalling the discussion of Section 7-8 regarding coupling, it is noted that the type of coupling—static, dynamic, or static and dynamic—depends on the coordinates chosen and on the method of analysis.

EXAMPLE 7-3 Obtain the principal-mode solution to the equations of motion for the case of static coupling above.

SOLUTION In order to simplify the equations, set $J_G = mr^2$ where r is the radius of gyration with respect to G, and let

$$\frac{k_1 + k_2}{m} = \beta \qquad \frac{k_2 b_2 - k_1 b_1}{m} = \gamma \qquad \frac{k_1 b_1^2 + k_2 b_2^2}{J_G} = \eta$$

The differential equations of motion then become

$$\ddot{x}_G + \beta x + \gamma\theta = 0$$

$$\ddot{\theta} + \frac{\gamma}{r^2} + \eta\theta = 0$$

Substituting in the harmonic solutions

$$x = A \sin (\omega t + \phi), \qquad \theta = B \sin (\omega t + \phi)$$

setting the determinant of the resulting amplitude equations to zero, and so on, yields the principal modes as follows:
First mode:

$$\omega_1 = \sqrt{\frac{\eta + \beta}{2} - \sqrt{\left(\frac{\eta - \beta}{2}\right)^2 + \left(\frac{\gamma}{r}\right)^2}}$$

$$\frac{A}{B} = \frac{\gamma}{(\eta - \beta)/2 - \sqrt{[(\eta - \beta)/2]^2 + (\gamma/r)^2}}$$

Second mode:

$$\omega_2 = \sqrt{\frac{\eta + \beta}{2} + \sqrt{\left(\frac{\eta - \beta}{2}\right)^2 + \left(\frac{\gamma}{r}\right)^2}}$$

$$\frac{A}{B} = \frac{\gamma}{(\eta - \beta)/2 + \sqrt{[(\eta - \beta)/2]^2 + (\gamma/r)^2}}$$

The nodes can be obtained from the amplitude ratios and simple geometry. In general, for the first mode the node will be outside the springs, and for the second mode it will be located between the springs.

EXAMPLE 7-4   Consider Fig. 7-10 to represent the equivalent system for a compact automobile. The data include the following:

$$b_1 = 50 \text{ in.} \qquad W = 1930 \text{ lb}$$
$$b_2 = 30 \text{ in.} \qquad k_1 = 120 \text{ lb/in.}$$
$$r = 31.62 \text{ in.} \qquad k_2 = 90 \text{ lb/in.}$$

Determine the principal modes, including the locations of the nodes.

SOLUTION

$$\beta = \frac{k_1 + k_2}{m} = \frac{120 + 90}{1930/386} = \frac{210}{5} = 42$$

$$\gamma = \frac{k_2 b_2 - k_1 b_1}{m} = \frac{90 \times 30 - 120 \times 50}{5} = -660$$

$$\eta = \frac{k_1 b_1^2 + k_2 b_2^2}{mr^2} = \frac{120 \times (50)^2 + 90 \times (30)^2}{5 \times (31.62)^2} = 76.2$$

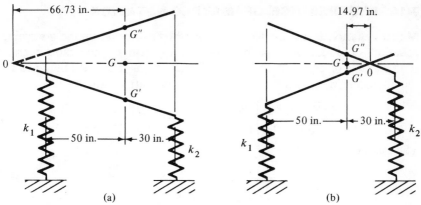

Figure 7-14

Then

$$\omega^2 = \frac{\eta + \beta}{2} \mp \sqrt{\left(\frac{\eta - \beta}{2}\right)^2 + \left(\frac{\gamma}{r}\right)^2}$$

$$= \frac{76.2 + 42}{2} \mp \sqrt{\left(\frac{76.2 - 42}{2}\right)^2 + \left(\frac{-660}{31.62}\right)^2}$$

$$= 59.10 \mp 26.99$$

and the amplitude ratios are

$$\frac{A}{B} = \frac{\gamma}{(\eta - \beta)/2 \mp \sqrt{[(\eta - \beta)/2]^2 + (\gamma/r)^2}} = \frac{-660}{17.10 \mp 26.99}$$

For the first mode, shown in Fig. 7-14(a),

$$\omega_1^2 = 59.10 - 26.99 = 32.11 \quad \text{and} \quad \omega_1 = 5.667 \text{ rad/sec}$$

$$\frac{A_1}{B_1} = \frac{-660}{17.10 - 26.99} = 66.73 \text{ in./rad} = 1.165 \text{ in./deg}$$

For the second mode, represented in Fig. 7-14(b),

$$\omega_2^2 = 59.10 + 26.99 = 86.09 \quad \text{and} \quad \omega_2 = 9.279 \text{ rad/sec}$$

$$\frac{A_2}{B_2} = \frac{-660}{17.10 + 26.99} = -14.97 \text{ in./rad} = -0.2612 \text{ in./deg}$$

The node locations may be obtained by considering the tangent defined for a small angular amplitude such as 0.01 rad. The tangent of the angle will then also be 0.01. Thus for the first mode, if $B_1 = 0.01$ rad, then $A_1 = 0.6673$ in. and the distance $OG = 0.6673$ in./0.01 $= 66.73$ in., which is the same as the amplitude ratio A_1/B_1. Similarly, for the second mode, the distance $OG = -14.97$ in., the same as the amplitude ratio.

7-10. FURTHER USE OF MATRIX NOTATION

Matrix notation is not limited by the method of setting up the governing equations of motion, and may be used for any system of two or more degrees of freedom. For example, the differential equations (Eq. 7-49) obtained by the displacement method may be expressed in matrix form as

$$\begin{bmatrix} a_{11} & a_{12} \\ a_{21} & a_{22} \end{bmatrix} \begin{bmatrix} m_1 & 0 \\ 0 & m_2 \end{bmatrix} \begin{Bmatrix} \ddot{x}_1 \\ \ddot{x}_2 \end{Bmatrix} + \begin{Bmatrix} x_1 \\ x_2 \end{Bmatrix} = \begin{Bmatrix} 0 \\ 0 \end{Bmatrix} \tag{7-81}$$

The matrix

$$\begin{bmatrix} a_{11} & a_{12} \\ a_{21} & a_{22} \end{bmatrix}$$

is called the *flexibility* matrix, and the other matrices are designated the same as in Section 7-6. It will be shown in Chapter 8 that $a_{12} = a_{21}$, so that the flexibility matrix is symmetric. The matrix form of the harmonic solutions is

$$\begin{Bmatrix} x_1 \\ x_2 \end{Bmatrix} = \begin{Bmatrix} A_1 \\ A_2 \end{Bmatrix} \sin(\omega t + \phi)$$

which has the second time derivative

$$\begin{Bmatrix} \ddot{x}_1 \\ \ddot{x}_2 \end{Bmatrix} = \begin{Bmatrix} A_1 \\ A_2 \end{Bmatrix} (-\omega^2) \sin(\omega t + \phi)$$

Substitution into Eq. 7-81 and canceling the common scalar time function results in

$$-\omega^2 \begin{bmatrix} a_{11} & a_{12} \\ a_{21} & a_{22} \end{bmatrix} \begin{bmatrix} m_1 & 0 \\ 0 & m_2 \end{bmatrix} \begin{Bmatrix} A_1 \\ A_2 \end{Bmatrix} + \begin{Bmatrix} A_1 \\ A_2 \end{Bmatrix} = \begin{Bmatrix} 0 \\ 0 \end{Bmatrix} \tag{7-82}$$

Multiplying the coefficient matrices and rearranging gives

$$\begin{bmatrix} (1 - a_{11} m_1 \omega^2) & -a_{12} m_2 \omega^2 \\ -a_{21} m_1 \omega^2 & (1 - a_{22} m_2 \omega^2) \end{bmatrix} \begin{Bmatrix} A_1 \\ A_2 \end{Bmatrix} = \begin{Bmatrix} 0 \\ 0 \end{Bmatrix} \tag{7-83}$$

representing the matrix form of the amplitude equations, which can be pursued further to obtain the principal modes as in Section 7-7.

In like manner for other systems, such as that of Fig. 7-10, the differential equations for the analyses of Section 7-9 involving mixed coordinates may be written in matrix form.

7-11. FREE VIBRATION OF DAMPED SYSTEM

Damping occurs in systems having more than one degree of freedom. An example of a two-degree-of-freedom system with viscous damping is shown in Fig. 7-15(a). Assuming x_1 and $\dot{x}_1$ to be greater than x_2 and $\dot{x}_2$, respec-

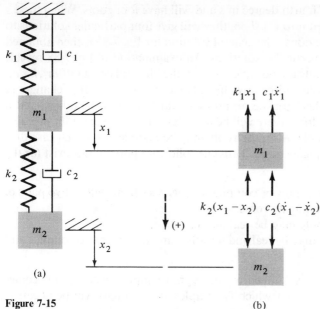

Figure 7-15

(b)

tively, the equation of motion will be

$$m_1\ddot{x}_1 = -k_1 x_1 - k_2(x_1 - x_2) - c_1\dot{x}_1 - c_2(\dot{x}_1 - \dot{x}_2)$$

$$m_2\ddot{x}_2 = k_2(x_1 - x_2) + c_2(\dot{x}_1 - \dot{x}_2) \tag{7-84}$$

which can be rearranged as

$$m_1\ddot{x}_1 + (c_1 + c_2)\dot{x}_1 + (k_1 + k_2)x_1 - c_2\dot{x}_2 - k_2 x_2 = 0$$

$$-c_2\dot{x}_1 - k_2 x_1 + m_2\ddot{x}_2 + c_2\dot{x}_2 + k_2 x_2 = 0 \tag{7-85}$$

Due to the presence of the first-derivative terms as well as the second derivatives, the solution form is taken as

$$x_1 = C_1 e^{st}$$

$$x_2 = C_2 e^{st} \tag{7-86}$$

where s, C_1, and C_2 are undefined constants which may be complex. Substituting these into the differential equations gives the auxiliary equations

$$[m_1 s^2 + (c_1 + c_2)s + (k_1 + k_2)]C_1 - (c_2 s + k_2)C_2 = 0$$

$$-(c_2 s + k_2)C_1 + (m_2 s^2 + c_2 s + k_2)C_2 = 0 \tag{7-87}$$

Setting the determinant to zero and expanding results in the following characteristic equation:

$$m_1 m_2 s^4 + [m_1 c_2 + m_2(c_1 + c_2)]s^3$$

$$+ [m_1 k_2 + m_2(k_1 + k_2) + c_1 c_2]s^2 + (k_1 c_2 + k_2 c_1)s + k_1 k_2 = 0 \tag{7-88}$$

This equation is of fourth degree in s and will have four roots. When these s roots are substituted into Eq. 7-86, they will give four particular solutions to the differential equations. The general solution for Eq. 7-85 is then obtained by combining the particular solutions. An examination of Eq. 7-88 reveals that all the coefficients are positive, since all the physical constants m_1, k_1, c_1, and so on, are considered positive here. Hence there are no sign changes in the relation, and there can be no positive real roots. On the other hand if $(-s)$ is substituted for s, there will be four sign changes; hence four negative real roots are possible. Also, the roots may be complex, in which case they must occur in conjugate pairs. Thus the following possibilities exist for the four roots:

1. All four roots may be complex. In this case, there will be two pairs of complex conjugate roots.
2. All four roots may be real and negative.
3. Two roots may be real and negative, and the other two complex and conjugate.

Since the system is damped, x_1 and x_2 must approach zero as t becomes large. Hence for any root which is complex the real part will be negative. Then for the first case listed, the roots can be expressed as

$$s_1 = -p_1 + iq_1$$
$$s_2 = -p_1 - iq_1$$
$$s_3 = -p_2 + iq_2$$
$$s_4 = -p_2 - iq_2 \qquad (7\text{-}89)$$

wherein p_1, p_2, q_1, and q_2 are positive. The first two roots will yield the following solutions:

$$x_1 = C_{11}e^{(-p_1+iq_1)t} + C_{12}e^{(-p_1-iq_1)t}$$
$$= e^{-p_1 t}(C_{11}e^{iq_1 t} + C_{12}e^{-iq_1 t})$$
$$= A_{11}e^{-p_1 t}\sin(q_1 t + \phi_{11}) \qquad (7\text{-}90a)$$

and

$$x_2 = C_{21}e^{(-p_1+iq_1)t} + C_{22}e^{(-p_1-iq_1)t}$$
$$= A_{21}e^{-p_1 t}\sin(q_1 t + \phi_{21}) \qquad (7\text{-}90b)$$

These represent the case of exponential decay of oscillatory motion, which is much the same as occurred for the single-degree-of-freedom damped system. The rate of decay is defined by the p_1 real part of the root, and the frequency of vibration is specified by the q_1 imaginary part. The ratio C_{21}/C_{11} can be obtained by substitution of s_1 into Eq. 7-87, and C_{22}/C_{12} can be similarly determined from s_2. From these the ratio A_{21}/A_{11} can be found, and ϕ_{21} and ϕ_{11} can be related. In like manner, roots s_3 and s_4 will give

$$x_1 = A_{12}e^{-p_2 t} \sin (q_2 t + \phi_{12})$$

$$x_2 = A_{22}e^{-p_2 t} \sin (q_2 t + \phi_{22}) \tag{7-91}$$

with the ratio A_{22}/A_{12} and the relation between ϕ_{22} and ϕ_{12} being known. These also express the case of exponentially decaying vibratory motion. The general solution is then

$$x_1 = A_{11}e^{-p_1 t} \sin (q_1 t + \phi_{11}) + A_{12}e^{-p_2 t} \sin (q_2 t + \phi_{12})$$

$$x_2 = A_{21}e^{-p_1 t} \sin (q_1 t + \phi_{21}) + A_{22}e^{-p_2 t} \sin (q_2 t + \phi_{22}) \tag{7-92}$$

wherein A_{11} and A_{21}, A_{12}, and A_{22}, ϕ_{11} and ϕ_{21}, and ϕ_{12} and ϕ_{22} are related as noted above, thus reducing the number of these constants to four, which can then be evaluated by the four initial conditions of motion, as specified by x_{1_0}, x_{2_0}, $\dot{x}_{1_0}$ and $\dot{x}_{2_0}$.

If the damping is slight, the values of p_1 and p_2 will be small, and q_1 and q_2 will be approximately the same as ω_1 and ω_2, the frequencies for the system without damping. Except for the slight decay, the motion will approximate that of the undamped system.

If all the roots of Eq. 7-88 are real and negative, then the specific solutions are all of the decaying exponential form. The complete motion of each mass, as defined by the general solution, is then aperiodic, being of the following form:

$$x_1 = C_{11}e^{s_1 t} + C_{12}e^{s_2 t} + C_{13}e^{s_3 t} + C_{14}e^{s_4 t}$$

$$x_2 = C_{21}e^{s_1 t} + C_{22}e^{s_2 t} + C_{23}e^{s_3 t} + C_{24}e^{s_4 t} \tag{7-93}$$

where all the s exponents are negative. The respective C coefficients (that is, C_{11} and C_{21}, C_{12} and C_{22}, and so on) can be related by substituting, separately, the roots s_1, s_2, s_3, and s_4 into either auxiliary equation. The four coefficients which then remain can be evaluated from the initial motion conditions for the two masses. The solutions here represent the case of large damping. When disturbed, the system returns without oscillation to its equilibrium configuration in a general manner similar to that discussed in Section 3-5.

Finally, the case occurs wherein two of the roots of Eq. 7-88 are real and negative, and the remaining two form a complex conjugate pair. These may be expressed as

$$-p + iq, \ -p - iq, \ -\sigma_3, \ -\sigma_4 \tag{7-94}$$

where p, q, σ_3, and σ_4 are positive. The general solutions then become

$$x_1 = C_{11}e^{(-p+iq)t} + C_{12}e^{(-p-iq)t} + C_{13}e^{-\sigma_3 t} + C_{14}e^{-\sigma_4 t}$$

$$x_2 = C_{21}e^{(-p+iq)t} + C_{22}e^{(-p-iq)t} + C_{23}e^{-\sigma_3 t} + C_{24}e^{-\sigma_4 t} \tag{7-95}$$

which may be written in the form

$$x_1 = A_{11} e^{-pt} \sin(qt + \phi_{11}) + C_{13} e^{-\sigma_3 t} + C_{14} e^{-\sigma_4 t}$$

$$x_2 = A_{21} e^{-pt} \sin(qt + \phi_{21}) + C_{23} e^{-\sigma_3 t} + C_{24} e^{-\sigma_4 t} \qquad (7\text{-}96)$$

Here, A_{11} and A_{21}, ϕ_{11} and ϕ_{21}, C_{13} and C_{23}, and C_{14} and C_{24} can be related by substituting into the auxiliary equations (Eq. 7-87) the roots listed as Eq. 7-94. The resulting four constants can be determined by the initial conditions of motion. Equation 7-96 represents, for each mass of the system, an aperiodic-motion part superposed on a damped vibration. The initial conditions can be such that C_{13} and C_{14} are zero, in which case it can be readily shown that C_{23} and C_{24} will also become zero. Thus the aperiodic component will vanish, and the motion will be entirely a decaying oscillation for each mass. Similarly, the damped vibration can be absent, so that the motion is aperiodic only. Particular initial conditions, of course, are required for this.

Although typical motion curves have not been shown for the preceding cases, it should be possible to visualize, without difficulty, the representative motion in each instance. Since the relative values of the various physical constants were not set down, the analyses here are necessarily rather general and may appear somewhat vague. In this connection it should be pointed out that if all the physical constants are known, then the methods outlined will yield solutions that define explicitly the motion of each mass.

7-12. FORCED VIBRATION OF TWO-DEGREE-OF-FREEDOM SYSTEM WITHOUT DAMPING

Forcing conditions occur in systems having more than one degree of freedom. An undamped system of this nature is shown in Fig. 7-16(a). The forcing function is harmonic, as indicated, being defined by $P = P_0 \sin \omega_f t$ acting on mass element m_1. From the attendant free-body diagrams of Fig. 7-16(b), where x_1 was assumed as greater than x_2, the equations of motion are

$$m_1 \ddot{x}_1 = -k_1 x_1 - k_2 (x_1 - x_2) + P$$

$$m_2 \ddot{x}_2 = k_2 (x_1 - x_2) \qquad (7\text{-}97)$$

whence

$$m_1 \ddot{x}_1 + (k_1 + k_2) x_1 - k_2 x_2 = P_0 \sin \omega_f t$$

$$m_2 \ddot{x}_2 - k_2 x_1 + k_2 x_2 = 0 \qquad (7\text{-}98)$$

The solution to the left-hand side of these equations has already been discussed. The main concern here is to obtain the particular solution resulting from the presence of the right-hand-side term. It is reasonable to take a

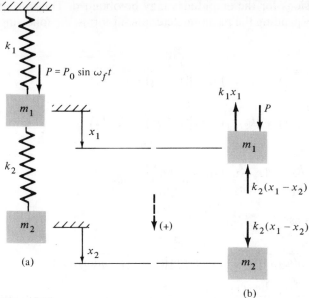

Figure 7-16

particular solution of the same form as that of the forcing function. Thus the following are assumed:

$$x_1 = X_1 \sin \omega_f t$$

$$x_2 = X_2 \sin \omega_f t \tag{7-99}$$

Convenient substitutions include

$\omega_1 = \sqrt{k_1/m_1}$ = natural frequency of system number 1 alone
$\omega_2 = \sqrt{k_2/m_2}$ = natural frequency of system number 2 alone
$X_0 = P_0/k_1$ = static displacement of m_1 that would result from the steady force P_0
$r_1 = \omega_f/\omega_1$ = forced-frequency ratio (for system number 1)
$r_2 = \omega_f/\omega_2$ = forced-frequency ratio (for system number 2)
$b = \omega_2/\omega_1$ = natural-frequency ratio
$\mu = m_2/m_1$ = mass ratio

Substituting these, as well as the assumed particular solutions, into the differential equations of motion reduces them to

$$\left(1 + \frac{k_2}{k_1} - r_1^2\right) X_1 - \left(\frac{k_2}{k_1}\right) X_2 = X_0$$

$$(-1)X_1 + (1 - r_2^2)X_2 = 0 \tag{7-100}$$

From these, expressions for the amplitudes may be obtained. Thus, using Cramer's rule and expanding the resulting determinant forms, the following are obtained:

$$\frac{X_1}{X_0} = \frac{(1 - r_2^2)}{[1 + (k_2/k_1) - r_1^2](1 - r_2^2) - (k_2/k_1)}$$

$$\frac{X_2}{X_0} = \frac{1}{[1 + (k_2/k_1) - r_1^2](1 - r_2^2) - (k_2/k_1)} \tag{7-101}$$

By setting

$$\frac{k_2}{k_1} = \frac{m_2}{m_1} \cdot \frac{k_2}{m_2} \cdot \frac{m_1}{k_1} = \mu \left(\frac{\omega_2}{\omega_1}\right)^2 = \mu b^2$$

these can be written as

$$\frac{X_1}{X_0} = \frac{1 - r_2^2}{[1 + \mu(\omega_2/\omega_1)^2 - (\omega_2/\omega_1)^2 r_2^2](1 - r_2^2) - \mu(\omega_2/\omega_1)^2}$$

$$= \frac{1 - r_2^2}{b^2 r_2^4 - [1 + (1 + \mu)b^2]r_2^2 + 1} \tag{7-102}$$

and

$$\frac{X_2}{X_0} = \frac{1}{b^2 r_2^4 - [1 + (1 + \mu)b^2]r_2^2 + 1} \tag{7-103}$$

The steady-state motions are given by Eq. 7-99, wherein X_1 and X_2 are defined by Eq. 7-101, or by 7-102 and 7-103.

Ordinarily the main consideration here is not the form of the steady-state motion but the amplitude. This can be determined from Eqs. 7-102 and 7-103. For $r_2 = 1$ (thus $\omega_f = \omega_2$), the numerator of Eq. 7-102 vanishes but the denominator does not and has a finite value; hence X_1 becomes zero. For this same condition, it is seen that

$$X_2 = \frac{-X_0}{\mu b^2} = -\frac{k_1}{k_2} \cdot X_0 = -\frac{P_0}{k_2} \tag{7-104}$$

Also, for $\omega_f = 0$, the amplitudes are $X_1 = X_0$ and $X_2 = X_0$; and when ω_f becomes very large, both X_1/X_0 and X_2/X_0 will be small. On the other hand, if the denominator vanishes, both X_1/X_0 and X_2/X_0 become infinite (since this cannot occur at $\omega_f = \omega_2$). The value of r_2, and thence ω_f, at which this occurs can be found by setting the denominator to zero. Thus

$$b^2 r_2^4 - [1 + (1 + \mu)b^2]r_2^2 + 1 = 0 \tag{7-105}$$

For a given system, this is a quadratic equation in r_2^2. Hence there could be two finite values of r_2 which would satisfy this, and thus two forcing frequencies for which there would be a resonant-type infinite amplitude. Note that

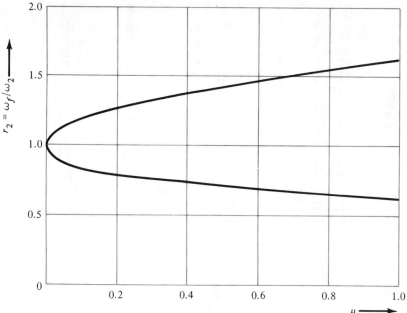

Figure 7-17

the two ω_f frequencies so determined will depend on the values of the physical constants k_1, k_2, m_1, and m_2 of the system. The two resonant frequencies can be determined for any assigned values of b and μ. In order to examine this further, Fig. 7-17 shows r_2 plotted against μ for the case in which $b = 1$ [so that Eq. 7-105 becomes $r_2^4 - (2 + \mu)r_2^2 + 1 = 0$]. For any mass ratio μ other than zero, two r_2 resonant frequency-ratio values are determined, one of these being less than unity and the other being greater than 1.

Returning now to the steady-state amplitudes, it is useful to plot magnification curves of X_1/X_0 and X_2/X_0 against the forced frequency or the frequency ratio. For equal masses and equal springs, $\omega_2 = \omega_1$, $\mu = 1$, and $r_1 = r_2 = r$. The resulting curves, obtained from Eqs. 7-102 and 7-103, are shown in Fig. 7-18. The solid lines indicate that the steady motion is in phase with the forcing function; for the dashed line, the motion is of opposite phase to the forcing condition.

It is significant that even though the forcing condition acts directly on only one mass, the double resonant situation results for both masses. Similar studies could be made of this case in which the forcing condition is applied to m_2 instead of m_1, and for the problem in which forcing functions act on both masses.

EXAMPLE 7-5 A forced two-degree-of-freedom system of the type shown in Fig. 7-16(a) consists of equal masses and equal springs. Determine the amplitudes, as measured by X_1/X_0 and X_2/X_0, for r_2 frequency ratios of

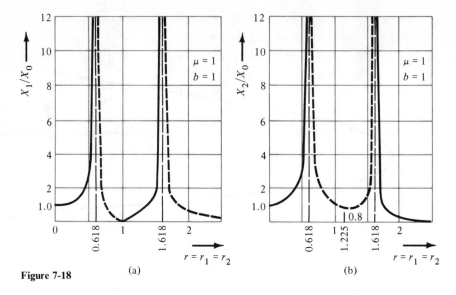

Figure 7-18

(a)   (b)

0.6, 0.63, 0.9, 1.6, 1.65, and 2. Also obtain the value of r_2 for the two resonant conditions.

SOLUTION   For the case of equal masses and equal springs, $\mu = 1$ and $b = 1$ (also $r_1 = r_2 = r$). The required amplitude values can be obtained from Eqs. 7-102 and 7-103, resulting in the following:

$$r_2 = \quad 0.6 \qquad 0.63 \qquad 0.9 \qquad 1.6 \qquad 1.65 \qquad 2$$
$$X_1/X_0 = 12.90 \quad -18.16 \quad -0.2455 \quad 12.34 \quad -7.045 \quad -0.600$$
$$X_2/X_0 = 20.16 \quad -30.12 \quad -1.2921 \quad -7.911 \quad 4.090 \quad 0.200$$

Equation 7-102 becomes $r_2^4 - 3r_2^2 + 1 = 0$, having roots $r_2^2 = 0.3820, 2.618$. Then for resonance, $r_2 = 0.6180, 1.618$.

7-13. UNDAMPED VIBRATION ABSORBER

In the preceding section the general conditions for the steady-state motion of a two-degree-of-freedom undamped system subjected to a harmonic forcing condition were discussed. The relations and analysis there may also be used as the basis for the undamped vibration absorber. First, consider a single-degree-of-freedom system, consisting of mass m_1 suspended from spring k_1 and subjected to the harmonic force $P = P_0 \sin \omega_f t$, such as that shown in Fig. 4-1(a). Under many conditions the steady-state amplitude will be appreciable or even large (as discussed in Section 4-3). However, if the mass m_2 is fastened to the mass m_1 by means of a spring k_2, then the system becomes identical to that of Fig. 7-16(a), and the relations and ideas of the preceding section will apply. It is now possible to reduce the steady amplitude X_1 of the mass m_1 to zero. This is done by setting $r_2 = 1$ in Eq. 7-102. For this condition, $\omega_f/\omega_2 = 1$ or $\omega_2 = \omega_f$ and

$$\sqrt{\frac{k_2}{m_2}} = \omega_f \qquad (7\text{-}106)$$

Thus by proper selection of the spring k_2 and mass m_2 for the absorber, the steady motion of the main mass m_1 can be eliminated. The absorber amplitude is then obtained by setting $r_2 = 1$ in Eq. 7-101, which gives

$$\frac{X_2}{X_0} = -\frac{k_1}{k_2}$$

whence

$$X_2 = -X_0 \frac{k_1}{k_2} = -\frac{P_0}{k_1} \frac{k_1}{k_2}$$

$$= -\frac{P_0}{k_2} \qquad \text{or} \qquad k_2 X_2 = -P_0 \qquad (7\text{-}107)$$

where the negative sign shows that X_2 is of opposite phase to the impressed force P_0. Actually, X_1 is zero because the force $(-k_2 X_2)$ exerted by spring number 2 on the mass m_1 is equal and opposite to the force P_0.

The foregoing can be used for designing an absorber, that is, selecting k_2 and m_2. It is presumed that ω_f and P_0 are known, and X_2 may be assumed or determined from an allowable displacement of the absorber mass, being dependent on practical limitations, such as space. The absorber spring constant can then be determined by Eq. 7-107, disregarding the negative sign; that is, by

$$k_2 = \frac{P_0}{X_2} \qquad (7\text{-}108)$$

The absorber mass may then be obtained from Eq. 7-106, since ω_f is known. In general, for this type of vibration absorber, it is desirable to use an absorber spring k_2 and a mass m_2 which are small compared to k_1 and m_1 of the main system. This results in a rather large amplitude X_2 for the absorber. This is illustrated by the magnification curves shown plotted against the absorber-frequency ratio r_2 in Fig. 7-19 for $\mu = 0.2$ and $\omega_2 = \omega_1$. The curves can be readily verified by Eqs. 7-102 and 7-103. For the operating condition of $r_2 = 1$, where $X_1 = 0$, it is found that $X_2 = 5X_0$.

There are certain inherent difficulties in this type of absorber. First, it is effective at only one frequency, and if ω_f varies appreciably from the expected value, the absorber may cause serious trouble because the range between the two resonant frequencies is rather small. On a fixed-speed machine the absorber will be satisfactory in this regard. However, when the machine is started, it is necessary to go through the lower resonant

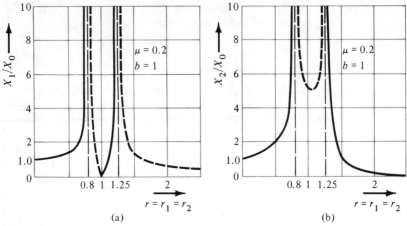

Figure 7-19

frequency in order to reach the operating frequency, and this is an undesirable condition. Also, if loss of load or some other factor results in a gain in speed, difficulty can be encountered at the upper resonant frequency.

For a torsional system composed of a flywheel or disk on a shaft, an absorber consisting of an additional disk and shaft can be employed. The foregoing analysis can then be applied simply by substituting the mass moment of inertia I of the disks for m of the mass elements and using the torsional spring constant K_T for the shaft parts in place of k for the rectilinear spring constants. Such a torsional absorber works satisfactorily, but it is subject to the same limitations as noted above for the rectilinear-system absorber.

EXAMPLE 7-6   A machine weighing 50 lb is supported by a set of four springs each of which has a spring modulus of 15 lb/in. It is subjected to a harmonic force having an amplitude of 3 lb and a frequency of 180 cycles/min. Design a vibration absorber for this system, determining the absorber spring constant and the mass. Clearance limits the absorber amplitude to 0.8 in., and standard springs are available with constants in 1 lb/in. multiples.

SOLUTION   The preliminary calculation for k_2 is

$$k_2 = \frac{P_0}{X_2} = \frac{3 \text{ lb}}{0.8 \text{ in.}} = 3.75 \text{ lb/in.}$$

Accordingly, select a spring so that

$$k_2 = 4 \text{ lb/in.}$$

Then

$$X_2 = \frac{P_0}{k_2} = \frac{3 \text{ lb}}{4 \text{ lb/in.}} = 0.75 \text{ in.} < 0.8 \text{ in.}$$

$$m_2 = \frac{k_2}{\omega_f^2} = \frac{4}{\left(\frac{180}{60} \times 2\pi\right)^2} = 0.011\,257\,9 \text{ lb sec/in.}$$

$$W_2 = m_2 g = 0.011\,257\,9 \times 386 = 4.3455 \text{ lb}$$

EXAMPLE 7-7 Consider an undamped vibration absorber for which $\omega_2 = \omega_1$ and $\mu = 0.2$. Determine the operating range of the r_2 frequency for which $|X_1/X_0| \le 0.5$.

SOLUTION   Equation 7-102 becomes

$$\pm 0.5 = \frac{1 - r_2^2}{r_2^4 - 2.2r_2^2 + 1}$$

As indicated by Fig. 7-19(a), the positive sign will give the upper limiting value for r_2, and the negative sign will yield the lower limit of r_2, for the range extending from somewhat below $r_2 = 1$ to slightly above this value.

For plus 0.5, the above expression can be rearranged as

$$r_2^4 - 0.2r_2^2 - 1 = 0$$

whence

$$r_2^2 = 1.105, \qquad r_2 = 1.051$$

For minus 0.5, the relation becomes

$$r_2^4 - 4.2r_2^2 + 3 = 0$$

whence

$$r_2^2 = 0.9126, \qquad r_2 = 0.9553$$

(There is another root here, for which $r_2 = 1.813$. This defines $X_1/X_0 = -0.5$ at r_2 above the upper resonance point.) Thus the range is less than 10% of the frequency value for which the absorber was designed ($\omega_f = \omega_2$ or $r_2 = 1$).

7-14. VIBRATION OF FORCED DAMPED SYSTEM

Forcing conditions act on damped systems of more than one degree of freedom, for example, as represented in Fig. 7-20. Although a damping element is shown only between the two mass elements, this does not greatly reduce the generality of the conditions or problem. The force P is harmonic

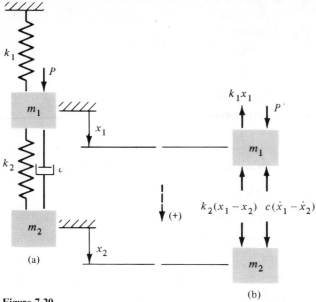

Figure 7-20

and could be defined by a trigonometric function. However, it can also be expressed as

$$P = P_0 e^{i\omega_f t} \tag{7-109}$$

where it is understood that the harmonic force P is represented by the real part of the expression. This form is used in order to simplify the solution procedure.

The differential equations are

$$m_1 \ddot{x}_1 = -k_1 x_1 - k_2(x_1 - x_2) - c(\dot{x}_1 - \dot{x}_2) + P$$
$$m_2 \ddot{x}_2 = k_2(x_1 - x_2) + c(\dot{x}_1 - \dot{x}_2) \tag{7-110}$$

which can be rearranged as

$$m_1 \ddot{x}_1 + c\dot{x}_1 + (k_1 + k_2)x_1 - c\dot{x}_2 - k_2 x_2 = P$$
$$-c\dot{x}_1 - k_2 x_1 + m_2 \ddot{x}_2 + c\dot{x}_2 + k_2 x_2 = 0 \tag{7-111}$$

The solution will consist of two parts, one of which is obtained by the solution for the left side of the equations, with the right side set equal to zero. This part will represent the transient or free vibration for the system. The solution will be similar to that explained in Section 7-11. The motion of this part will eventually die out, and hence it is not of as great concern as the steady state defined by the particular solution. The part which remains to be determined must satisfy the complete differential equations. Since this involves the harmonic driving function P, it is reasonable to select a solution of the same form. Experience with the similar single-degree-of-freedom forced case indicates that the motion would exhibit phase lag with respect to the forcing condition. Hence the steady-state solution is set down as

$$x_1 = \bar{X}_1 e^{i\omega_f t}$$

$$x_2 = \bar{X}_2 e^{i\omega_f t} \tag{7-112}$$

where the complex amplitudes are given by

$$\bar{X}_1 = X_1 e^{-i\psi}$$

$$\bar{X}_2 = X_2 e^{-i\psi} \tag{7-113}$$

It is understood here that the harmonic motion is defined by the real parts of Eq. 7-112.

Substituting Eqs. 7-109 and 7-112 into the differential equations results in

$$(-m_1 \omega_f^2 + k_1 + k_2 + ic\omega_f)\bar{X}_1 - (k_2 + ic\omega_f)\bar{X}_2 = P_0$$

$$-(k_2 + ic\omega_f)\bar{X}_1 + (-m_2 \omega_f^2 + k_2 + ic\omega_f)\bar{X}_2 = 0 \tag{7-114}$$

Using determinants and Cramer's rule, the following can be set down:

$$\bar{X}_1 = \frac{D_{\bar{X}_1}}{D} = \frac{\begin{vmatrix} P_0 & -(k_2 + ic\omega_f) \\ 0 & (-m_2 \omega_f^2 + k_2 + ic\omega_f) \end{vmatrix}}{\begin{vmatrix} (-m_1 \omega_f^2 + k_1 + k_2 + ic\omega_f) & -(k_2 + ic\omega_f) \\ -(k_2 + ic\omega_f) & (-m_2 \omega_f^2 + k_2 + ic\omega_f) \end{vmatrix}} \tag{7-115}$$

Expanding and simplifying gives

$$\bar{X}_1 = \frac{P_0[(k_2 - m_2 \omega_f^2) + ic\omega_f]}{[(k_1 - m_1 \omega_f^2)(k_2 - m_2 \omega_f^2) - m_2 \omega_f^2 k_2] + ic\omega_f(-m_1 \omega_f^2 + k_1 - m_2 \omega_f^2)} \tag{7-116}$$

Before proceeding, some additional relations for complex numbers will be developed. From Fig. 2-6 and Eq. 2-30,

$$a + ib = re^{i\theta}$$

$$= \sqrt{a^2 + b^2}\, e^{i\theta} \tag{7-117}$$

and

$$\tan \theta = \frac{b}{a} \tag{7-118}$$

Then a complex-number ratio can be transformed as follows:

$$\frac{A + iB}{C + iD} = \frac{A + iB}{C + iD} \cdot \frac{C - iD}{C - iD} = \frac{(AC + BD) + i(BC - AD)}{C^2 + D^2}$$

$$= \frac{\sqrt{(AC + BD)^2 + (BC - AD)^2}}{C^2 + D^2} e^{i\eta}$$

$$= \frac{\sqrt{(A^2 + B^2)(C^2 + D^2)}}{C^2 + D^2} e^{i\eta}$$

$$= \sqrt{\frac{A^2 + B^2}{C^2 + D^2}} \cdot e^{i\eta} \qquad (7\text{-}119)$$

where

$$\tan \eta = \frac{BC - AD}{AC + BD} \qquad (7\text{-}120)$$

Equation 7-116 can now be written as

$$X_1 e^{-i\psi} = \frac{P_0 \sqrt{(k_2 - m_2 \omega_f^2)^2 + (c\omega_f)^2} \cdot e^{i\eta}}{\sqrt{\begin{array}{l}[(k_1 - m_1\omega_f^2)(k_2 - m_2\omega_f^2) - m_2 k_2 \omega_f^2]^2 \\ \qquad + (c\omega_f)^2(-m_1\omega_f^2 + k_1 - m_2\omega_f^2)^2\end{array}}}$$

whence $\psi = -\eta$, and $\qquad (7\text{-}121)$

$$X_1 = \frac{P_0 \sqrt{(k_2 - m_2 \omega_f^2)^2 + (c\omega_f)^2}}{\sqrt{\begin{array}{l}[(k_1 - m_1\omega_f^2)(k_2 - m_2\omega_f^2) - m_2 k_2 \omega_f^2]^2 \\ \qquad + (c\omega_f)^2(m_1\omega_f^2 - k_1 + m_2\omega_f^2)^2\end{array}}} \qquad (7\text{-}122)$$

Similarly,

$$\bar{X}_2 = \frac{D_{\bar{x}_2}}{D} = \frac{\begin{vmatrix} (-m_1\omega_f^2 + k_1 + k_2 + ic\omega_f) & P_0 \\ -(k_2 + ic\omega_f) & 0 \end{vmatrix}}{D} \qquad (7\text{-}123)$$

and

$$X_2 = \frac{P_0 \sqrt{k_2^2 + (c\omega_f)^2}}{\sqrt{\begin{array}{l}[(k_1 - m_1\omega_f^2)(k_2 - m_2\omega_f^2) - m_2 k_2 \omega_f^2]^2 \\ \qquad + (c\omega_f)^2(m_1\omega_f^2 - k_1 + m_2\omega_f^2)^2\end{array}}} \qquad (7\text{-}124)$$

The forcing function and the steady-state solutions can be written in trigonometric form as

$$P = P_0 \cos \omega_f t$$

$$x_1 = X_1 \cos (\omega_f t - \psi_1)$$

$$x_2 = X_2 \cos (\omega_f t - \psi_2) \qquad (7\text{-}125)$$

wherein X_1 and X_2 are defined by Eqs. 7-122 and 7-124.

The phase angles have been designated by different subscripts. This is because they will have different values, as can be seen from Eq. 7-120, since the factor A will not be the same for the two cases. Thus the x_1 motion and the x_2 motion will lag the force P by different amounts, which means that x_1 and x_2 will not be in phase. This is generally not important; hence the relations that define ψ_1 and ψ_2 have not been developed here.

The steady-state motion consists of a continuous harmonic oscillation of each mass. The displacement amplitude of each mass is a significant measurement and will be investigated. In order to do so the amplitude relations will be reduced to nondimensional form, and for this purpose the relations for $\omega_1, \omega_2, \ldots, r_1, r_2, \ldots$, and so on, listed in Section 7-12, and the following additional substitutions will be used:

$$c_c = 2m_2 \omega_1 = \text{critical damping constant}$$

$$\zeta = \frac{c}{c_c} = \text{damping factor}$$

On the right-hand side of Eq. 7-122, both the numerator and the denominator are then multiplied by $(1/k_1) \cdot (m_1/m_2 k_1)$. Rearranging and making the indicated substitutions results in

$$\frac{X_1}{X_0} = \frac{\sqrt{(r_1^2 - b^2)^2 + (2\zeta r_1)^2}}{\sqrt{(2\zeta r_1)^2 (r_1^2 - 1 + \mu r_1^2)^2 + [\mu b^2 r_1^2 - (r_1^2 - 1)(r_1^2 - b^2)]^2}} \qquad (7\text{-}126)$$

A similar treatment of Eq. 7-124 gives

$$\frac{X_2}{X_0} = \frac{\sqrt{b^4 + (2\zeta r_1)^2}}{\sqrt{(2\zeta r_1)^2 (r_1^2 - 1 + \mu r_1^2)^2 + [\mu b^2 r_1^2 - (r_1^2 - 1)(r_1^2 - b^2)]^2}} \qquad (7\text{-}127)$$

These relations express the amplitude, relative to the reference X_0, as a function of the four nondimensional parameters μ, ζ, r_1 and b. In studying this, it is useful to plot the amplitude against the forced-frequency ratio. This necessitates selecting values for the remaining parameters. For equal masses and equal springs, $\mu = 1, b = 1$, and $r_1 = r_2 = r$. Equations 7-126 and 7-127 then become

$$\frac{X_1}{X_0} = \frac{\sqrt{(r^2 - 1)^2 + (2\zeta r)^2}}{\sqrt{(2\zeta r)^2 (2r^2 - 1)^2 + (r^4 - 3r^2 + 1)^2}} \qquad (7\text{-}128)$$

$$\frac{X_2}{X_0} = \frac{\sqrt{1 + (2\zeta r)^2}}{\sqrt{(2\zeta r)^2 (2r^2 - 1)^2 + (r^4 - 3r^2 + 1)^2}} \qquad (7\text{-}129)$$

The form of the curve obtained from these depends on damping. Two limiting cases are significant—for $\zeta = 0$ and for $\zeta = \infty$. Placing $\zeta = 0$ in the above expressions reduces these to the same form as for the undamped case of Section 7-12 and results in curves identical to those in Fig. 7-18. These

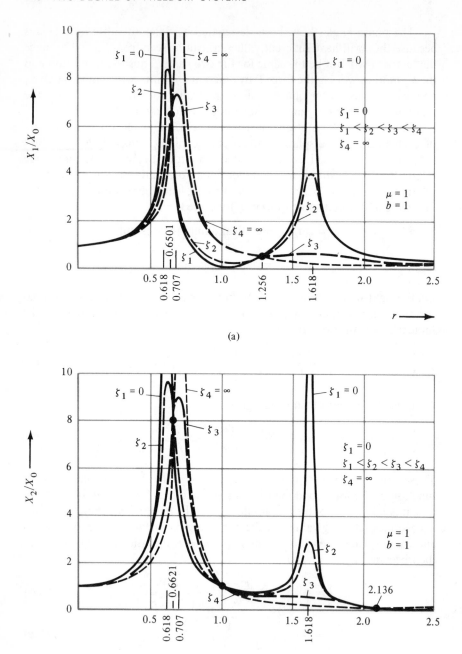

(a)

(b)

Figure 7-21

have been repeated in Fig. 7-21. For $\zeta = \infty$, Eqs. 7-128 and 7-129 become

$$\frac{X_1}{X_0} = \frac{\pm 1}{1 - 2r^2} = \frac{X_2}{X_0} \qquad (7\text{-}130)$$

In effect, the two masses are locked, but with a single spring acting, so that $\omega = \sqrt{k/2m}$ and $r = \sqrt{2r_1}$, and Eq. 7-130 thus checks Eqs. 4-16 and 4-17. Resonance then occurs at $r = 0.707$. Curves for this case are also shown in Fig. 7-21. For small damping, the curve should be similar to that for $\zeta_1 = 0$, but with both resonant amplitudes reduced to finite values. This is shown by the curve for ζ_2. Similarly, for moderately large damping, the curve should be similar to that for $\zeta_4 = \infty$, but with the resonant amplitude reduced to a finite amount. A curve of this type is represented by that for ζ_3. Curves such as those for ζ_2 and ζ_3 can be determined from Eqs. 7-128 and 7-129. All the curves have common crossover points, as shown by the circled points in the figures.

The preceding discussions and relations can be used as the basis for the damped vibration absorber since the arrangement contains the same system elements. A thorough analysis of such an absorber is lengthy and detailed and has not been included here.†

7-15. EXCITATION AND STABILITY OF SYSTEMS

For single-degree-of-freedom systems subjected to some form of self-excitation, it is relatively simple to determine stability from the physical constants of the system, as explained in Section 4-19. However, the problem of determining stability becomes more involved with increase in the number of degrees of freedom, and accordingly, it is desirable to ascertain the criteria for stability of such systems.

Consider a damped two-degree-of-freedom system to be subjected to self-excitation which is a linear function of the velocities, that is, for forcing functions of the form $P_1\dot{x}_1$, $P_{12}(\dot{x}_1 - \dot{x}_2)$ and $P_2\dot{x}_2$. Also assume that no other forcing conditions act on the system. The self-exciting functions may be combined with the damping terms, resulting in differential equations of the form

$$\ddot{x}_1 + \alpha_{11}\dot{x}_1 + \alpha_{12}\dot{x}_2 + \beta_{11}x_1 + \beta_{12}x_2 = 0$$

$$\ddot{x}_2 + \alpha_{21}\dot{x}_1 + \alpha_{22}\dot{x}_2 + \beta_{21}x_1 + \beta_{22}x_2 = 0 \qquad (7\text{-}131)$$

Note that the signs of the α coefficients are not defined here. Substituting in the solutions $x_1 = C_1 e^{st}$ and $x_2 = C_2 e^{st}$ and expanding the determinant of

† See J. P. Den Hartog, *Mechanical Vibrations*, 4th ed. (New York: McGraw-Hill, 1956), pp. 97–105. See also S. P. Timoshenko, D. H. Young, and W. Weaver, Jr., *Vibration Problems in Engineering*, 4th ed. (New York: Wiley, 1974), Example, pp. 273–278.

the resulting auxiliary equations will produce a characteristic equation of the following general form:

$$s^4 + a_3 s^3 + a_2 s^2 + a_1 s + a_0 = 0 \qquad (7\text{-}132)$$

In this relation the coefficients a_0, a_1, a_2, and a_3 are real since they result from physical constants of the system. For a two-degree-of-freedom system, in general, such a quartic equation is obtained. If there is a condition of degeneracy, then a cubic equation will result; that is, the general form will be

$$s^3 + a_2 s^2 + a_1 s + a_0 = 0 \qquad (7\text{-}133)$$

This latter equation will be investigated with a view to determining the conditions for stability. In the first place, if all the coefficients a_0, a_1, and a_2 are positive, then there are no sign changes and there can be no positive real roots for s. Hence for this condition the solutions cannot be increasing exponentials, representing diverging nonoscillatory motion. On the other hand, if all the a coefficients are positive, substituting $-s$ for $+s$ results in three sign changes, and the roots can be real and negative. For such roots, the solutions would represent converging nonoscillatory motion. From the foregoing it is concluded that one condition for stability is that all the coefficients a_0, a_1, and a_2 of Eq. 7-133 must be positive.

An additional criterion will now be obtained, based on the prospect that the roots for s may be complex, in which case they must occur in conjugate pairs. Accordingly the roots can be expressed as

$$s_1 = p_1$$
$$s_2 = p_2 + iq_2$$
$$s_3 = p_2 - iq_2 \qquad (7\text{-}134)$$

Equation 7-133 can now be written as

$$(s - s_1) \cdot (s - s_2) \cdot (s - s_3) = 0$$

and expanding gives

$$s^3 - (s_1 + s_2 + s_3)s^2 + (s_1 s_2 + s_1 s_3 + s_2 s_3)s - s_1 s_2 s_3 = 0 \qquad (7\text{-}135)$$

Comparing this with Eq. 7-133 shows that

$$a_2 = -(s_1 + s_2 + s_3)$$
$$a_1 = s_1 s_2 + s_1 s_3 + s_2 s_3$$
$$a_0 = -s_1 s_2 s_3$$

and substituting in Eq. 7-134 gives

$$a_2 = -(p_1 + 2p_2)$$
$$a_1 = 2p_1 p_2 + p_2^2 + q_2^2$$
$$a_0 = -p_1(p_2^2 + q_2^2) \qquad (7\text{-}136)$$

The solution represented by the roots for Eq. 7-133 will be of the form

$$x_1 = C_{11}e^{p_1t} + e^{p_2t}C_{12}\sin(q_2t + \phi_{11})$$

$$x_2 = C_{21}e^{p_1t} + e^{p_2t}C_{22}\sin(q_2t + \phi_{12}) \tag{7-137}$$

For the roots expressed in Eq. 7-134, the q_2 coefficient of the imaginary part represents the frequency, and the real part p_2 determines the rate of change of amplitude of the oscillatory portion of the motion. The real part p_1 determines the rate of change of the exponential, nonoscillatory, part of the motion. For dynamic stability, both p_1 and p_2 must be negative. Examination of Eq. 7-136 reveals that for p_1 and p_2 negative, all the coefficients a_0, a_1, and a_2 will be positive. However, the a coefficients being positive is not sufficient to ensure that both p_1 and p_2 are negative. The third relation shows that if a_0 is positive then p_1 will be negative. However, a_0, a_1, and a_2 being positive does not require p_2 to be negative. For p_2, the limiting condition between stability and instability is that p_2 be zero in passing from a negative to a positive value. Accordingly, p_2 is set to zero in Eq. 7-136, resulting in

$$a_2 = -p_1$$

$$a_1 = q_2^2$$

$$a_0 = -p_1 q_2^2 \tag{7-138}$$

Eliminating p_1 and q_2 between these gives

$$a_1 a_2 = a_0 \tag{7-139}$$

This holds for the border line between stability and instability. For the condition of stability, it is necessary to determine whether $a_1 a_2$ is greater or less than a_0. It will now be shown that if p_2, as well as p_1, is negative, then

$$a_1 a_2 > a_0 \tag{7-140}$$

Substituting the expressions from Eq. 7-136 into Eq. 7-140 results in

$$(2p_1 p_2 + p_2^2 + q_2^2) \cdot [-(p_1 + 2p_2)] > [-p_1(p_2^2 + q_2^2)]$$

Expanding gives

$$-2p_1 p_2(p_1 + 2p_2) - p_1(p_2^2 + q_2^2) - 2p_2(p_2^2 + q_2^2) > -p_1(p_2^2 + q_2^2)$$

and canceling results in

$$-2p_1 p_2(p_1 + 2p_2) - 2p_2(p_2^2 + q_2^2) > 0$$

Since $-2p_2$ is positive, dividing by this will not alter the order of the equality. Thus

$$p_1(p_1 + 2p_2) + (p_2^2 + q_2^2) > 0$$

and

$$p_1^2 + 2p_1 p_2 + p_2^2 + q_2^2 > 0 \tag{7-141}$$

Since p_1 and p_2 were to be taken as negative here, all parts of Eq. 7-141 are positive, and the inequality is fulfilled. Thus for stability, Eq. 7-140 must hold.

Finally, it can be stated that if all the coefficients a_0, a_1, and a_2 are positive and meet the requirement $a_1 a_2 > a_0$, the system defined by the cubic characteristic equation (Eq. 7-133) will be dynamically stable. If any of these conditions are not met by the coefficients, the system will be dynamically unstable.

An analysis similar to the foregoing, but more lengthy, can be made for the fourth-degree characteristic equation (Eq. 7-132). From this the criteria† for stability of the system defined by Eq. 7-132 are that all the coefficients a_0, a_1, a_2, and a_3 must be positive and also meet the requirement

$$a_1 a_2 a_3 > a_1^2 + a_3^2 a_0 \tag{7-142}$$

EXAMPLE 7-8 For the two-degree-of-freedom system shown in Fig. 7-22(a), determine the conditions for dynamic stability, discuss briefly the physical significance of these conditions, and interpret these for the case of equal masses. Consider that $F_0 > 0$.

SOLUTION The free-body diagrams are shown in Fig. 7-22(b) and from these the differential equations are

$$m_1 \ddot{x}_1 = F_0 \dot{x}_1 - k(x_1 - x_2)$$
$$m_2 \ddot{x}_2 = k(x_1 - x_2) - c\dot{x}_2$$

which may be rearranged as

$$m_1 \ddot{x}_1 - F_0 \dot{x}_1 + kx_1 - kx_2 = 0$$
$$-kx_1 + m_2 \ddot{x}_2 + c\dot{x}_2 + kx_2 = 0$$

Assuming the solutions $x_1 = C_1 e^{st}$ and $x_2 = C_2 e^{st}$ and substituting results in

$$(m_1 s^2 - F_0 s + k)C_1 - kC_2 = 0$$
$$-kC_1 + (m_2 s^2 + cs + k)C_2 = 0$$

The characteristic equation is then

$$s\left\{ s^3 + \left(\frac{m_1 c - m_2 F_0}{m_1 m_2} \right) s^2 + \left[\frac{(m_1 + m_2)k - cF_0}{m_1 m_2} \right] s + \frac{(c - F_0)k}{m_1 m_2} \right\} = 0$$

The root $s = 0$ shows that there is a degeneracy and the characteristic equation becomes cubic. For all coefficients to be positive requires that

a. $c > F_0$
b. $(m_1 + m_2)k > cF_0$
c. $m_1 c > m_2 F_0$

† See J. P. Den Hartog, *Mechanical Vibrations*, pp. 288–289.

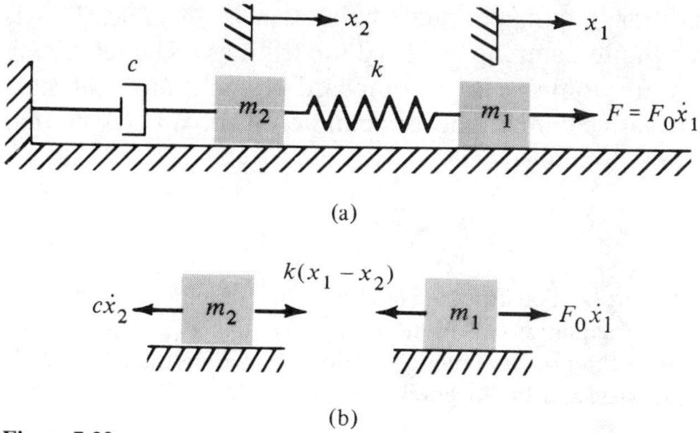

(a)

(b)

Figure 7-22

The condition of Eq. 7-140 requires that

$$\left(\frac{m_1 c - m_2 F_0}{m_1 m_2}\right)\left[\frac{(m_1 + m_2)k - cF_0}{m_1 m_2}\right] > \frac{(c - F_0)k}{m_1 m_2}$$

which simplifies to

d. $(m_1 c - m_2 F_0)[(m_1 + m_2)k - cF_0] > (c - F_0)km_1 m_2$

It is not surprising to find the condition (a), which merely states that the damping constant must be greater than the negative damping coefficient. The importance of the other criteria is that additional requirements involving the size of the masses and the spring constant must be met for the system to be stable.

For equal masses the four conditions reduce to the following three: $c > F_0$, $2mk > cF_0$, and $mk > cF_0$. Since the second of these will be satisfied if the third one is, the conditions become

$$c > F_0 \qquad \text{and} \qquad mk > cF_0$$

The significance of the first requirement is apparent, but that of the second is not. In effect, this latter condition states that the mass and the spring constant must be large compared to the velocity coefficients. The requirements can be partially verified and interpreted by considering the following two cases:

1. Assume k to be infinite, but the other constants to be finite. Then the second requirement, $mk > cF_0$, is met. However, if $c < F_0$, the system is unstable. Physically, if k is infinite, the two masses are locked together and move as one, so that the system reduces to one degree of freedom having a mass $2m$. Stability then would depend on the comparative values of c and F_0.

2. Assume c to be infinite, but the remaining constants to be finite. Then the first requirement, $c > F_0$, is fulfilled; but $mk < cF_0$, so the

second condition is not met, and the system is unstable. This is because the infinite damping has fixed the left mass. The remaining one-degree-of-freedom system is composed of k and the right mass m with the negative damping force acting on it. It can be seen that the system would be unstable.

PROBLEMS

7-1. For the system shown: (a) Construct the free-body diagram for each body. (b) Write the differential equations of motion. (c) Assuming equal masses and equal springs, obtain the principal-mode solutions, including the modal form and natural frequencies, and the location of the nodes.

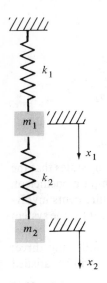

Problem 7-1

7-2. For the system of Prob. 7-1, derive the expressions for the flexibility coefficients. Check these against Eqs. 7-51 through 7-54 by setting $k_3 = 0$.

7-3. Using the flexibility coefficients determined in Prob. 7-2, carry out the inverse method of analysis and obtain the solution for the system of Prob. 7-1. (a) Construct the free-body diagrams for the masses alone and write Newton's relation for these. (b) Make the deformation diagram for the spring system and write the displacement relations for the spring arrangement. (c) By combining the relations of parts (a) and (b), express the differential equations that govern the motion. (d) Assuming equal masses and equal springs, obtain the principal-mode solutions, including the modal form and natural frequencies, and the location of the nodes. (e) Compare the results with those for Prob. 7-1.

7-4. Two masses are arranged on a stretched cord, as shown. The displacements are to be assumed as small, so that the cord tension Q does not change appreciably due to motion of the system. (a) Construct the free-body diagram for each mass. (b) Write the differential equations of motion. (c) Assuming equal masses and equal cord lengths, obtain the principal-mode solutions, including the modal form and natural frequencies, and the location of the nodes.

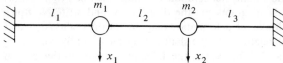

Problem 7-4

7-5. For the system of Prob. 7-4, show that the flexibility coefficients are defined as follows:

$$a_{11} = \frac{l_1(l_2 + l_3)}{Ql_0}, \qquad a_{12} = a_{21} = \frac{l_1 l_3}{Ql_0}, \qquad a_{22} = \frac{l_3(l_1 + l_2)}{Ql_0}$$

where $l_0 = l_1 + l_2 + l_3$.

7-6. Using the flexibility coefficients from Prob. 7-5, carry out the inverse method of analysis for the system of Prob. 7-4. (a) Construct the free-body diagrams for the masses alone and write Newton's relation for these. (b) Make the deformation diagram for the cord system and write the displacement relations for the cord. (c) By combining the relations of parts (a) and (b), express the differential equations that govern the motion. (d) Assuming equal masses and equal cord lengths, obtain the principal-mode solutions, including the modal form and natural frequencies, and the location of the nodes. (e) Compare the results with those of Prob. 7-4.

7-7. The two disks shown in the accompanying figure, having mass moments of inertia I_1 and I_2, are fastened to a continuous shaft, both ends of which are fixed. The torsional stiffness of the different portions of the shaft are defined by the respective k indicated. (a) Construct the free-body diagram for each disk. (b) Write the differential equations of motion. (c) Assuming that $I_1 = I$, $I_2 = 2I$, and $k_1 = k_2 = k_3 = k$, obtain the principal-mode solutions, including the modal form and natural frequencies, and the location of the nodes.

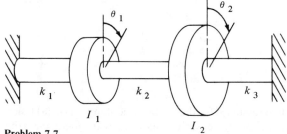

Problem 7-7

7-8. For the system of Prob. 7-7, show that the flexibility coefficients are defined as follows:

$$a_{11} = \frac{k_2 + k_3}{k_0}, \qquad a_{12} = a_{21} = \frac{k_2}{k_0}, \qquad a_{22} = \frac{k_1 + k_2}{k_0}$$

where $k_0 = k_1 k_2 + k_1 k_3 + k_2 k_3$. Note that a is the angular displacement for unit torque.

7-9. Using the flexibility coefficients determined in Prob. 7-8, carry out the inverse method of analysis and obtain the solution for the system of Prob. 7-7. (a) Construct the free-body diagrams for the disks alone and write Newton's relation for these. (b) Make the deformation diagram for the shaft system and write the displacement relations for the shaft arrangement. (c) By combining the relations of parts (a) and (b), express the differential equations that govern the motion. (d) Assuming that $I_1 = I$, $I_2 = 2I$, and $k_1 = k_2 = k_3 = k$, obtain the principal-mode solutions, including the modal form and natural frequencies, and the location of the nodes. (e) Compare the results with those of Prob. 7-7.

7-10. The two pendulums shown in the accompanying figure are firmly fastened to shaft AB. The torsional stiffness of the shaft portion is defined by the corresponding k. The shaft is fixed at end A and is free to turn in a smooth bearing at B. (a) Construct the free-body diagrams and write the differential equations of motion. (b) Obtain the principal-mode solutions, including the modal form and natural frequencies, and the location of the nodes, for the case of equal masses, equal pendulum lengths, and equal torsional spring constants.

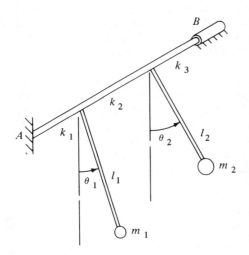

Problem 7-10

7-11. Small amplitudes are to be assumed for the double pendulum shown. (a) Using the angular coordinates θ_1 and θ_2, construct the free-body diagrams and write the differential equations, using A and B as reference points for the motion of m_1 and m_2, respectively. (Observe that B is a general reference for the plane motion of the lower pendulum.) (b) Assuming equal masses and equal pendulum lengths, obtain the principal-mode solutions. (c) Plot the principal modes, showing specifically the location of the nodes.

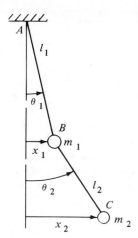

Problem 7-11

7-12. Two simple pendulums are connected by a spring k, as shown. The entire system lies in the plane of the page. Determine the principal modes and natural frequencies for the case of equal masses and equal pendulum lengths.

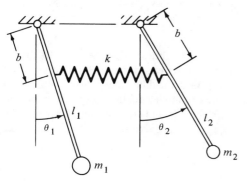

Problem 7-12

7-13. Determine the principal modes and natural frequencies for the system shown. Explain the significance of the first mode and the corresponding frequency.

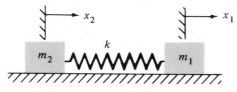

Problem 7-13

7-14. The differential equations of motion for a 2-degree-of-freedom system are expressed as

$$\ddot{x}_1 + a_1 x_1 + a_2 x_2 = 0$$

$$\ddot{x}_2 + b_1 x_1 + b_2 x_2 = 0$$

Show that for a degenerate mode to exist, the condition $a_1/a_2 = b_1/b_2$ must be fulfilled.

7-15. A 9000-lb trailer is connected to a 4000-lb car by means of a flexible hitch having a modulus of 750 lb/in. While traveling on a level highway, the frequency of oscillatory movement of the trailer is to be determined, (a) assuming that the car is coasting so that both vehicles move freely on the roadway, and (b) considering that the car is operating under power so that its drive train is firmly engaged and the rear wheels do not slip on the paving.

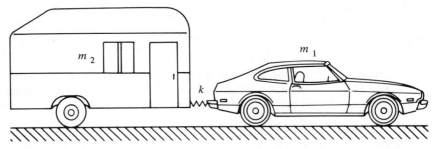

Problem 7-15

7-16. The arrangement shown consists of a weightless rigid member AB that is pivoted at O and contains small mass m_1 at end A. Cord BC connects small mass m_2 to the bar, as indicated. (a) Write the differential equations for small motion of the system and obtain the amplitude equations, expressing these in matrix form. (b) For the case of $a = l$, $b = 2l$, and $m_1 = m = m_2$, determine the principal modes.

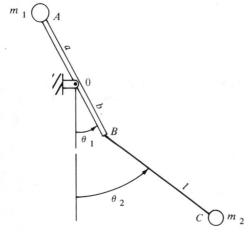

Problem 7-16

7-17. For the system of Prob. 7-16, determine the following: (a) For the case of $a = l = b$ and $m_1 = m = m_2$, obtain the principal modes and discuss the kinds of motion they represent. (b) Considering that $a = 2l$, $b = l$, and $m_1 = m = m_2$, obtain the principal modes and discuss the kinds of motion and conditions they represent.

7-18. The equivalent system for an arrangement of machine parts is shown. Slender bars AB (of length l and mass m) and CD (of length $2l$ and mass m) pivot freely at A and C, respectively, and are fastened to springs at ends B and D. Determine the principal modes and frequencies for small angular motion of the bars.

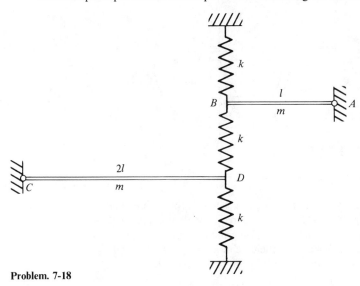

Problem. 7-18

7-19. For the system shown, the body having mass m_1 moves freely on frictionless rollers on a horizontal surface and is fastened to spring k. It also contains the simple pendulum with mass m_2, as shown. (a) Determine the differential equations for small motion of the system. (b) Obtain the amplitude equations, expressing these in matrix form. (c) For the case of $m_1 = m = m_2$ and $kl = mg$, determine the principal modes.

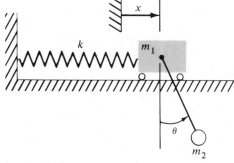

Problem 7-19

7-20. A circular disk is centered on and fastened to a shaft having a torsional spring constant K. The mass moment of inertia of the disk about its center is I. A pendulum of length l with a bob mass m pivots about point B on a vertical centerline of the disk at distance b from the disk center. (a) Write the differential equations for small motion of the system. (b) Obtain the amplitude equations and express these in matrix form.

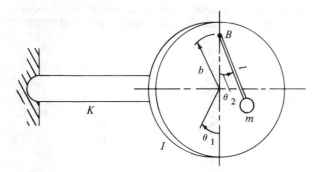

Problem 7-20

7-21. Three springs lie in the x-y plane, and m is constrained so that it can move only in this plane. (a) Construct the free-body diagram for m. (b) Write the differential equations of motion. (c) Determine the principal modes and corresponding natural frequencies for the case of $k_1 = k_2 = k_3 = k$ and $\alpha_1 = 45°$, $\alpha_2 = 135°$, $\alpha_3 = -45°$ (hence $\beta_1 = -45°$, $\beta_2 = 45°$, $\beta_3 = 225°$).

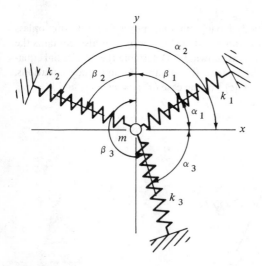

Problem 7-21

7-22. The uniform cantilever beam is of negligible mass. Show that the influence coefficients are defined as follows:

$$a_{11} = \frac{1}{3}\frac{l^3}{EI}, \qquad a_{12} = a_{21} = \frac{5}{48}\frac{l^3}{EI}, \qquad a_{22} = \frac{1}{24}\frac{l^3}{EI}$$

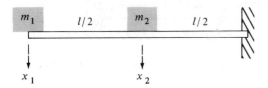

Problem 7-22

7-23. Using the influence coefficients from Prob. 7-22, carry out the inverse analysis for the system and obtain the differential equations of motion. Then assuming that $m_1 = m = m_2$, obtain the principal modes and natural frequencies and plot the mode forms.

7-24. The slender uniform bar AB is supported by springs k_1 and k_2 as shown. The bar has a total length l, its mass is m, and the moment of inertia about mass center G is $J_G = \frac{1}{12}ml^2$. Horizontal movement of the bar is restrained. (a) Obtain differential equations that govern the motion of the bar. (b) Taking $k_1 = k = k_2$ and $a = \frac{1}{3}l$, $b = \frac{1}{6}l$, determine the principal modes and natural frequencies.

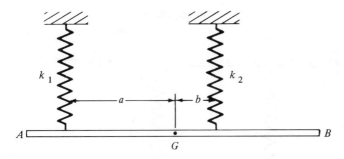

Problem 7-24

7-25. The equivalent spring systems for the main landing gear and the nose gear of an airplane are shown. The airplane mass is m and its moment of inertia is J_G about mass center axis at G. (a) Express the differential equations of motion for the system. (b) Taking $b = 2a$, $k_2 = k$ and $k_1 = 10k$, and $J_G = 100ma^2$, obtain the principal modes and frequencies of vibration.

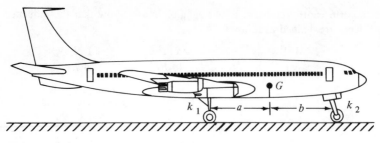

Problem 7-25

7-26 through 7-28. In each case, do the following: (a) Construct the appropriate free-body diagrams and write the differential equations of motion. (b) Solve the differential equations of motion, expressing the principal modes and corresponding natural frequencies. (c) Determine the location of the node for each mode. (d) Plot the mode diagrams.

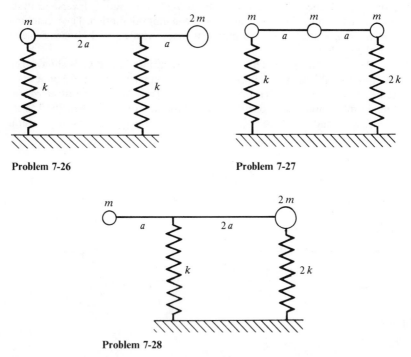

Problem 7-26   **Problem 7-27**

Problem 7-28

7-29. An airfoil section is shown arranged for a wind-tunnel test. The member is supported at B by a rectilinear spring k and a torsional spring k_T. The section has a mass m, and its mass center G is located a distance b from the support B. The moment of inertia of the mass of the section about an axis at B is given as I_B. (a) Determine the differential equations of motion for the system. (b) Obtain the frequency equation.

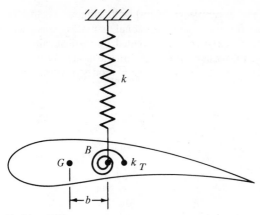

Problem 7-29

7-30. The system shown consists of a weightless slender cylindrical rod AB which is fixed at end A and has a horizontal weightless rigid crossbar fastened at end B. The masses m_1 and m_2 are fastened to the ends of the crossbar. Rod AB has vertical bending freedom (as a cantilever beam) with stiffness coefficient $k = 3EI/l^3$ and torsional freedom with torsional stiffness coefficient $K_T = GJ/l$. Lateral movement of the rod is prevented by vertical rails (not shown). Using coordinate x for the vertical displacement of point B and θ for the angular displacement of the crossbar, obtain the differential equations that govern small motion of the masses.

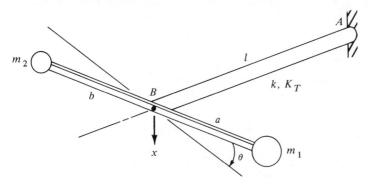

Problem 7-30

7-31. For the system of Prob. 7-30, obtain (a) the amplitude equations and the frequency equation. (b) For the case of $a = 20$ in., $b = 10$ in., $m_1 = m_2 = 0.02$ lb sec²/in., $k = 50$ lb/in., $K_T = 12\,600$ lb in./rad, determine the natural frequencies and mode shapes for the principal modes.

7-32. The arrangement shown consists of a weightless slender cylindrical rod AB which is fixed at end A, and a rigid crossbar fastened at end B. For the crossbar, the mass is m and the moment of inertia about axis AB is J_0. Rod AB has vertical bending freedom (as a cantilever beam) with stiffness coefficient $k = 3EI/l^3$, and torsional freedom with torsional stiffness coefficient $K_T = GJ/l$. Lateral movement of the rod is prevented by vertical rails (not shown). (a) Obtain the differential equations that govern the motion of the bar. (b) Determine the natural frequencies for the principal modes of vibration.

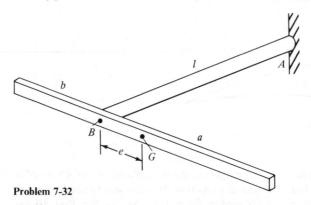

Problem 7-32

7-33. A vertical plane segment of a building frame is composed of the columns and rigid girders shown. The columns are constrained against rotation at the upper and lower ends. The weights of the girders, including floor or roof, and so on, are indicated, and the section and material property for the columns are designated. The columns are to be considered as weightless. (a) Establish the differential equations for horizontal motion of the girders. (b) For $W_1 = W = W_2$, $I_1 = 2I$, and $I_2 = I$, determine the principal modes and natural frequencies.

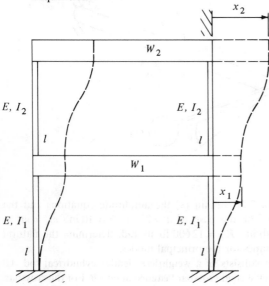

Problem 7-33

Note: In Probs. 7-34 through 7-37, consider the system of Fig. 7-1 but with equal masses and equal springs. The principal modes are defined (refer to Eqs. 7-12 and 7-14) by

Mode 1:
$$\left. \begin{array}{l} x_1 = A \sin (\omega_1 t + \phi_1) \\ x_2 = A \sin (\omega_1 t + \phi_1) \end{array} \right\} \quad \text{where } \omega_1 = \sqrt{\frac{k}{m}}$$

Mode 2:
$$\left. \begin{array}{l} x_1 = B \sin (\omega_2 t + \phi_2) \\ x_2 = -B \sin (\omega_2 t + \phi_2) \end{array} \right\} \quad \text{where } \omega_2 = \sqrt{\frac{3k}{m}}$$

The following are to be determined for the initial conditions listed in each problem:
(a) Evaluate the arbitrary constants A, B, ϕ_1, and ϕ_2. (b) Write the general solution for the motion as defined by x_1 and x_2.

7-34. $x_1 = 0$, $x_2 = 0$, and $\dot{x}_1 = \dot{x}_0$, $\dot{x}_2 = \dot{x}_0$; at $t = 0$.

7-35. $x_1 = x_0$, $x_2 = -x_0$, and $\dot{x}_1 = -\dot{x}_0$, $\dot{x}_2 = \dot{x}_0$; at $t = 0$.

7-36. $x_1 = x_0$, $x_2 = -x_0$, and $\dot{x}_1 = \dot{x}_0$, $\dot{x}_2 = \dot{x}_0$; at $t = 0$.

7-37. $x_1 = x_0$, $x_2 = x_0$, and $\dot{x}_1 = \dot{x}_0$, $\dot{x}_2 = -\dot{x}_0$; at $t = 0$.

7-38. A 2-degree-of-freedom system of the form shown in Fig. 7-16(a) is excited by a harmonic force having a maximum value of 9 lb. The physical constants for the system are $W_1 = 19.3$ lb, $W_2 = 9.65$ lb, $k_1 = 45$ lb/in., and $k_2 = 8.1$ lb/in. Determine the forced amplitude of each mass for an impressed frequency of 121.53 cycles/min.

7-39. For the system described in Prob. 7-38, determine the forced amplitude of each mass for an impressed frequency of 154.7 cycles/min.

7-40. For the system described in Prob. 7-38, determine the forced amplitude of each mass for an impressed frequency of 343.8 cycles/min.

7-41. For the system described in Prob. 7-38, determine the forced frequencies at which the amplitude of the masses would tend to become infinite.

7-42. For the damped two-degree-of-freedom system of Fig. 7-15, take $k_1 = 76.46$ lb/in., $k_2 = 11.77$ lb/in., $W_1 = W_2 = 38.6$ lb, $c_1 = 0.0034$ lb sec/in., and $c_2 = 0.0283$ lb sec/in. Obtain (a) the auxiliary equations and (b) the characteristic equation.

7-43. For Prob. 7-42, (a) verify the roots of the characteristic equation as $-0.1 + 10i$, $-0.1 - 10i$, $-0.2 + 30i$, and $-0.2 - 30i$. (b) Express the equations that define the motion of the mass elements.

7-44. A weight of 77.2 lb is suspended by a spring having a modulus of 80 lb/in. A harmonic force having a frequency of 152.8 cycles/min acts on the mass, resulting in a steady motion with an amplitude of 1.111 in. This amplitude is to be reduced to zero by a suspended mass-spring-type absorber. (a) Determine the nearest integral value of the absorber spring constant k_2 so that the mass ratio is approximately 0.1. (b) Calculate the resulting amplitude of the absorber mass.

7-45. For the system and vibration described in Prob. 7-44, select the absorber spring and mass so that the absorber amplitude will be limited to 2 in.

7-46. A damped 2-degree-of-freedom system of the form shown in Fig. 7-20(a) is excited by a harmonic force having a maximum value of 9 lb. The physical constants of the system are $W_1 = 19.3$ lb, $W_2 = 9.65$ lb, $k_1 = 45$ lb/in., $k_2 = 8.1$ lb/in., and the damping factor is 0.1. Determine the forced amplitude of each mass for an impressed frequency of 114.6 cycles/min.

7-47. For the system described in Prob. 7-46, determine the forced amplitude of each mass for an impressed frequency of 573 cycles/min.

7-48. For the system shown, determine the conditions for dynamic stability and discuss briefly the corresponding physical conditions, noting how these would be fulfilled. Consider that $P_0 > 0$.

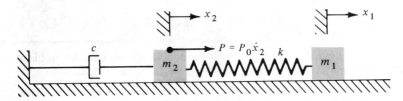

Problem 7-48

7-49. For the system shown, determine the conditions for dynamic stability, discuss briefly the physical significance of these conditions, and interpret these for the case of equal masses. Consider that $P_0 > 0$.

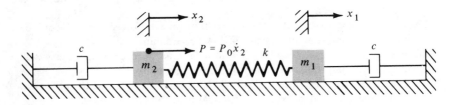

Problem 7-49

chapter
8

Multidegree-of-Freedom
Systems

8-1. INTRODUCTION

The case of two degrees of freedom is part of the general problem of systems
having many degrees of freedom. The discussion in the preceding chapter
was limited to two degrees of freedom, mainly so that the reader could
become acquainted with the analysis and solution, and with related ideas for
the problem of several degrees of freedom in its simplest form. It is natural to
presume that the methods of analysis for two degrees could be extended to
many-degree-of-freedom systems but to wonder what complications might
arise. Most of the difficulties will probably be anticipated, but certain
significant ideas and principles are not apparent and need to be developed.

With an increase in the number of degrees of freedom and consequent
increase in the number of variables, coefficients and terms involved, the
writing of the various relations and equations becomes increasingly more
cumbersome. Accordingly, the use of matrix notation greatly simplifies the
writing of such expressions and will be used to a considerable extent. This
introduces new concepts and principles which require explanation and
example.

The limitations of the classical method of solution are also discussed in
the present chapter. As a consequence of these limitations, the need for other
methods of solution becomes apparent, and various methods of numerical
analysis are subsequently presented in Chapter 9.

8-2. FORCE ANALYSIS AND SOLUTION

Consider the general case of a multidegree-of-freedom system, such as that represented by Fig. 8-1. From the corresponding free-body diagrams, by Newton's second law the differential equations of motion are

$$m_1 \ddot{x}_1 = -k_1 x_1 + k_2(x_2 - x_1)$$
$$m_2 \ddot{x}_2 = -k_2(x_2 - x_1) + k_3(x_3 - x_2)$$
$$\cdots$$
$$m_i \ddot{x}_i = -k_i(x_i - x_{i-1}) + k_{i+1}(x_{i+1} - x_i)$$
$$\cdots$$
$$m_n \ddot{x}_n = -k_n(x_n - x_{n-1}) - k_{n+1} x_n \tag{8-1}$$

These may be rearranged and written in matrix form as follows:

$$
\begin{bmatrix}
m_1 & 0 & 0 & \cdots & 0 \\
0 & m_2 & 0 & \cdots & 0 \\
0 & 0 & m_3 & \cdots & 0 \\
& & & \cdots & \\
0 & & & \cdots & m_n
\end{bmatrix}
\begin{Bmatrix}
\ddot{x}_1 \\ \ddot{x}_2 \\ \ddot{x}_3 \\ \vdots \\ \ddot{x}_n
\end{Bmatrix}
$$
$$
+
\begin{bmatrix}
(k_1 + k_2) & -k_2 & 0 & \cdots & 0 \\
-k_2 & (k_2 + k_3) & -k_3 & \cdots & 0 \\
0 & -k_3 & & \cdots & 0 \\
& & & \cdots & \\
& & & \cdots & (k_{n+1} + k_n)
\end{bmatrix}
\begin{Bmatrix}
x_1 \\ x_2 \\ x_3 \\ \vdots \\ x_n
\end{Bmatrix}
=
\begin{Bmatrix}
0 \\ 0 \\ 0 \\ \vdots \\ 0
\end{Bmatrix} \tag{8-2}
$$

This is a particular case of the general equation

$$
\begin{bmatrix}
m_{11} & m_{12} & m_{13} & \cdots & m_{1n} \\
m_{21} & m_{22} & m_{23} & \cdots & m_{2n} \\
m_{31} & m_{32} & m_{33} & \cdots & m_{3n} \\
& & & \cdots & \\
m_{n1} & & & \cdots & m_{nn}
\end{bmatrix}
\begin{Bmatrix}
\ddot{x}_1 \\ \ddot{x}_2 \\ \ddot{x}_3 \\ \vdots \\ \ddot{x}_n
\end{Bmatrix}
$$
$$
+
\begin{bmatrix}
k_{11} & k_{12} & k_{13} & \cdots & k_{1n} \\
k_{21} & k_{22} & k_{23} & \cdots & k_{2n} \\
k_{31} & k_{32} & k_{33} & \cdots & k_{3n} \\
& & & \cdots & \\
k_{n1} & & & \cdots & k_{nn}
\end{bmatrix}
\begin{Bmatrix}
x_1 \\ x_2 \\ x_3 \\ \vdots \\ x_n
\end{Bmatrix}
=
\begin{Bmatrix}
0 \\ 0 \\ 0 \\ \vdots \\ 0
\end{Bmatrix} \tag{8-3}
$$

which may be written more simply as

$$[m]\{\ddot{x}\} + [k]\{x\} = \{0\} \tag{8-4}$$

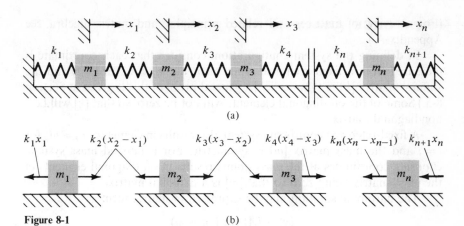

Figure 8-1
(b)

where

$$[m] = \begin{bmatrix} m_{11} & m_{12} & m_{13} & \cdots & m_{1n} \\ m_{21} & m_{22} & m_{23} & \cdots & m_{2n} \\ m_{31} & m_{32} & m_{33} & \cdots & m_{3n} \\ \cdot & \cdot & \cdot & \cdots & \cdot \\ m_{n1} & \cdot & \cdot & \cdots & m_{nn} \end{bmatrix}$$

$$= \textit{mass matrix}; \text{ a square scalar matrix}$$

(8-5)

$$[k] = \begin{bmatrix} k_{11} & k_{12} & k_{13} & \cdots & k_{1n} \\ k_{21} & k_{22} & k_{23} & \cdots & k_{2n} \\ k_{31} & k_{32} & k_{33} & \cdots & k_{3n} \\ \cdot & \cdot & \cdot & \cdots & \cdot \\ k_{n1} & \cdot & \cdot & \cdots & k_{nn} \end{bmatrix}$$

$$= \textit{stiffness matrix}; \text{ a square scalar matrix}$$

(8-6)

$$\{x\} = \begin{Bmatrix} x_1 \\ x_2 \\ x_3 \\ \cdot \\ x_n \end{Bmatrix} = \textit{displacement vector matrix}; \text{ a column matrix}$$

(8-7)

$$\{\ddot{x}\} = \begin{Bmatrix} \ddot{x}_1 \\ \ddot{x}_2 \\ \ddot{x}_3 \\ \cdot \\ \ddot{x}_n \end{Bmatrix} = \textit{acceleration vector matrix}; \text{ a column matrix}$$

(8-8)

$$\{0\} = \begin{Bmatrix} 0 \\ 0 \\ 0 \\ \cdot \\ 0 \end{Bmatrix} = \textit{null column matrix}$$

(8-9)

(For a review of matrices and related principles, and matrix algebra, see Appendix A.)

In defining the system configuration, consider that each coordinate is measured from a fixed reference. Then elements $k_{ij} = k_{ji}$ for $k \neq j$ and the stiffness matrix $[k]$ is symmetric. (A general proof of this is given in Section 8-5.) Some of the off-diagonal elements will not be zero, so that $[k]$ will be a nondiagonal matrix.

A fixed-reference coordinate system also results in elements $m_{ij} = m_{ji}$ for $i \neq j$ and the mass matrix $[m]$ is symmetric. For a lumped mass system, choosing coordinates at the mass points results in off-diagonal elements of the mass matrix being zero so that $[m]$ is a diagonal matrix.

The harmonic solution may be expressed in matrix form as

$$\{x\} = \{A\} \sin (\omega t + \phi) \tag{8-10}$$

where

$$\{A\} = \begin{Bmatrix} A_1 \\ A_2 \\ A_3 \\ \cdot \\ A_n \end{Bmatrix} = amplitude \text{ column matrix} \tag{8-11}$$

From this the second time derivative is

$$\{\ddot{x}\} = \{A\}(-\omega^2) \sin (\omega t + \phi) \tag{8-12}$$

Substitution into Eq. 8-4 results in

$$-\omega^2[m]\{A\} + [k]\{A\} = \{0\} \tag{8-13}$$

representing the matrix form of the amplitude equations, which may be also written as

$$([k] - \xi[m])\{A\} = \{0\} \tag{8-14}$$

or

$$[C]\{A\} = \{0\} \tag{8-15}$$

where $[C]$ is the *characteristic matrix* defined by

$$[C] = [k] - \xi[m] \tag{8-16}$$

and

$$\omega^2 = \xi = \text{direct frequency factor} \tag{8-17}$$

For the nontrivial solution, the determinant of $[C]$ must vanish. Thus the determinant of the characteristic matrix is set to zero.

$$|[k] - \xi[m]| = 0 \tag{8-18}$$

The *characteristic determinant* may be written as

$$\begin{vmatrix} (k_{11} - \xi m_{11}) & (k_{12} - \xi m_{12}) & \cdots & (k_{1n} - \xi m_{1n}) \\ (k_{21} - \xi m_{21}) & (k_{22} - \xi m_{22}) & \cdots & (k_{2n} - \xi m_{2n}) \\ (k_{31} - \xi m_{31}) & \cdot & \cdots & \cdot \\ \cdot & \cdot & \cdots & \cdot \\ (k_{n1} - \xi m_{n1}) & \cdot & \cdots & (k_{nn} - \xi m_{nn}) \end{vmatrix} = 0 \qquad (8\text{-}19)$$

Expanding this nth-order determinant results in the polynomial *characteristic equation*

$$\xi^n + \alpha_1 \xi^{n-1} + \alpha_2 \xi^{n-2} + \cdots + \alpha_{n-1} \xi + \alpha_n = 0 \qquad (8\text{-}20)$$

where $\alpha_1, \alpha_2, \ldots, \alpha_n$ are numerical coefficients. This will have n positive real roots[†] $\xi_1, \xi_2, \ldots, \xi_n$, where $\xi_1 < \xi_2 < \xi_3 \cdots < \xi_n$, which are called the *eigenvalues* or characteristic numbers. From these the n natural frequencies $\omega_1, \omega_2, \ldots, \omega_n$ can be determined. If n is small, it may be possible to solve for the ξ roots readily and conveniently, but for a large number of degrees of freedom it would require some form of numerical procedure. Substituting an eigenvalue ξ_i into the matrix amplitude equation (Eq. 8-14) determines the corresponding mode shape $\{A\}_i$, which is called the *eigenvector*. There will be n eigenvalues and n associated eigenvectors for the n-degree-of-freedom system. The general solution is then the sum of these principal-mode solutions.

The eigenvectors can also be obtained from the adjoint matrix for the system (see Appendix A-9). Recalling the characteristic matrix $[C]$ as defined by Eq. 8-16, the inverse is written as follows:

$$[C]^{-1} = \frac{\text{adj } [C]}{|C|} \qquad (8\text{-}21)$$

Premultiplying by $|C|[C]$ results in

$$|C|[C][C]^{-1} = |C|[C] \frac{\text{adj } [C]}{|C|} \qquad (8\text{-}22)$$

whence

$$|C|[I] = [C] \text{ adj } [C] \qquad (8\text{-}23)$$

where $[I]$ is the identity matrix.

Replacing $[C]$ according to its definition (Eq. 8-16) gives

$$|[k] - \xi[m]|[I] = ([k] - \xi[m]) \text{ adj } ([k] - \xi[m])$$

Now for the eigenvalue ξ_i, the determinant $|[k] - \xi_i[m]|$ is zero, and the left side of the above equation is then zero. That is,

$$0 = ([k] - \xi_i[m]) \text{ adj } ([k] - \xi_i[m]) \qquad (8\text{-}24)$$

† See H. Lamb, *Higher Mechanics*, p. 222.

which holds for all n values of ξ_i and hence represents n equations. From Eq. 8-14 the amplitude equation for mode i is

$$([k] - \xi_i[m])\{A\}_i = 0 \tag{8-25}$$

Comparison of this and the preceding relation leads to the conclusion that

$$\{A\}_i = \text{adj } ([k] - \xi_i[m]) \tag{8-26}$$

Thus the adjoint matrix must be composed of columns each one of which is the eigenvector $\{A\}_i$ multiplied by a constant.

EXAMPLE 8-1 Consider the case of three masses on a stretched massless elastic cable or cord as shown in Fig. 8-2(a). Assume that the dynamic displacements are small, so that the tension F in the cord does not change appreciably. (a) Write the differential equations of motion and express these in matrix form. (b) Taking $l_1 = l_2 = l_3 = l_4 = l$ and $m_1 = m$, $m_2 = 3m$, and $m_3 = 2m$ obtain the solution. (c) Define the principal modes.

SOLUTION

a. The free-body diagrams for a motion configuration are shown in Fig. 8-2(b), where the x's have been measured from static equilibrium positions. Since three independent coordinates are needed, there are three degrees of freedom. By considering the vertical components of cord tension F, the equations of motion may be written as

$$m_1 \ddot{x}_1 = -F\frac{x_1}{l_1} + F\left(\frac{x_2 - x_1}{l_2}\right)$$

$$m_2 \ddot{x}_2 = -F\left(\frac{x_2 - x_1}{l_2}\right) - F\left(\frac{x_2 - x_3}{l_3}\right)$$

$$m_3 \ddot{x}_3 = F\left(\frac{x_2 - x_3}{l_3}\right) - F\frac{x_3}{l_4}$$

In these expressions, the coefficients F/l_1, F/l_2, F/l_3, and F/l_4 can be thought of as equivalent spring constants for a rectilinear system composed of a set of springs k_1, k_2, k_3, and k_4 arranged alternately with the masses. These equations may be rearranged and expressed in matrix form as

$$
\begin{bmatrix} m_1 & 0 & 0 \\ 0 & m_2 & 0 \\ 0 & 0 & m_3 \end{bmatrix}
\begin{Bmatrix} \ddot{x}_1 \\ \ddot{x}_2 \\ \ddot{x}_3 \end{Bmatrix}
+
\begin{bmatrix} \left(\frac{F}{l_1} + \frac{F}{l_2}\right) & -\frac{F}{l_2} & 0 \\ -\frac{F}{l_2} & \left(\frac{F}{l_2} + \frac{F}{l_3}\right) & -\frac{F}{l_3} \\ 0 & -\frac{F}{l_3} & \left(\frac{F}{l_3} + \frac{F}{l_4}\right) \end{bmatrix}
\begin{Bmatrix} x_1 \\ x_2 \\ x_3 \end{Bmatrix}
=
\begin{Bmatrix} 0 \\ 0 \\ 0 \end{Bmatrix}
$$

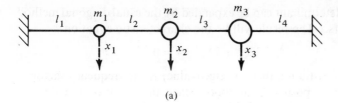

(a)

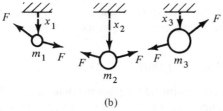

(b)

Figure 8-2

b. The harmonic solutions and their second time derivatives may be expressed as

$$\begin{Bmatrix} x_1 \\ x_2 \\ x_3 \end{Bmatrix} = \begin{Bmatrix} A_1 \\ A_2 \\ A_3 \end{Bmatrix} \sin(\omega t + \phi) \qquad \text{and} \qquad \begin{Bmatrix} \ddot{x}_1 \\ \ddot{x}_2 \\ \ddot{x}_3 \end{Bmatrix} = -\omega^2 \begin{Bmatrix} A_1 \\ A_2 \\ A_3 \end{Bmatrix} \sin(\omega t + \phi)$$

Substituting these and the assigned lengths and masses gives the amplitude equations as

$$\begin{bmatrix} \left(\dfrac{2F}{l} - m\omega^2 \right) & -\dfrac{F}{l} & 0 \\[2mm] -\dfrac{F}{l} & \left(\dfrac{2F}{l} - 3m\omega^2 \right) & -\dfrac{F}{l} \\[2mm] 0 & -\dfrac{F}{l} & \left(\dfrac{2F}{l} - 2m\omega^2 \right) \end{bmatrix} \begin{Bmatrix} A_1 \\ A_2 \\ A_3 \end{Bmatrix} = \begin{Bmatrix} 0 \\ 0 \\ 0 \end{Bmatrix}$$

This can be simplified somewhat by multiplying by $(-l/F)$ and setting $(lm/F)\omega^2 = \xi$. Then

$$\begin{bmatrix} (\xi - 2) & 1 & 0 \\ 1 & (3\xi - 2) & 1 \\ 0 & 1 & (2\xi - 2) \end{bmatrix} \begin{Bmatrix} A_1 \\ A_2 \\ A_3 \end{Bmatrix} = \begin{Bmatrix} 0 \\ 0 \\ 0 \end{Bmatrix}$$

where ξ represents a direct frequency factor. For other than the trivial case of the A's being zero, the determinant must vanish. Thus

$$\begin{vmatrix} (\xi - 2) & 1 & 0 \\ 1 & (3\xi - 2) & 1 \\ 0 & 1 & (2\xi - 2) \end{vmatrix} = 0$$

This third-order determinant can be expanded by the usual diagonal method or by the cofactors.† The result will be the characteristic equation

$$6\xi^3 - 22\xi^2 + 21\xi - 4 = 0$$

where the characteristic number (or eigenvalue) is the frequency factor ξ. This will have three positive real roots. Since the equation is cubic, its solution is inconvenient, although it can be accomplished by Newton's and other methods. In general, the roots will not be integers and they are likely to be irrational numbers, so that their determination will involve rounding-off errors. The roots of the characteristic equation for this case can be verified as

$$\xi = 0.252\,82 \ldots, \ 1.1809 \ldots, \ 2.2329 \ldots$$

Substitution of these into the amplitude equations will define the three principal modes. The frequency factor can also be used to express the natural frequencies.

The principal modes are found to be as follows:

Mode 1:

$$\omega_1^2 = 0.252\,82\,\frac{F}{lm} \qquad \frac{A_2}{A_1} = 1.7472, \qquad \frac{A_3}{A_1} = 1.1692$$

$$(A_1 = 1, \quad A_2 = 1.7472, \quad A_3 = 1.1692)$$

Mode 2:

$$\omega_2^2 = 1.1809\,\frac{F}{lm} \qquad \frac{A_2}{A_1} = 0.819\,10, \qquad \frac{A_3}{A_1} = -2.2636$$

$$(A_1 = 1, \quad A_2 = 0.819\,10, \quad A_3 = -2.2636)$$

Mode 3:

$$\omega_3^2 = 2.2329\,\frac{F}{lm} \qquad \frac{A_2}{A_1} = -0.232\,90, \qquad \frac{A_3}{A_1} = 0.094\,327$$

$$(A_1 = 1, \quad A_2 = -0.232\,90, \quad A_3 = 0.094\,327)$$

Instead of obtaining amplitude ratios, in any mode one of the amplitudes may be arbitrarily assigned the value of unity, and the other amplitudes can then be determined. This represents a procedure of normalizing the amplitudes. These are shown in parentheses for the present case. For each mode, the amplitude ratios or normalized amplitude values can be determined from any two of the amplitude equations. The remaining amplitude equation is redundant and in such a determination can be used only as a check. Slight rounding-off errors in the natural frequencies obtained will cause

† See Appendix A-6.

larger errors in the amplitudes determined. In the present problem the amplitudes were calculated from the first and second amplitude equations. Small errors would be found by checking these in the third amplitude equation.

The three modes are shown in Fig. 8-3. The actual amplitudes in any mode depend on the initial conditions. Comparison of the magnitude of the amplitudes for one mode to those in another is meaningless. The position of the nodes can be readily determined from the geometry of the figure.

EXAMPLE 8-2 Using the adjoint matrix, determine the eigenvectors for Example 8-1.

SOLUTION The characteristic matrix for Example 8-1 is

$$(\xi_i[m] - [k]) = \begin{bmatrix} (\xi - 2) & 1 & 0 \\ 1 & (3\xi - 2) & 1 \\ 0 & 1 & (2\xi - 2) \end{bmatrix}_i$$

(Note that sign of the matrix was reversed in the solution process for Example 8-1. This is of no consequence.) Then the adjoint is

$$\text{adj} \ (\xi_i[m] - [k]) = \begin{bmatrix} \begin{vmatrix} (3\xi - 2) & 1 \\ 1 & (2\xi - 2) \end{vmatrix} & -\begin{vmatrix} 1 & 0 \\ 1 & (2\xi - 2) \end{vmatrix} & \begin{vmatrix} 1 & 0 \\ (3\xi - 2) & 1 \end{vmatrix} \\ -\begin{vmatrix} 1 & 1 \\ 0 & (2\xi - 2) \end{vmatrix} & \begin{vmatrix} (\xi - 2) & 0 \\ 0 & (2\xi - 2) \end{vmatrix} & -\begin{vmatrix} (\xi - 2) & 0 \\ 1 & 1 \end{vmatrix} \\ \begin{vmatrix} 1 & (3\xi - 2) \\ 0 & 1 \end{vmatrix} & -\begin{vmatrix} (\xi - 2) & 1 \\ 0 & 1 \end{vmatrix} & \begin{vmatrix} (\xi - 2) & 1 \\ 1 & (3\xi - 2) \end{vmatrix} \end{bmatrix}_i$$

$$= \begin{bmatrix} (6\xi^2 - 10\xi + 3) & (2 - 2\xi) & 1 \\ (2 - 2\xi) & (2\xi^2 - 6\xi + 4) & (2 - \xi) \\ 1 & (2 - \xi) & (3\xi^2 - 8\xi + 3) \end{bmatrix}_i$$

Substituting the eigenvalue $\xi_1 = 0.252\,82$ from Eq. 8-1 gives

$$\begin{bmatrix} 0.855\,308 & 1.494\,36 & 1 \\ 1.494\,36 & 2.610\,92 & 1.747\,18 \\ 1 & 1.747\,18 & 1.169\,19 \end{bmatrix}_1$$

Normalizing each column to unity in the first row results in the first eigenvector as

$$\begin{bmatrix} 1 & 1 & 1 \\ 1.7472 & 1.7472 & 1.7472 \\ 1.1692 & 1.1692 & 1.1692 \end{bmatrix} \quad \text{whence } \{A\}_1 = \begin{Bmatrix} 1 \\ 1.7472 \\ 1.1692 \end{Bmatrix}$$

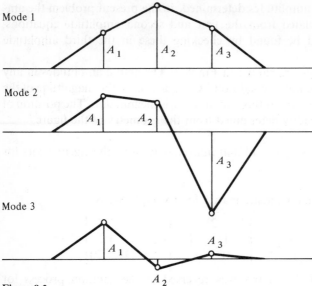

Figure 8-3

Similarly, substituting the second eigenvalue of $\xi_2 = 1.1809$ gives the corresponding eigenvector as

$$\{A\}_2 = \left\{ \begin{matrix} 1 \\ 0.819\,10 \\ -2.2636 \end{matrix} \right\}$$

Finally, the eigenvalue $\xi_3 = 2.2329$ results in

$$\{A\}_3 = \left\{ \begin{matrix} 1 \\ -0.232\,90 \\ 0.094\,327 \end{matrix} \right\}$$

8-3. THE STIFFNESS MATRIX, AND AN ALTERNATIVE ANALYSIS

In the preceding section the general stiffness matrix was defined by Eq. 8-6 as

$$[k] = \begin{bmatrix} k_{11} & k_{12} & k_{13} & \cdots & k_{1n} \\ k_{21} & k_{22} & \cdot & \cdots & \cdot \\ k_{31} & \cdot & \cdot & \cdots & \cdot \\ \cdot & \cdot & \cdot & \cdots & \cdot \\ k_{n1} & \cdot & \cdot & \cdots & k_{nn} \end{bmatrix} \tag{8-27}$$

The k_{ij} elements are known as the *stiffness coefficients* or stiffness influence coefficients. To define k_{ij} explicitly, consider station j of the system to be given a unit displacement with all other stations restrained to zero displacement. The corresponding force required (a holding-type force) at station i is designated as k_{ij}. Thus if station 1 is displaced by unity with all other station displacements of zero (that is, $x_1 = 1$, $x_2 = 0$, and $x_3 = 0$, and so on) the force required at station 1 is k_{11}, at station 2 it is k_{21}, at station 3 it is k_{31}, and so on. Then for $x_1 = 0$, $x_2 = 1$, $x_3 = 0$, $x_4 = 0$, and so on, the holding force required at station 1 is k_{12}, at station 2 it is k_{22}, at station 3 it is k_{32}, and so on.

The elements of the stiffness matrix may be established by columns. To obtain the elements of any column, set the corresponding station j displacement to unity and all other displacements to zero and determine the force required at each i station of the system.

Once the elements of the stiffness matrix have been established, the matrix equation of motion can be written as

$$[m]\{\ddot{x}\} + [k]\{x\} = 0 \tag{8-28}$$

The elements of the mass matrix depend on the coordinate system chosen. If coordinates are chosen at the mass points, then the acceleration of each mass is a function of a single coordinate and $m_{ij} = 0$ for $i \neq j$. The nondiagonal terms of the mass matrix are zero and $m_{11}, m_{22}, m_{33}, \ldots$ may be written as $m_1, m_2, m_3, \ldots$, respectively. Thus

$$[m] = \begin{bmatrix} m_1 & 0 & 0 & \cdots & 0 \\ 0 & m_2 & 0 & \cdots & 0 \\ 0 & 0 & m_3 & \cdots & 0 \\ \cdot & \cdot & \cdot & \cdots & \cdot \\ 0 & \cdot & \cdot & \cdots & m_n \end{bmatrix} \tag{8-29}$$

EXAMPLE 8-3   For the three-mass stretched-cord system of Example 8-1, determine (a) the stiffness coefficients or elements of $[k]$, and (b) write the matrix form of the equations of motion.

SOLUTION   Diagrams for the appropriate displacement arrangements and forces are shown in Fig. 8-4.

a. Referring to the diagram (a), by equilibrium

$$k_{11} = F \cdot \frac{1}{l_1} + F \cdot \frac{1}{l_2}$$

$$k_{21} = -F \cdot \frac{1}{l_2} \qquad k_{31} = 0$$

From diagram (b),

$$k_{12} = -F \cdot \frac{1}{l_2} \qquad k_{32} = -F \cdot \frac{1}{l_3}$$

$$k_{22} = F \cdot \frac{1}{l_2} + F \cdot \frac{1}{l_3}$$

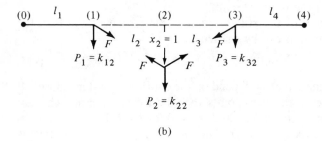

(a)

(b)

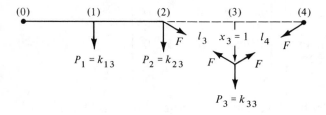

(c)

Figure 8-4

Also, from diagram (c),

$$k_{13} = 0 \qquad k_{23} = -F \cdot \frac{1}{l_3}$$

$$k_{33} = F \cdot \frac{1}{l_3} + F \cdot \frac{1}{l_4}$$

Then

$$[k] = \begin{bmatrix} \left(\dfrac{F}{l_1} + \dfrac{F}{l_2}\right) & \dfrac{-F}{l_2} & 0 \\[2mm] \dfrac{-F}{l_2} & \left(\dfrac{F}{l_2} + \dfrac{F}{l_3}\right) & \dfrac{-F}{l_3} \\[2mm] 0 & \dfrac{-F}{l_3} & \left(\dfrac{F}{l_3} + \dfrac{F}{l_4}\right) \end{bmatrix}$$

b. Since the displacement of each mass is defined by a single coordinate, then

$$[m] = \begin{bmatrix} m_1 & 0 & 0 \\ 0 & m_2 & 0 \\ 0 & 0 & m_3 \end{bmatrix}$$

and the matrix equation of motion is

$$[m]\{\ddot{x}\} + [k]\{x\} = 0$$

wherein the mass and stiffness matrices are those established above. This is the same as was obtained in Example 8-1.

8-4. THE FLEXIBILITY AND STIFFNESS MATRICES

The flexibility matrix in which the elements are flexibility influence coefficients or *flexibility coefficients*, a_{ij}, can be used to establish the displacement equations that govern the motion of a multidegree-of-freedom system. The flexibility coefficient a_{ij} defines the displacement at station i due to a unit force acting on the system at station j. Letting P_j represent the force on the elastic system at station j, the displacement relations are then

$$x_1 = a_{11}P_1 + a_{12}P_2 + a_{13}P_3 + \cdots + a_{1n}P_n$$

$$x_2 = a_{21}P_1 + a_{22}P_2 + a_{23}P_3 + \cdots + a_{2n}P_n$$

$$x_3 = a_{31}P_1 + a_{32}P_2 + a_{33}P_3 + \cdots + a_{3n}P_n$$

$$\vdots$$

$$x_n = a_{n1}P_1 + a_{n2}P_2 + a_{n3}P_3 + \cdots + a_{nn}P_n \tag{8-30}$$

These may be written in matrix form as

$$\begin{Bmatrix} x_1 \\ x_2 \\ x_3 \\ \cdot \\ x_n \end{Bmatrix} = \begin{bmatrix} a_{11} & a_{12} & a_{13} & \cdots & a_{1n} \\ a_{21} & a_{22} & a_{23} & \cdots & a_{2n} \\ a_{31} & a_{32} & a_{33} & \cdots & a_{3n} \\ \cdot & \cdot & \cdot & \cdots & \cdot \\ a_{n1} & \cdot & \cdot & \cdots & a_{nn} \end{bmatrix} \begin{Bmatrix} P_1 \\ P_2 \\ P_3 \\ \cdot \\ P_n \end{Bmatrix} \tag{8-31}$$

or more briefly as

$$\{x\} = [a]\{P\} \tag{8-32}$$

where

$$[a] = \begin{bmatrix} a_{11} & a_{12} & a_{13} & \cdots & a_{1n} \\ a_{21} & a_{22} & a_{23} & \cdots & a_{2n} \\ a_{31} & a_{32} & a_{33} & \cdots & a_{3n} \\ \cdot & \cdot & \cdot & \cdots & \cdot \\ a_{n1} & \cdot & \cdot & \cdots & a_{nn} \end{bmatrix} = flexibility \text{ matrix; a square matrix}$$

$$\tag{8-33}$$

$$\{x\} = \begin{Bmatrix} x_1 \\ x_2 \\ x_3 \\ \cdot \\ x_n \end{Bmatrix} = \text{displacement vector matrix; a column matrix} \qquad (8\text{-}34)$$

$$\{P\} = \begin{Bmatrix} P_1 \\ P_2 \\ P_3 \\ \cdot \\ P_n \end{Bmatrix} = \text{force vector matrix} \qquad (8\text{-}35)$$

It will now be shown that the stiffness matrix and flexibility matrix bear a reciprocal or inverse relation to each other. From the definition of the stiffness matrix, the following relation is established for the forces acting on the system at the various stations in terms of the displacements:

$$\begin{Bmatrix} P_1 \\ P_2 \\ P_3 \\ \cdot \\ P_n \end{Bmatrix} = \begin{bmatrix} k_{11} & k_{12} & k_{13} & \cdots & k_{1n} \\ k_{21} & k_{22} & k_{23} & \cdots & k_{2n} \\ k_{31} & k_{32} & k_{33} & \cdots & k_{3n} \\ \cdot & \cdot & \cdot & \cdots & \cdot \\ k_{n1} & k_{n2} & k_{n3} & \cdots & k_{nn} \end{bmatrix} \begin{Bmatrix} x_1 \\ x_2 \\ x_3 \\ \cdot \\ x_n \end{Bmatrix} \qquad (8\text{-}36)$$

or

$$\{P\} = [k]\{x\} \qquad (8\text{-}37)$$

Substituting Eq. 8-37 into Eq. 8-32 results in

$$\{x\} = [a][k]\{x\} \qquad (8\text{-}38)$$

from which it is concluded that

$$[a][k] = [I] \qquad (8\text{-}39)$$

where $[I]$ is the identity matrix having all diagonal elements of unity and all off-diagonal elements of zero. Substituting $[I] = [a][a]^{-1}$ gives

$$[a][k] = [a][a]^{-1}$$

whence

$$[k] = [a]^{-1} \qquad (8\text{-}40)$$

Similarly, substituting $[I] = [k]^{-1}[k]$ into Eq. 8-39 results in

$$[a][k] = [k]^{-1}[k]$$

which determines that

$$[a] = [k]^{-1} \qquad (8\text{-}41)$$

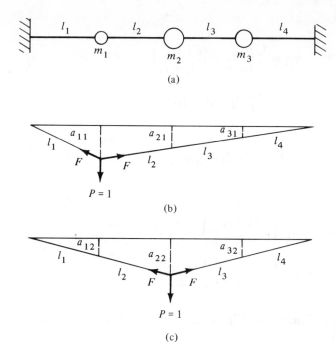

Figure 8-5

Thus the flexibility matrix can be obtained from the stiffness matrix, and vice versa. This may be a useful procedure. In many cases the flexibility coefficients are needed for a structural system, but may be more difficult to determine than the stiffness coefficients. They can, however, be obtained by the inverse matrix, although this procedure in itself is not always simple.

The inverse matrix may be obtained by several methods. One method (see Appendix A-11) follows the procedure indicated by

$$[k]^{-1} = \frac{\text{adj } [k]}{|k|} \tag{8-42}$$

where adj $[k]$ is the adjoint of matrix $[k]$ and $|k|$ is its determinant.

EXAMPLE 8-4   Determine the flexibility matrix for the stretched-cord system of Example 8-1, using appropriate displacement unit-load diagrams and principles of equilibrium.

SOLUTION   Placing a unit load on the cord at the number 1 position results in the free-body configuration of Fig. 8-5(b). By equilibrium,

$$1 = F \frac{a_{11}}{l_1} + F \frac{a_{11}}{l_2 + l_3 + l_4}$$

whence

$$a_{11} = \frac{l_1}{F}\left(\frac{l_2 + l_3 + l_4}{l_1 + l_2 + l_3 + l_4}\right)$$

$$= \frac{l_1}{l_0 F}(l_2 + l_3 + l_4)$$

where

$$l_0 = l_1 + l_2 + l_3 + l_4$$

Then by geometry,

$$a_{21} = a_{11}\left(\frac{l_3 + l_4}{l_2 + l_3 + l_4}\right) = \frac{l_1}{l_0 F}(l_3 + l_4)$$

$$a_{31} = a_{11}\left(\frac{l_4}{l_2 + l_3 + l_4}\right) = \frac{l_1}{l_0 F}l_4$$

From the free-body diagram of Fig. 8-5(c), showing a unit force at the number 2 location,

$$1 = F\frac{a_{22}}{l_1 + l_2} + F\frac{a_{22}}{l_3 + l_4}$$

and

$$a_{22} = \frac{1}{l_0 F}(l_1 + l_2)(l_3 + l_4)$$

$$a_{12} = a_{22}\frac{l_1}{l_1 + l_2} = \frac{l_1}{l_0 F}(l_3 + l_4)$$

$$a_{32} = a_{22}\frac{l_4}{l_3 + l_4} = \frac{l_4}{l_0 F}(l_1 + l_2)$$

Similarly,

$$a_{33} = \frac{l_4}{l_0 F}(l_1 + l_2 + l_3)$$

$$a_{23} = \frac{l_4}{l_0 F}(l_1 + l_2)$$

$$a_{13} = \frac{l_4}{l_0 F}l_1$$

The flexibility matrix is then

$$[a] = \frac{1}{l_0 F}\begin{bmatrix} l_1(l_2 + l_3 + l_4) & l_1(l_3 + l_4) & l_1 l_4 \\ l_1(l_3 + l_4) & (l_1 + l_2)(l_3 + l_4) & l_4(l_1 + l_2) \\ l_1 l_4 & l_4(l_1 + l_2) & l_4(l_1 + l_2 + l_3) \end{bmatrix}$$

EXAMPLE 8-5 Using the stiffness matrix established in Example 8-3 for the stretched-cord case, obtain the flexibility matrix and check this against the results of Example 8-4.

SOLUTION For the 3-degree-of-freedom system, the stiffness matrix was established as

$$[k] = \begin{bmatrix} \left(\dfrac{F}{l_1} + \dfrac{F}{l_2}\right) & -\dfrac{F}{l_2} & 0 \\ -\dfrac{F}{l_2} & \left(\dfrac{F}{l_2} + \dfrac{F}{l_3}\right) & -\dfrac{F}{l_3} \\ 0 & -\dfrac{F}{l_3} & \left(\dfrac{F}{l_3} + \dfrac{F}{l_4}\right) \end{bmatrix}$$

Then the adjoint of $[k]$ is

$$\text{adj } [k] = \begin{bmatrix} \begin{vmatrix} \left(\dfrac{F}{l_2} + \dfrac{F}{l_3}\right) & -\dfrac{F}{l_3} \\ -\dfrac{F}{l_3} & \left(\dfrac{F}{l_3} + \dfrac{F}{l_4}\right) \end{vmatrix} & -\begin{vmatrix} -\dfrac{F}{l_2} & 0 \\ -\dfrac{F}{l_3} & \left(\dfrac{F}{l_3} + \dfrac{F}{l_4}\right) \end{vmatrix} & \begin{vmatrix} -\dfrac{F}{l_2} & 0 \\ \left(\dfrac{F}{l_2} + \dfrac{F}{l_3}\right) & -\dfrac{F}{l_3} \end{vmatrix} \\ -\begin{vmatrix} -\dfrac{F}{l_2} & -\dfrac{F}{l_3} \\ 0 & \left(\dfrac{F}{l_3} + \dfrac{F}{l_4}\right) \end{vmatrix} & \begin{vmatrix} \left(\dfrac{F}{l_1} + \dfrac{F}{l_2}\right) & 0 \\ 0 & \left(\dfrac{F}{l_3} + \dfrac{F}{l_4}\right) \end{vmatrix} & -\begin{vmatrix} \left(\dfrac{F}{l_1} + \dfrac{F}{l_2}\right) & 0 \\ -\dfrac{F}{l_2} & -\dfrac{F}{l_3} \end{vmatrix} \\ \begin{vmatrix} -\dfrac{F}{l_2} & \left(\dfrac{F}{l_2} + \dfrac{F}{l_3}\right) \\ 0 & -\dfrac{F}{l_3} \end{vmatrix} & -\begin{vmatrix} \left(\dfrac{F}{l_1} + \dfrac{F}{l_2}\right) & 0 \\ 0 & -\dfrac{F}{l_3} \end{vmatrix} & \begin{vmatrix} \left(\dfrac{F}{l_1} + \dfrac{F}{l_2}\right) & -\dfrac{F}{l_2} \\ -\dfrac{F}{l_2} & \left(\dfrac{F}{l_2} + \dfrac{F}{l_3}\right) \end{vmatrix} \end{bmatrix}$$

$$= F^2 \begin{bmatrix} \left(\dfrac{l_2 + l_3 + l_4}{l_2 l_3 l_4}\right) & \left(\dfrac{l_3 + l_4}{l_2 l_3 l_4}\right) & \dfrac{1}{l_2 l_3} \\ \left(\dfrac{l_3 + l_4}{l_2 l_3 l_4}\right) & \dfrac{(l_1 + l_2)(l_3 + l_4)}{l_1 l_2 l_3 l_4} & \left(\dfrac{l_1 + l_2}{l_1 l_2 l_3}\right) \\ \dfrac{1}{l_2 l_3} & \left(\dfrac{l_1 + l_2}{l_1 l_2 l_3}\right) & \left(\dfrac{l_1 + l_2 + l_3}{l_1 l_2 l_3}\right) \end{bmatrix}$$

The determinant of $[k]$ is

$$
|k| = \begin{vmatrix} \left(\dfrac{F}{l_1}+\dfrac{F}{l_2}\right) & -\dfrac{F}{l_2} & 0 \\[2mm] -\dfrac{F}{l_2} & \left(\dfrac{F}{l_2}+\dfrac{F}{l_3}\right) & -\dfrac{F}{l_3} \\[2mm] 0 & -\dfrac{F}{l_3} & \left(\dfrac{F}{l_3}+\dfrac{F}{l_4}\right) \end{vmatrix}
$$

$$
= F^3 \frac{l_1 + l_2 + l_3 + l_4}{l_1 l_2 l_3 l_4}
$$

Then

$$
[k]^{-1} = \frac{\text{adj } [k]}{|k|} = \frac{1}{l_0 F} \begin{bmatrix} l_1(l_2 + l_3 + l_4) & l_1(l_3 + l_4) & l_1 l_4 \\[2mm] l_1(l_3 + l_4) & (l_1 + l_2)(l_3 + l_4) & l_4(l_1 + l_2) \\[2mm] l_1 l_4 & l_4(l_1 + l_2) & l_4(l_1 + l_2 + l_3) \end{bmatrix}
$$

where $l_0 = l_1 + l_2 + l_3 + l_4$. This is in agreement with the flexibility matrix determined in Example 8-4.

8-5. SYMMETRY OF FLEXIBILITY AND STIFFNESS MATRICES

It has been noted that both the flexibility and stiffness matrices are symmetric. The reasons for this will now be shown. For a linearly elastic system, consider stations i and j and forces P_i and P_j applied at these positions, respectively. Then, irrespective of what other loads may be acting on the system, by the principle of superposition both the final configuration and the work done on the system are independent of the order in which forces P_i and P_j are applied.

Consider P_i to be applied first. As P_i is applied, the displacement produced at i is $P_i a_{ii}$ and the work done is $\frac{1}{2}P_i(P_i a_{ii})$. Applying P_j next causes the further displacement $P_j a_{ij}$ at i and displacement $P_j a_{jj}$ at j, and the work done is $P_i(P_j a_{ij}) + \frac{1}{2}P_j(P_j a_{jj})$. The total work is then

$$
U' = \tfrac{1}{2}P_i(P_i a_{ii}) + P_i(P_j a_{ij}) + \tfrac{1}{2}P_j(P_j a_{jj}) \tag{8-43}
$$

Similarly, if load P_j is applied first and then P_i, the work done is

$$
U'' = \tfrac{1}{2}P_j(P_j a_{jj}) + P_j(P_i a_{ji}) + \tfrac{1}{2}P_i(P_i a_{ii}) \tag{8-44}
$$

Equating U' and U'' leads to

$$
a_{ij} = a_{ji} \qquad i \neq j \tag{8-45}
$$

and it is concluded that the flexibility matrix $[m]$ is symmetric. (Equation 8-45 represents a form of Maxwell's law of reciprocal deflections.)

Again considering the loads P_i and P_j and designating the corresponding deflections as x_i and x_j, the following can be written:

$$\begin{Bmatrix} P_i \\ P_j \end{Bmatrix} = \begin{bmatrix} k_{ii} & k_{ij} \\ k_{ji} & k_{jj} \end{bmatrix} \begin{Bmatrix} x_i \\ x_j \end{Bmatrix} \quad \text{and} \quad \begin{Bmatrix} x_i \\ x_j \end{Bmatrix} = \begin{bmatrix} a_{ii} & a_{ij} \\ a_{ji} & a_{jj} \end{bmatrix} \begin{Bmatrix} P_i \\ P_j \end{Bmatrix} \tag{8-46}$$

where $[k]$ and $[a]$ are the stiffness and flexibility matrices, respectively. Recalling from Section 8-4, that $[a] = [k]^{-1}$, then

$$\begin{bmatrix} a_{ii} & a_{ij} \\ a_{ji} & a_{jj} \end{bmatrix} = \begin{bmatrix} k_{ii} & k_{ij} \\ k_{ji} & k_{jj} \end{bmatrix}^{-1} = \frac{\text{adj } [k]}{|k|} \tag{8-47}$$

$$= \frac{\begin{bmatrix} k_{jj} & -k_{ij} \\ -k_{ji} & k_{ii} \end{bmatrix}}{|k|} \tag{8-48}$$

From this it is concluded that

$$a_{ij} = \frac{-k_{ij}}{|k|} \quad \text{and} \quad a_{ji} = \frac{-k_{ji}}{|k|} \tag{8-49}$$

Since $a_{ij} = a_{ji}$, then

$$\frac{-k_{ij}}{|k|} = \frac{-k_{ji}}{|k|} \tag{8-50}$$

and

$$k_{ij} = k_{ji} \qquad i \neq j \tag{8-51}$$

which establishes that the stiffness matrix $[k]$ is symmetric.

8-6. DISPLACEMENT ANALYSIS—AN ALTERNATIVE METHOD

By the direct analysis of Section 8-2, the general form of the equation of motion was established as

$$[m]\{\ddot{x}\} + [k]\{x\} = \{0\} \tag{8-52}$$

Premultiplying this by the flexibility matrix $[a]$ gives

$$[a][m]\{\ddot{x}\} + [a][k]\{x\} = \{0\} \tag{8-53}$$

and since from Eq. 8-41, $[a] = [k]^{-1}$, then

$$[a][m]\{\ddot{x}\} + [k]^{-1}[k]\{x\} = \{0\}$$

whence

$$[a][m]\{\ddot{x}\} + \{x\} = \{0\} \tag{8-54}$$

The matrices of this equation are as previously defined. Writing the equation in detail for the general case,

$$
\begin{bmatrix}
a_{11} & a_{12} & a_{13} & \cdots & a_{1n} \\
a_{21} & a_{22} & a_{23} & \cdots & a_{2n} \\
a_{31} & a_{32} & a_{33} & \cdots & a_{3n} \\
\cdot & \cdot & \cdot & \cdots & \cdot \\
a_{n1} & \cdot & \cdot & \cdots & a_{nn}
\end{bmatrix}
\begin{bmatrix}
m_{11} & m_{12} & m_{13} & \cdots & m_{1n} \\
m_{21} & m_{22} & m_{23} & \cdots & m_{2n} \\
m_{31} & m_{32} & m_{33} & \cdots & m_{3n} \\
\cdot & \cdot & \cdot & \cdots & \cdot \\
m_{n1} & \cdot & \cdot & \cdots & m_{nn}
\end{bmatrix}
\begin{Bmatrix}
\ddot{x}_1 \\
\ddot{x}_2 \\
\ddot{x}_3 \\
\cdot \\
\ddot{x}_n
\end{Bmatrix}
+
\begin{Bmatrix}
x_1 \\
x_2 \\
x_3 \\
\cdot \\
x_n
\end{Bmatrix}
=
\begin{Bmatrix}
0 \\
0 \\
0 \\
\cdot \\
0
\end{Bmatrix}
$$

(8-55)

The manner of expressing the matrix product $[a][m]$ should be carefully noted. Thus

$$
[a][m] =
\begin{bmatrix}
(a_{11}m_{11} + a_{12}m_{21} + a_{13}m_{31} + \cdots) & (a_{11}m_{12} + a_{12}m_{22} + a_{13}m_{32} + \cdots) & \cdots \\
(a_{21}m_{11} + a_{22}m_{21} + a_{23}m_{31} + \cdots) & (a_{21}m_{12} + a_{22}m_{22} + a_{23}m_{32} + \cdots) & \cdots \\
(a_{31}m_{11} + a_{32}m_{21} + a_{33}m_{31} + \cdots) & (a_{31}m_{12} + a_{32}m_{22} + a_{33}m_{32} + \cdots) & \cdots \\
\cdots & \cdots & \cdots \\
\cdots & \cdots & \cdots
\end{bmatrix}
$$

(8-56)

All of the terms of the flexibility matrix $[a]$ will be present. Then if the mass matrix is also nondiagonal (that is, if nondiagonal terms are present), expanding the matrix product $[a][m]$ can be lengthy and cumbersome. However, if coordinates are chosen so that a single coordinate defines the displacement of each mass point, then $m_{ij} = 0$ for $i \neq j$ and $[m]$ is a diagonal matrix, and the matrix product $[a][m]$ is reasonably simple. The diagonal mass matrix may then be expressed as

$$
[m] =
\begin{bmatrix}
m_1 & 0 & 0 & \cdots & 0 \\
0 & m_2 & 0 & \cdots & 0 \\
0 & 0 & m_3 & \cdots & 0 \\
\cdot & \cdot & \cdot & \cdots & \cdot \\
0 & \cdot & \cdot & \cdots & m_n
\end{bmatrix}
$$

(8-57)

and the displacement equation is written as

$$
\begin{bmatrix}
a_{11} & a_{12} & a_{13} & \cdots & a_{1n} \\
a_{21} & a_{22} & a_{23} & \cdots & a_{2n} \\
a_{31} & \cdots & \cdots & \cdots & a_{3n} \\
\cdot & \cdot & \cdot & \cdots & \cdot \\
a_{n1} & \cdot & \cdot & \cdots & a_{nn}
\end{bmatrix}
\begin{bmatrix}
m_1 & 0 & 0 & \cdots & 0 \\
0 & m_2 & 0 & \cdots & 0 \\
0 & 0 & m_3 & \cdots & 0 \\
\cdot & \cdot & \cdot & \cdots & \cdot \\
0 & \cdot & \cdot & \cdots & m_n
\end{bmatrix}
\begin{Bmatrix}
\ddot{x}_1 \\
\ddot{x}_2 \\
\ddot{x}_3 \\
\cdot \\
\ddot{x}_n
\end{Bmatrix}
+
\begin{Bmatrix}
x_1 \\
x_2 \\
x_3 \\
\cdot \\
x_n
\end{Bmatrix}
=
\begin{Bmatrix}
0 \\
0 \\
0 \\
\cdot \\
0
\end{Bmatrix}
$$

(8-58)

This is the same as Eq. 8-55, but with the off-diagonal terms of $[m]$ equal to zero.

Substituting the harmonic function

$$
\{x\} = \{A\} \sin(\omega t + \phi)
$$

(8-59)

into Eq. 8-54 results in

$$-\omega^2[a][m]\{A\} + [I]\{A\} = \{0\} \tag{8-60}$$

This is the amplitude equation, which can then also be written as

$$([a][m] - \lambda[I])\{A\} = \{0\} \tag{8-61}$$

where $1/\omega^2 = \lambda$ is the inverse frequency factor and

$$[a][m] - \lambda[I] = [E] \text{ is the characteristic matrix.} \tag{8-62}$$

The determinant of the characteristic matrix must vanish, so that

$$|[a][m] - \lambda[I]| = 0 \tag{8-63}$$

This characteristic determinant would be expressed in detail as

$$\left| \begin{bmatrix} a_{11} & a_{12} & a_{13} & \cdots \\ a_{21} & a_{22} & \cdot & \cdots \\ a_{31} & \cdot & \cdot & \cdots \\ \cdot & \cdot & \cdot & \cdots \end{bmatrix} \begin{bmatrix} m_{11} & m_{12} & m_{13} & \cdots \\ m_{21} & m_{22} & \cdot & \cdots \\ m_{31} & \cdot & \cdot & \cdots \\ \cdot & \cdot & \cdot & \cdots \end{bmatrix} - \lambda \begin{bmatrix} 1 & 0 & 0 & \cdots & \cdots \\ 0 & 1 & 0 & \cdots & \cdots \\ 0 & 0 & 1 & 0 & \cdots \\ \cdot & \cdot & \cdot & \cdot & \cdots \end{bmatrix} \right| = 0 \tag{8-64}$$

Expansion of this determinant will yield a polynomial in which the highest-order term is λ^n, having the n roots $\lambda_1, \lambda_2, \lambda_3, \ldots, \lambda_n$ representing the n eigenvalues. For several degrees of freedom these roots cannot be obtained by factoring but could be determined by a numerical procedure. The natural frequencies can be determined from the eigenvalues.

The n eigenvectors $\{A\}_i$ can be obtained by substitution of the eigenvalues into the amplitude matrix equation (Eq. 8-61), or from the adjoint of the characteristic matrix.

EXAMPLE 8-6   For the stretched-cord case of Example 8-1, (a) write the matrix form of the displacement equations that govern the motion of the system, using the flexibility coefficients determined in Example 8-4. (b) Determine the principal modes for the case of equal cord lengths and $m_1 = m$, $m_2 = 3m$, $m_3 = 2m$.

SOLUTION   Figure 8-6(a) represents a dynamic loading diagram of the cord alone, in which P_i is the force acting on the cord at station i. The displacement relations are expressed as

$$x_1 = a_{11}P_1 + a_{12}P_2 + a_{13}P_3$$

$$x_2 = a_{21}P_1 + a_{22}P_2 + a_{23}P_3$$

$$x_3 = a_{31}P_1 + a_{32}P_2 + a_{33}P_3$$

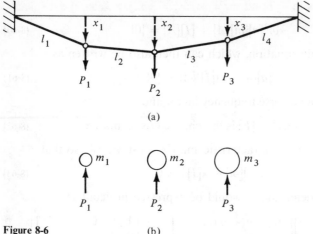

Figure 8-6   (b)

Figure 8-6(b) shows free-body diagrams of each mass alone. Here P_i is the force at i acting on mass m_i and is due to the cord connection at that position. Newton's second law can be expressed for each of these as follows:

$$m_1 \ddot{x}_1 = -P_1 \qquad m_2 \ddot{x}_2 = -P_2 \qquad m_3 \ddot{x}_3 = -P_3$$

Substituting these into the displacement relations and rearranging results in

$$a_{11} m_1 \ddot{x}_1 + a_{12} m_2 \ddot{x}_2 + a_{13} m_3 \ddot{x}_3 + x_1 = 0$$

$$a_{21} m_1 \ddot{x}_1 + a_{22} m_2 \ddot{x}_2 + a_{23} m_3 \ddot{x}_3 + x_2 = 0$$

$$a_{31} m_1 \ddot{x}_1 + a_{32} m_2 \ddot{x}_2 + a_{33} m_3 \ddot{x}_3 + x_3 = 0$$

These can be written in matrix form as

$$\begin{bmatrix} a_{11} & a_{12} & a_{13} \\ a_{21} & a_{22} & a_{23} \\ a_{31} & a_{32} & a_{33} \end{bmatrix} \begin{bmatrix} m_1 & 0 & 0 \\ 0 & m_2 & 0 \\ 0 & 0 & m_3 \end{bmatrix} \begin{Bmatrix} \ddot{x}_1 \\ \ddot{x}_2 \\ \ddot{x}_3 \end{Bmatrix} + \begin{Bmatrix} x_1 \\ x_2 \\ x_3 \end{Bmatrix} = \begin{Bmatrix} 0 \\ 0 \\ 0 \end{Bmatrix}$$

This equation could have been set down immediately by considering Eq. 8-54 or Eq. 8-58 for this case.

Substituting in the flexibility coefficients or matrix obtained in Example 8-4 for this system gives

$$\frac{1}{l_0 F} \begin{bmatrix} l_1(l_2 + l_3 + l_4) & l_1(l_3 + l_4) & l_1 l_4 \\ l_1(l_3 + l_4) & (l_1 + l_2)(l_3 + l_4) & l_4(l_1 + l_2) \\ l_1 l_4 & l_4(l_1 + l_2) & l_4(l_1 + l_2 + l_3) \end{bmatrix}$$

$$\times \begin{bmatrix} m_1 & 0 & 0 \\ 0 & m_2 & 0 \\ 0 & 0 & m_3 \end{bmatrix} \begin{Bmatrix} \ddot{x}_1 \\ \ddot{x}_2 \\ \ddot{x}_3 \end{Bmatrix} + \begin{Bmatrix} x_1 \\ x_2 \\ x_3 \end{Bmatrix} = \begin{Bmatrix} 0 \\ 0 \\ 0 \end{Bmatrix}$$

(b) For $l_1 = l_2 = l_3 = l_4 = l$ and $m_1 = m$, $m_2 = 3m$, $m_3 = 2m$, the displacement equation simplifies to

$$\frac{ml}{F} \begin{bmatrix} \frac{3}{4} & \frac{1}{2} & \frac{1}{4} \\ \frac{1}{2} & 1 & \frac{1}{2} \\ \frac{1}{4} & \frac{1}{2} & \frac{3}{4} \end{bmatrix} \begin{bmatrix} 1 & 0 & 0 \\ 0 & 3 & 0 \\ 0 & 0 & 2 \end{bmatrix} \begin{Bmatrix} \ddot{x}_1 \\ \ddot{x}_2 \\ \ddot{x}_3 \end{Bmatrix} + \begin{Bmatrix} x_1 \\ x_2 \\ x_3 \end{Bmatrix} = \begin{Bmatrix} 0 \\ 0 \\ 0 \end{Bmatrix}$$

Multiplying the two square matrices gives

$$\frac{ml}{F} \begin{bmatrix} \frac{3}{4} & \frac{3}{2} & \frac{1}{2} \\ \frac{1}{2} & 3 & 1 \\ \frac{1}{4} & \frac{3}{2} & \frac{3}{2} \end{bmatrix} \begin{Bmatrix} \ddot{x}_1 \\ \ddot{x}_2 \\ \ddot{x}_3 \end{Bmatrix} + \begin{Bmatrix} x_1 \\ x_2 \\ x_3 \end{Bmatrix} = \begin{Bmatrix} 0 \\ 0 \\ 0 \end{Bmatrix}$$

Substituting in the harmonic function $\{x\} = \{A\} \sin(\omega t + \phi)$, multiplying by $(-F/ml\omega^2)$, and setting $F/ml\omega^2 = \lambda$ results in the amplitude equation

$$\begin{bmatrix} (3/4 - \lambda) & 3/2 & 1/2 \\ 1/2 & (3 - \lambda) & 1 \\ 1/4 & 3/2 & (3/2 - \lambda) \end{bmatrix} \begin{Bmatrix} A_1 \\ A_2 \\ A_3 \end{Bmatrix} = \begin{Bmatrix} 0 \\ 0 \\ 0 \end{Bmatrix}$$

The determinant of the characteristic matrix is then set to zero,

$$\begin{vmatrix} (3/4 - \lambda) & 3/2 & 1/2 \\ 1/2 & (3 - \lambda) & 1 \\ 1/4 & 3/2 & (3/2 - \lambda) \end{vmatrix} = 0$$

Expanding gives the characteristic or frequency equation as

$$4\lambda^3 - 21\lambda^2 + 22\lambda - 6 = 0$$

The roots of this can be verified as

$$\lambda_i = 3.9554, \quad 0.846\,74, \quad 0.447\,85$$

from which

$$\omega^2 = 0.252\,82 \frac{F}{ml}, \quad 1.1809 \frac{F}{ml}, \quad 2.2329 \frac{F}{ml}$$

as obtained in Example 8-1. These will define the same eigenvectors or modal configurations as were determined by the direct method of Example 8-1.

EXAMPLE 8-7 For the beam and lumped-mass system of Fig. 8-7, determine the flexibility coefficients, express the flexibility matrix, and write the mass matrix. Also obtain the matrix form of the displacement equations for the system.

$$m_1 = m, \quad m_2 = 2m, \quad m_3 = 3m$$

SOLUTION The influence coefficients can be obtained by placing a unit load at a load position, as shown and determining the deflection at all stations. The deflections may then be obtained from the moment diagrams by the moment-area method (Fig. 8-8).

a. a_{11} is equal to the moment of the M diagram taken about (1) and divided by EI. Thus

$$a_{11} = \left(\frac{1}{2}3l \cdot 3l\right) \cdot \frac{2}{3}3l \cdot \frac{1}{EI} = \frac{9l^3}{EI}$$

Similarly, a_{21} is obtained by the moment of the area between (2) and (4) taken about (2).

$$a_{21} = \left[\left(\frac{1}{2}2l \cdot l\right)\left(\frac{1}{3}2l\right) + \left(\frac{1}{2}2l \cdot 3l\right) \cdot \left(\frac{2}{3}2l\right)\right]\frac{1}{EI} = \frac{14}{3}\frac{l^3}{EI}$$

$$a_{31} = \left[\left(\frac{1}{2}l \cdot 2l\right)\frac{1}{3}l + \left(\frac{1}{2}l \cdot 3l\right) \cdot \left(\frac{2}{3}l\right)\right]\frac{1}{EI} = \frac{4}{3}\frac{l^3}{EI}$$

b. Similarly,

$$a_{12} = \left(\frac{1}{2}2l \cdot 2l\right) \cdot \left(\frac{2}{3}2l + l\right)\frac{1}{EI} = \frac{14}{3}\frac{l^3}{EI}$$

$$a_{22} = \left(\frac{1}{2}2l \cdot 2l\right) \cdot \left(\frac{2}{3}2l\right)\frac{1}{EI} = \frac{8}{3}\frac{l^3}{EI}$$

$$a_{32} = \left[\left(\frac{1}{2}l \cdot l\right) \cdot \left(\frac{1}{3}l\right) + \left(\frac{1}{2}l \cdot 2l\right) \cdot \left(\frac{2}{3}l\right)\right]\frac{1}{EI} = \frac{5}{6}\frac{l^3}{EI}$$

c. In like manner,

$$a_{13} = \left(\frac{1}{2}l \cdot l\right)\left(\frac{2}{3}l + 2l\right)\frac{1}{EI} = \frac{4}{3}\frac{l^3}{EI}$$

$$a_{23} = \left(\frac{1}{2}l \cdot l\right)\left(\frac{2}{3}l + l\right)\frac{1}{EI} = \frac{5}{6}\frac{l^3}{EI}$$

$$a_{33} = \left(\frac{1}{2}l \cdot l\right)\left(\frac{2}{3}l\right)\frac{1}{EI} = \frac{1}{3}\frac{l^3}{EI}$$

(Note that $a_{12} = a_{21}$, $a_{13} = a_{31}$, $a_{23} = a_{32}$.) Then

$$[a] = \frac{l^3}{EI}\begin{bmatrix} 9 & 14/3 & 4/3 \\ 14/3 & 8/3 & 5/6 \\ 4/3 & 5/6 & 1/3 \end{bmatrix}$$

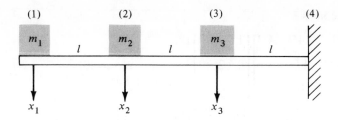

Figure 8-7

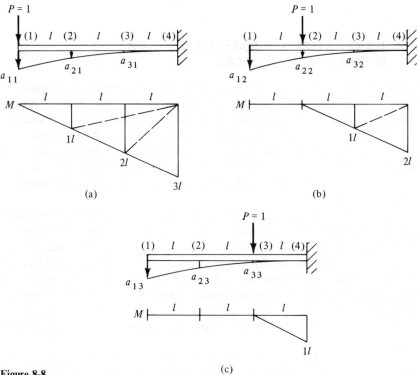

Figure 8-8

Using displacement coordinates at the load position results in

$$[m] = m \begin{bmatrix} 1 & 0 & 0 \\ 0 & 2 & 0 \\ 0 & 0 & 3 \end{bmatrix}$$

The governing displacement equation then is

$$[a][m]\{\ddot{x}\} + [k]\{x\} = 0$$

which in detail becomes

$$\frac{ml^3}{EI}\begin{bmatrix} 9 & 14/3 & 4/3 \\ 14/3 & 8/3 & 5/6 \\ 4/3 & 5/6 & 1/3 \end{bmatrix}\begin{bmatrix} 1 & 0 & 0 \\ 0 & 2 & 0 \\ 0 & 0 & 3 \end{bmatrix}\begin{Bmatrix} \ddot{x}_1 \\ \ddot{x}_2 \\ \ddot{x}_3 \end{Bmatrix} + \begin{Bmatrix} x_1 \\ x_2 \\ x_3 \end{Bmatrix} = \{0\}$$

$$\frac{ml^3}{EI}\begin{bmatrix} 9 & 28/3 & 4 \\ 14/3 & 16/3 & 5/2 \\ 4/3 & 5/3 & 1 \end{bmatrix}\begin{Bmatrix} \ddot{x}_1 \\ \ddot{x}_2 \\ \ddot{x}_3 \end{Bmatrix} + \begin{Bmatrix} x_1 \\ x_2 \\ x_3 \end{Bmatrix} = \{0\}$$

8-7. GENERALIZED COORDINATES AND GENERALIZED FORCES

In the dynamic analysis of a mass system, a set of coordinates is required. Often different sets of coordinates may be available, any of which would adequately describe the configuration of the system. Such sets of coordinates may be linear or angular, or a combination of these. When starting an analysis, it is necessary to select a set of coordinates; and for this purpose, it is desirable to choose a set which will be suitable and efficient. This involves an understanding of generalized coordinates and the idea of the independence of coordinates.

Generalized coordinates may be defined as a set of space or position parameters which express the configuration of a system and which meet the requirement of being independent of each other. They are also independent of any conditions of constraint. Most coordinate systems that might be used are generalized coordinates. The only coordinates that do not come within this classification are those having more coordinates than there are degrees of freedom. Generalized coordinates may be lengths, angles, areas, or any other set of numbers which define uniquely the dynamic configuration of the system.

In order to illustrate the foregoing, consider the double pendulum arrangement of Fig. 8-9, having the fixed lengths l_1 and l_2, with pivots at A and B, and containing the masses m_1 and m_2. The configuration can be specified by the four coordinates x_1, y_1, x_2, and y_2. However, these coordinates are not independent of each other. For example, changing x_1 alters y_1, and changing x_2 alters either y_2 or x_1 and y_1. The coordinates are constrained by the pendulums, and from this the equations of constraint can be written as follows:

$$x_1^2 + y_1^2 = l_1^2$$

$$(x_2 - x_1)^2 + (y_2 - y_1)^2 = l_2^2 \tag{8-65}$$

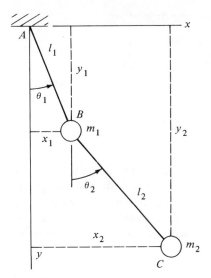

Figure 8-9

From the conditions previously stated, it is then concluded that this set of coordinates is not generalized. It can be observed that the number of coordinates (four) minus the number of equations of constraint (two) is equal to the number of degrees of freedom (two) here. If there had been no constraints—that is, if the pendulums were removed so that m_1 and m_2 were free to move anywhere in the x-y plane—then there would have been four degrees of freedom requiring four coordinates such as x_1, y_1, x_2, and y_2, which would then be generalized coordinates.

For the double pendulum the coordinates θ_1 and θ_2 shown would properly define the configuration of the mass system. They are independent of each other (that is, they can be changed independently) and are independent of the constraints, which are simply that l_1 and l_2 are constant. Hence θ_1 and θ_2 form a set of generalized coordinates. The number of these generalized coordinates (two) is equal to the number of degrees of freedom (two) for this case.

The n generalized coordinates of a system may be designated by

$$q_1, q_2, \ldots, q_k, \ldots, q_n \qquad (8\text{-}66)$$

Next, it is assumed that an arrangement of forces acts on the system under consideration. For a small displacement δq_1 of the generalized coordinate q_1, consider the work done to be U_1. Then the generalized force Q_1 is defined as

$$Q_1 = \frac{U_1}{\delta q_1} \qquad (8\text{-}67a)$$

Similarly, the remaining generalized forces are defined by

$$Q_2 = \frac{U_2}{\delta q_2}, \ldots, Q_k = \frac{U_k}{\delta q_k}, \ldots, Q_n = \frac{U_n}{\delta q_n} \tag{8-67b}$$

Note that if q_k is linear, then Q_k is a force; however, when q_k is angular, then Q_k is a moment.

8-8. LAGRANGE'S EQUATIONS

A notable simplification in the derivation of the equations of motion for a mass system can be obtained by using generalized coordinates and generalized forces. This makes use of the energy relations for the system. The potential energy V will be a function of the generalized coordinates of the system. Likewise, the kinetic energy T will usually be a function of the generalized velocities only. Thus

$$V = V(q_k)$$

$$T = T(\dot{q}_k) \tag{8-68}$$

The following governing relation can be derived.†

$$\frac{d}{dt}\left(\frac{\partial T}{\partial \dot{q}_k}\right) - \frac{\partial T}{\partial q_k} + \frac{\partial V}{\partial q_k} = Q'_k \qquad (k = 1, 2, \ldots, n) \tag{8-69}$$

where Q'_k is the generalized force which has no potential. This is known as Lagrange's equation (or equations). It is an advanced formulation based on Newton's second law and consideration of the energies and virtual work. For a conservative system, $Q'_k = 0$. Lagrange's equation is then

$$\frac{d}{dt}\left(\frac{\partial T}{\partial \dot{q}_k}\right) - \frac{\partial T}{\partial q_k} + \frac{\partial V}{\partial q_k} = 0 \qquad (k = 1, 2, \ldots, n) \tag{8-70}$$

In applying Eq. 8-70, it is written separately for each value of k from 1 to n. Thus n differential equations will result, corresponding to the n degrees of freedom defined by the n generalized coordinates. This enables the differential equations of motion to be derived quite simply, provided the energy functions V and T are not difficult to write.

Consider a system having n degrees of freedom defined by the n generalized coordinates $q_1, q_2, \ldots, q_k, \ldots, q_n$. Then, if the amplitudes are small so that the elastic system is linear, it is expected that V would be a second-degree function of the generalized coordinates and T would be a second-

† See Appendix B.

degree expression of the generalized velocities only. Thus

$$V = \tfrac{1}{2}[(k_{11}q_1^2 + k_{22}q_2^2 + k_{33}q_3^2 + \cdots + k_{nn}q_n^2)$$
$$+ 2(k_{12}q_1q_2 + k_{13}q_1q_3 + k_{14}q_1q_4 + \cdots + k_{1n}q_1q_n$$
$$+ k_{23}q_2q_3 + k_{24}q_2q_4 + k_{25}q_2q_5 + \cdots + k_{2n}q_2q_n$$
$$+ k_{34}q_3q_4 + k_{35}q_3q_5 + \cdots + k_{3n}q_3q_n + \cdots)] \tag{8-71}$$

The coefficients $k_{11}, k_{22}, \ldots, k_{12}, k_{13}, \ldots$ are defined by the elastic constants of the system. Also,

$$T = \tfrac{1}{2}[(m_{11}\dot{q}_1^2 + m_{22}\dot{q}_2^2 + m_{33}\dot{q}_3^2 + \cdots + m_{nn}\dot{q}_n^2)$$
$$+ 2(m_{12}\dot{q}_1\dot{q}_2 + m_{13}\dot{q}_1\dot{q}_3 + m_{14}\dot{q}_1\dot{q}_4 + \cdots + m_{1n}\dot{q}_1\dot{q}_n$$
$$+ m_{23}\dot{q}_2\dot{q}_3 + m_{24}\dot{q}_2\dot{q}_4 + m_{25}\dot{q}_2\dot{q}_5 + \cdots + m_{2n}\dot{q}_2\dot{q}_n$$
$$+ m_{34}\dot{q}_3\dot{q}_4 + m_{35}\dot{q}_3\dot{q}_5 + \cdots + m_{3n}\dot{q}_3\dot{q}_n + \cdots)] \tag{8-72}$$

The coefficients $m_{11}, m_{22}, \ldots, m_{12}, m_{13}, \ldots$ are defined by the mass or inertial elements of the system.

The Lagrange differential equations will then be of the form

$$m_{11}\ddot{q}_1 + m_{12}\ddot{q}_2 + m_{13}\ddot{q}_3 + \cdots + k_{11}q_1 + k_{12}q_2 + k_{13}q_3 + \cdots = 0$$

$$m_{21}\ddot{q}_1 + m_{22}\ddot{q}_2 + m_{23}\ddot{q}_3 + \cdots + k_{21}q_1 + k_{22}q_2 + k_{23}q_3 + \cdots = 0$$

$$m_{31}\ddot{q}_1 + m_{32}\ddot{q}_2 + m_{33}\ddot{q}_3 + \cdots + k_{31}q_1 + k_{32}q_2 + k_{33}q_3 + \cdots = 0$$

$$\vdots$$

$$m_{n1}\ddot{q}_1 + m_{n2}\ddot{q}_2 + m_{n3}\ddot{q}_3 + \cdots + k_{n1}q_1 + k_{n2}q_2 + k_{n3}q_3 + \cdots = 0 \tag{8-73}$$

The differential equations exhibit static coupling resulting from the presence of the cross-product coordinate terms $(q_1q_2, q_1q_3, \ldots, q_2q_3, \ldots)$ in the potential-energy function and dynamic coupling due to the occurrence of the cross-product velocity terms $(\dot{q}_1\dot{q}_2, \dot{q}_1\dot{q}_3, \ldots, \dot{q}_2\dot{q}_3, \ldots)$ in the kinetic-energy function. There are n such differential equations. In these, certain coefficients (for example, $k_{13}, k_{14}, \ldots$ and $m_{13}, m_{14}, \ldots$) may be missing.

Using matrix notation, Eqs. 8-73 would be written as

$$\begin{bmatrix} m_{11} & m_{12} & m_{13} & \cdots & m_{1n} \\ m_{21} & m_{22} & m_{23} & \cdots & m_{2n} \\ m_{31} & m_{32} & m_{33} & \cdots & m_{3n} \\ \cdot & \cdot & \cdot & \cdots & \cdot \\ m_{n1} & \cdot & \cdot & \cdots & m_{nn} \end{bmatrix} \begin{Bmatrix} \ddot{q}_1 \\ \ddot{q}_2 \\ \ddot{q}_3 \\ \cdot \\ \ddot{q}_n \end{Bmatrix}$$

$$+ \begin{bmatrix} k_{11} & k_{12} & k_{13} & \cdots & k_{n1} \\ k_{21} & k_{22} & k_{23} & \cdots & k_{2n} \\ k_{31} & k_{32} & k_{33} & \cdots & k_{3n} \\ \cdot & \cdot & \cdot & \cdots & \cdot \\ k_{n1} & \cdot & \cdot & \cdots & k_{nn} \end{bmatrix} \begin{Bmatrix} q_1 \\ q_2 \\ q_3 \\ \cdot \\ q_n \end{Bmatrix} = \begin{Bmatrix} 0 \\ 0 \\ 0 \\ \cdot \\ 0 \end{Bmatrix} \tag{8-74}$$

or more simply as

$$[m]\{\ddot{q}\} + [k]\{q\} = \{0\} \tag{8-75}$$

Observe that in Eqs. 8-74 and 8-75, $m_{12} = m_{21}$, since it comes from the same source, namely, the coefficient of $\dot{q}_1 \dot{q}_2$ in the kinetic-energy function T (Eq. 8-72). That is,

$$\frac{d}{dt}\left(\frac{\partial}{\partial \dot{q}_1}(m_{12}\dot{q}_1\dot{q}_2)\right) = m_{12}\ddot{q}_2 \tag{8-76a}$$

and

$$\frac{d}{dt}\left(\frac{\partial}{\partial \dot{q}_2}(m_{12}\dot{q}_1\dot{q}_2)\right) = m_{12}\ddot{q}_1 \tag{8-76b}$$

For the same reason, $m_{13} = m_{31}, \ldots, m_{23} = m_{32}, \ldots,$ and so on, so that $m_{ij} = m_{ji}$ for $i \neq j$ and $[m]$ is a symmetric matrix for the Lagrange equations. This applies as well to the mass matrix $[m]$ for the direct and the displacement analyses of Sections 8-2, 8-3, and 8-6, *providing* the coordinates are measured from a fixed reference system.

Also, in Eqs. 8-74 and 8-75, $k_{12} = k_{21}$ as both have the same source—the coefficient of $q_1 q_2$ in the potential-energy function V (Eq. 8-71), and so on—so that $k_{ij} = k_{ji}$ for $i \neq j$ and $[k]$ is symmetric here.

The form of Eq. 8-75 is the same as that of Eq. 8-4, so that the solution would be carried out in the same manner. This would yield the eigenvalues and eigenvectors—that is, the principal modes and corresponding natural frequencies.

EXAMPLE 8-8   For the torsional arrangement of Fig. 8-10, the angular displacements $\theta_1, \ldots, \theta_4$ are generalized coordinates for the four-degree-of-freedom system. The k's represent the torsional spring constants for the indicated shaft portions, and the I's are the mass moments of inertia of the disks. Write the kinetic and potential energy functions and obtain the Lagrange equations.

SOLUTION   The energy relations will be

a. $$T = \tfrac{1}{2}I_1\dot{\theta}_1^2 + \tfrac{1}{2}I_2\dot{\theta}_2^2 + \tfrac{1}{2}I_3\dot{\theta}_3^2 + \tfrac{1}{2}I_4\dot{\theta}_4^2$$

For a portion of a shaft in torsion, the potential energy is equal to the work done by the shaft as it returns from the dynamic configuration to the reference equilibrium position. Thus

$$V = \int_\theta^0 (-k\theta)\, d\theta = \frac{k\theta^2}{2}$$

Applying this to the present case gives the potential-energy function as

b. $V = \tfrac{1}{2}k_1(\theta_1 - \theta_2)^2 + \tfrac{1}{2}k_2(\theta_2 - \theta_3)^2 + \tfrac{1}{2}k_3(\theta_3 - \theta_4)^2 + \tfrac{1}{2}k_4\theta_4^2$

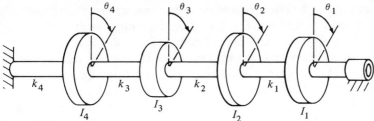

Figure 8-10

The Lagrange equations are now written. For subscript $k = 1$, the following are obtained:

$$\frac{\partial T}{\partial \dot\theta_1} = I_1 \dot\theta_1 \quad \text{and} \quad \frac{d}{dt}\left(\frac{\partial T}{\partial \dot\theta_1}\right) = I_1 \ddot\theta_1$$

$$\frac{\partial T}{\partial \theta_1} = 0$$

since T is independent of θ_1, and

$$\frac{\partial V}{\partial \theta_1} = k_1(\theta_1 - \theta_2)$$

whence

$$\frac{d}{dt}\left(\frac{\partial T}{\partial \dot\theta_1}\right) - \frac{\partial T}{\partial \theta_1} + \frac{\partial V}{\partial \theta_1} = 0$$

gives

$$I_1 \ddot\theta_1 + k_1(\theta_1 - \theta_2) = 0$$

Similarly, letting subscript $k = 2, 3, 4$ in Lagrange's relation yields

$$I_2 \ddot\theta_2 - k_1(\theta_1 - \theta_2) + k_2(\theta_2 - \theta_3) = 0$$
$$I_3 \ddot\theta_3 - k_2(\theta_2 - \theta_3) + k_3(\theta_3 - \theta_4) = 0$$
$$I_4 \ddot\theta_4 - k_3(\theta_3 - \theta_4) + k_4\theta_4 \qquad = 0$$

Rearranging and writing these in matrix form gives

c. $\begin{bmatrix} I_1 & 0 & 0 & 0 \\ 0 & I_2 & 0 & 0 \\ 0 & 0 & I_3 & 0 \\ 0 & 0 & 0 & I_4 \end{bmatrix} \begin{Bmatrix} \ddot\theta_1 \\ \ddot\theta_2 \\ \ddot\theta_3 \\ \ddot\theta_4 \end{Bmatrix}$

$$+ \begin{bmatrix} k_1 & -k_1 & 0 & 0 \\ -k_1 & (k_1 + k_2) & -k_2 & 0 \\ 0 & -k_2 & (k_2 + k_3) & -k_3 \\ 0 & 0 & -k_3 & (k_3 + k_4) \end{bmatrix} \begin{Bmatrix} \theta_1 \\ \theta_2 \\ \theta_3 \\ \theta_4 \end{Bmatrix} = \begin{Bmatrix} 0 \\ 0 \\ 0 \\ 0 \end{Bmatrix}$$

EXAMPLE 8-9 A system composed of two masses, on a stretched cord having tension F is shown in Fig. 8-11. The generalized coordinates selected are y_1 representing the absolute displacement of the mass m_1, and y_2 defining the displacement of the mass m_2 relative to the mass m_1. Express the kinetic and potential energy functions, and obtain Lagrange's equations of motion, observing the dynamic and static coupling which occurs. Then, for the case of $m_1 = m_2 = m$ and $l_1 = l_2 = l_3 = l$, obtain the solution to these differential equations, defining the principal modes and frequencies.

SOLUTION The energy relations will be

a.
$$T = \tfrac{1}{2}m_1\dot{y}_1^2 + \tfrac{1}{2}m_2(\dot{y}_1 + \dot{y}_2)^2$$
$$= \tfrac{1}{2}[(m_1 + m_2)\dot{y}_1^2 + m_2\dot{y}_2^2 + 2m_2\dot{y}_1\dot{y}_2]$$

b.
$$V = \tfrac{1}{2}F\frac{y_1}{l_1}y_1 - \tfrac{1}{2}F\frac{y_2}{l_2}y_1 + \tfrac{1}{2}F\frac{y_2}{l_2}(y_1 + y_2) + \tfrac{1}{2}F\left(\frac{y_1 + y_2}{l_3}\right)(y_1 + y_2)$$
$$= \frac{F}{2}\left[\left(\frac{1}{l_1} + \frac{1}{l_3}\right)y_1^2 + \left(\frac{1}{l_2} + \frac{1}{l_3}\right)y_2^2 + \frac{2}{l_3}y_1 y_2\right]$$

Note the cross-product terms in each energy function.
Lagrange's relation gives

c.
$$(m_1 + m_2)\ddot{y}_1 + m_2\ddot{y}_2 + F\left(\frac{1}{l_1} + \frac{1}{l_3}\right)y_1 + \frac{F}{l_3}y_2 = 0$$
$$m_2\ddot{y}_1 + m_2\ddot{y}_2 + \frac{F}{l_3}y_1 + F\left(\frac{1}{l_2} + \frac{1}{l_3}\right)y_2 = 0$$

These equations show both dynamic and static coupling. Note the way in which the coefficients $m_{11}, m_{12}, \dots$ and $k_{11}, k_{12}, \dots$ occur. Also, observe how $m_{12} = m_{21}$ and $k_{12} = k_{21}$.

Substituting $y_1 = A_1 \sin(\omega t + \phi)$, $y_2 = A_2 \sin(\omega t + \phi)$, and taking the case of equal masses and equal cord lengths results in

d.
$$(2\xi - 2)A_1 + (\xi - 1)A_2 = 0$$
$$(\xi - 1)A_1 + (\xi - 2)A_2 = 0$$

where $\xi = m\omega^2 l/F$. The frequency equation is

e.
$$\xi^2 - 4\xi + 3 = 0$$

having the roots $\xi = 1, 3$. This defines the principal modes as
   Mode 1: $\xi_1 = 1$

$$\omega_1 = \sqrt{\frac{F}{ml}} \qquad A_1 = 1, \quad A_2 = 0$$

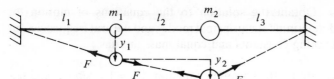

Figure 8-11

and

$$y_1 = A_1 \sin (\omega_1 t + \phi)$$

$$y_2 = 0$$

Mode 2: $\xi_2 = 3$

$$\omega_2 = \sqrt{\frac{3F}{ml}} \qquad A_1 = 1, \quad A_2 = -2$$

and

$$y_1 = A_1 \sin (\omega_2 t + \phi)$$

$$y_2 = -2A_1 \sin (\omega_2 t + \phi)$$

8-9. GENERAL DISCUSSION OF ANALYSIS AND SOLUTION

In preceding sections of this chapter, classical approaches were shown for solving the problem of several-degrees-of-freedom vibration systems which were without damping and were free of external forcing conditions. It was found that the three-degree-of-freedom problem was more inconvenient to solve than the two-degree case. For the four-degree-of-freedom case, an additional increase in difficulty would be encountered. It is reasonable to expect that a further advance in the number of degrees of freedom would be accompanied by greater difficulty of solution, until the classical method would eventually become impractical. The nature of these complications should be taken into account. They are encountered mainly after the amplitude equations have been obtained. In part, this is because the higher the order of a determinant, the more difficult it is to expand. The resulting characteristic equation is more inconvenient to combine and simplify, and its solution becomes more unwieldy. Furthermore, any inaccuracy in the roots is magnified in the resulting amplitude ratios calculated. However, an increase in the number of degrees of freedom does not complicate the initial stages of solving the problem. The analysis and derivation of the differential equations and the procedure for obtaining the amplitude equations are simple and straightforward, no matter how many degrees of freedom there may be. This should be kept in mind, since the amplitude equations become the basis for certain numerical methods which will be explained later.

EXAMPLE 8-10 Obtain the solution to the equations of motion for Example 8-8, and determine the principal modes and natural frequencies for the case of equal spring constants and equal mass inertias.

SOLUTION Substituting the harmonic functions $\{\theta\} = \{A\} \sin \omega t$, and setting $k_1 = k_2 = k_3 = k_4 = k$ and $I_1 = I_2 = I_3 = I_4 = I$, enables the amplitude equations to be written as

$$
\begin{bmatrix}
(\xi - 1) & 1 & 0 & 0 \\
1 & (\xi - 2) & 1 & 0 \\
0 & 1 & (\xi - 2) & 1 \\
0 & 0 & 1 & (\xi - 2)
\end{bmatrix}
\begin{Bmatrix}
A_1 \\
A_2 \\
A_3 \\
A_4
\end{Bmatrix}
=
\begin{Bmatrix}
0 \\
0 \\
0 \\
0
\end{Bmatrix}
$$

where $\xi = I\omega^2/k$. Setting the determinant of the A's to zero and expanding by the cofactors† gives the frequency equation

$$\xi^4 - 7\xi^3 + 15\xi^2 - 10\xi + 1 = 0$$

There are four positive real roots for this equation. These can be verified as

$$\xi = 0.120\,62, \quad 1, \quad 2.3473, \quad 3.5321$$

(The second root of 1 is exact; the others are irrational numbers for which the values involve rounding off.) Substitution into the amplitude equations determines the amplitudes and defines the modes as follows:

Mode 1:
$$\omega_1 = 0.347\,30 \sqrt{\frac{k}{I}}$$

$$A_1 = 1, \quad A_2 = 0.879\,39, \quad A_3 = 0.652\,70, \quad A_4 = 0.347\,30$$

Mode 2:
$$\omega_2 = \sqrt{\frac{k}{I}}$$

$$A_1 = 1, \quad A_2 = 0, \quad A_3 = -1, \quad A_4 = -1$$

Mode 3:
$$\omega_3 = 1.5321 \sqrt{\frac{k}{I}}$$

$$A_1 = 1, \quad A_2 = -1.3473, \quad A_3 = -0.532\,09, \quad A_4 = 1.5321$$

Mode 4:
$$\omega_4 = 1.8794 \sqrt{\frac{k}{I}}$$

$$A_1 = 1, \quad A_2 = -2.5321, \quad A_3 = 2.8794, \quad A_4 = -1.8794$$

† This is required here, as the diagonal method is incorrect for fourth and higher-order determinants. See Appendix A-6.

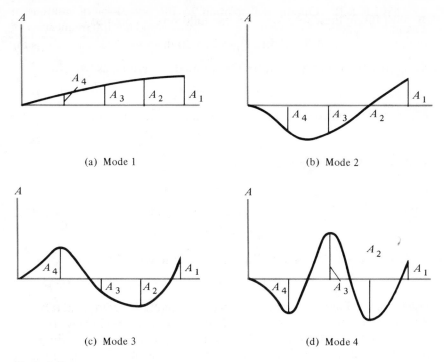

(a) Mode 1   (b) Mode 2

(c) Mode 3   (d) Mode 4

Figure 8-12

The amplitudes have been normalized by setting $A_1 = 1$, and the modal patterns are represented in Fig. 8-12.

8-10. ORTHOGONALITY OF THE PRINCIPAL MODES

In the solution of problems involving multidegree-of-freedom vibration, a useful fundamental relation exists between the principal modes. Consider any two principal modes of oscillation of a system of several degrees of freedom. Designate these as the rth and sth modes, and the corresponding eigenvalues as ξ_r and ξ_s. The amplitude equation for the rth mode can be written as (refer to Eq. 8-14)

$$[k]\{A\}_r = \xi_r[m]\{A\}_r \tag{8-77}$$

Premultiplying this by the transpose of principal mode s results in

$$\{A\}_s^T[k]\{A\}_r = \xi_r\{A\}_s^T[m]\{A\}_r \tag{8-78}$$

Considering that $[k]$ and $[m]$ are symmetric matrices, the following relations (see Appendix A-5) apply here

$$\{A\}_s^T[k]\{A\}_r = \{A\}_r^T[k]\{A\}_s$$
$$\{A\}_s^T[m]\{A\}_r = \{A\}_r^T[m]\{A\}_s \tag{8-79}$$

By substituting these into Eq. 8-78, the following is obtained

$$\{A\}_r^T[k]\{A\}_s = \xi_r\{A\}_r^T[m]\{A\}_s \qquad (8\text{-}80)$$

Next, the amplitude equation for the sth mode is written as

$$[k]\{A\}_s = \xi_s[m]\{A\}_s$$

This is premultiplied by the transpose of mode r resulting in

$$\{A\}_r^T[k]\{A\}_s = \xi_s\{A\}_r^T[m]\{A\}_s \qquad (8\text{-}81)$$

Subtracting this from Eq. 8-80, yields

$$0 = (\xi_r - \xi_s)\{A\}_r^T[m]\{A\}_s \qquad (8\text{-}82)$$

For different modes, $\xi_r \neq \xi_s$, and the preceding equation then requires that

$$\{A\}_r^T[m]\{A\}_s = 0 \qquad r \neq s \qquad (8\text{-}83)$$

If this is now substituted into Eq. 8-81, it also follows that

$$\{A\}_r^T[k]\{A\}_s = 0 \qquad r \neq s \qquad (8\text{-}84)$$

Equations 8-83 and 8-84 define the matrix form of the orthogonal relationships between principal modes of vibration. Equation 8-83 shows that the eigenvectors are orthogonal with respect to $\{m\}$ and Eq. 8-84 specifies that they are orthogonal with regard to $\{k\}$. Since $\{m\}$ is often a diagonal matrix and $\{k\}$ is not, it is usually simpler to write the orthogonality matrix with respect to $\{m\}$.

The orthogonality relation, Eq. 8-83, can be written in detail as

$$[A_1 A_2 A_3 \cdots A_n]_r \begin{bmatrix} m_{11} & m_{12} & m_{13} & \cdots & m_{1n} \\ m_{21} & m_{22} & m_{23} & \cdots & m_{2n} \\ m_{31} & m_{32} & m_{33} & \cdots & m_{3n} \\ & \cdot & \cdot & \cdots & \cdot \\ m_{n1} & \cdot & \cdot & \cdots & m_{nn} \end{bmatrix} \begin{Bmatrix} A_1 \\ A_2 \\ A_3 \\ \cdot \\ A_n \end{Bmatrix}_s = 0 \qquad r \neq s \qquad (8\text{-}85)$$

Using superscripts (r) and (s) to designate the modes, Eq. 8-85 can be multiplied out, giving the algebraic form of the orthogonality relation as

$$(m_{11} A_1^{(r)} A_1^{(s)} + m_{12} A_1^{(r)} A_2^{(s)} + m_{13} A_1^{(r)} A_3^{(s)} + \cdots$$

$$+ m_{21} A_2^{(r)} A_1^{(s)} + m_{22} A_2^{(r)} A_2^{(s)} + m_{23} A_2^{(r)} A_3^{(s)} + \cdots$$

$$+ m_{31} A_3^{(r)} A_1^{(s)} + m_{32} A_3^{(r)} A_2^{(s)} + m_{33} A_3^{(r)} A_3^{(s)} + \cdots$$

$$+ \cdots \qquad\qquad\qquad\qquad + \cdots$$

$$+ m_{n1} A_n^{(r)} A_1^{(s)} + m_{n2} A_n^{(r)} A_2^{(s)} + m_{n3} A_n^{(r)} A_3^{(s)} + \cdots) = 0 \qquad r \neq s \qquad (8\text{-}86)$$

which can be written in summation form as

$$\sum_{i=1}^{n} \sum_{j=1}^{n} m_{ij} A_i^{(r)} A_j^{(s)} = 0 \qquad r \neq s \qquad (8\text{-}87)$$

The orthogonality relation, as expressed by any of the forms Eq. 8-83 through 8-87, holds irrespective of the manner in which the differential equations of motion are obtained, *providing* $[k]$ and $[m]$ are symmetric matrices. Thus it applies if the direct approach of Section 8-3 or Section 8-4 is used or for the displacement analysis of Section 8-6, as well as if the procedure of writing Lagrange's equations is employed as in Section 8-8.

If the mass matrix $[m]$ is diagonal, the orthogonality relation would still be written as Eq. 8-85 but the nondiagonal terms $m_{12}, m_{13}, \ldots, m_{21}, \ldots$ would be zero.

The orthogonality relation for the principal modes of vibration is essentially a relation between the amplitudes of two principal modes. These are not necessarily successive modes but any two modes. Thus the relation can be written between modes 1 and 2, 1 and 3, 1 and 4, ..., 1 and n; between modes 2 and 3, 2 and 4, ..., and so on. If the amplitude ratios of mode r have been determined, then their substitution into Eq. 8-84 gives an expression which will be satisfied by the amplitude ratios of any other principal mode but not by those of the rth mode.

For most systems the term "orthogonality" has no geometrical interpretation. The use of the word comes from a case such as a mass suspended by springs, like that shown in Fig. 8-13. An analysis of this system of three degrees of freedom shows that the orthogonality relation here has the geometrical meaning that two principal modes of vibration take place along straight lines which are perpendicular to each other.†

EXAMPLE 8-11   Using the values obtained from the solution of Example 8-10 for the 4-disk torsional system, check the orthogonality relation between the principal modes.

SOLUTION   Writing Eq. 8-85 for the principal modes gives

$$\lfloor A_1 A_2 A_3 A_4 \rfloor_r \begin{bmatrix} I_1 & 0 & 0 & 0 \\ 0 & I_2 & 0 & 0 \\ 0 & 0 & I_3 & 0 \\ 0 & 0 & 0 & I_4 \end{bmatrix} \begin{Bmatrix} A_1 \\ A_2 \\ A_3 \\ A_4 \end{Bmatrix}_s = 0$$

Then for modes 1 and 2:

$$\lfloor 1 \quad 0.879\,39 \quad 0.652\,70 \quad 0.347\,30 \rfloor \begin{bmatrix} I & 0 & 0 & 0 \\ 0 & I & 0 & 0 \\ 0 & 0 & I & 0 \\ 0 & 0 & 0 & I \end{bmatrix} \begin{Bmatrix} 1 \\ 0 \\ -1 \\ -1 \end{Bmatrix} = 0$$

$$I(1 \cdot 1 + 0.879\,39 \cdot (0) + 0.652\,70 \cdot (-1) + 0.347\,30 \cdot (-1)) = 0.000\,00$$

---

† See T. von Karman and M. A. Biot, *Mathematical Methods in Engineering* (New York: McGraw-Hill, 1940), p. 176.

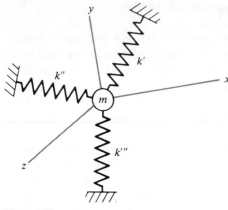

Figure 8-13

For modes 1 and 3:

$$1 \cdot 1 + 0.879\,39 \cdot (-1.3473) + 0.652\,70 \cdot (-0.532\,09)$$
$$+ 0.347\,30 \cdot (1.5321) = 0.000\,00$$

For modes 1 and 4:

$$1 \cdot 1 + 0.879\,39 \cdot (-2.5321) + 0.652\,70 \cdot (2.8794)$$
$$+ 0.347\,30 \cdot (-1.8794) = -0.000\,03$$

For modes 2 and 3:

$$1 \cdot 1 + (0) \cdot (-1.3473) + (-1)(-0.532\,09) + (-1)(1.5321) = -0.000\,01$$

For modes 2 and 4:

$$1 \cdot 1 + (0) \cdot (-2.5321) + (-1)(2.8794) + (-1)(-1.8794) = 0.000\,00$$

For modes 3 and 4:

$$1 \cdot 1 + (-1.3473)(-2.5321) + (-0.532\,09)(2.8794)$$
$$+ 1.5321 \cdot (-1.8794) = 0.000\,00$$

EXAMPLE 8-12 Using the results from Example 8-9, check the orthogonality relation for the two modes determined.

SOLUTION For the generalized coordinates selected, both dynamic coupling and static coupling occur. This is exhibited by the energy expressions as well as the differential equations. Thus the orthogonality relation is written as follows:

$$m_{11} A_1^{(1)} A_1^{(2)} + m_{12} A_1^{(1)} A_2^{(2)} + m_{21} A_2^{(1)} A_1^{(2)} + m_{22} A_2^{(1)} A_2^{(2)} = 0$$
$$2m \times 1 \times 1 + m \times 1 \times (-2) + m \times 0 \times 1 + m \times 0 \times (-2) = 0$$

8-11. EQUIVALENT SYSTEMS

So far, idealized systems (that is, systems composed of separate and distinct mass and elastic elements) have been contemplated. Such lumped-parameter systems may appear to be unrealistic. Most mechanical systems are not simple but are made up of gears, levers, linkages, and similar parts. The form of the members also may not be simple; they may overlap or in some manner not be separate, and in general, a member containing mass is also elastic. However, such an arrangement may be replaced by a system of lumped mass-and-elastic elements which would exhibit vibration character- istics approximately the same as those of the actual machine or structure this system represents. The method of determining the equivalent system is carried out in a rational manner and is usually based on equating the ener- gies of the actual and idealized systems. In some instances, force or moment may be equated or related for the systems.

Some examples of the determination of equivalent systems will be described below. In these the basic principles and general procedure should be followed. However, assumptions such as factors assumed in determining equivalent length should not be accepted as rules to be followed with great precision for all cases.

Lever System

Consider the overhead-valve system shown in Fig. 8-14(a), composed of valve A with stem AB, valve spring k_s, rocker arm BC, tappet rod CD, and cam D. In the beginning it should be assumed that all parts are elastic and have mass. Calculations can then be made to determine the comparative value of such physical constants for the parts. For example, from the defor- mation relation $\delta = Pl/AE$ for an axially loaded elastic rod (where P is the load, l is the length, A is the area, and E is the tension-compression modulus of elasticity), the spring constant k' of the rod is expressed by the force per unit deformation as

$$k' = \frac{P}{\delta} = \frac{AE}{l} \tag{8-88}$$

Since E is very large for a material such as steel, k' will be quite large for both the tappet rod and the valve stem, in comparison to k_s of the valve spring. Similarly, deflection calculations for the rocker arm will show that its equiv- alent spring constant is very large compared to k_s. Accordingly, the tappet rod, valve stem, and rocker arm can be considered as rigid; thus only the valve spring is elastic. The equivalent spring k_e for the system is then equal to k_s.

The mass elements move together and may be replaced by an effective mass. For this purpose the mass of the parts is transposed to a common

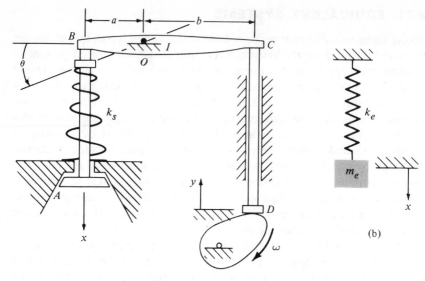

Figure 8-14   (a)

position. Since the motion of the valve is the important measure here, the line AB will be selected. The equivalent system is then rectilinear, as shown in Fig. 8-14(b). The displacement of the valve is designated by the coordinate x shown. Let m_t, m_v, and m_s represent the mass of the tappet rod, valve (including stem), and spring, respectively, and I be the moment of inertia of the rocker-arm mass about its rotation center O. The valve-spring mass is distributed along the member so that x does not define the displacement of all portions of m_s. Because of this, the effective mass may be taken as $m_s/3$, in accordance with Eq. 2-75 of Section 2-11. The kinetic energy of the actual system is expressed as

$$T = \frac{1}{2}\left(m_v + \frac{m_s}{3}\right)\dot{x}^2 + \tfrac{1}{2}I\dot{\theta}^2 + \tfrac{1}{2}m_t\dot{y}^2$$

$$= \frac{1}{2}\left(m_v + \frac{m_s}{3}\right)\dot{x}^2 + \tfrac{1}{2}I\left(\frac{\dot{x}}{a}\right)^2 + \tfrac{1}{2}m_t\left(\frac{b}{a}\dot{x}\right)^2 \qquad (8\text{-}89)$$

For the equivalent system, the kinetic energy is

$$T = \tfrac{1}{2}m_e\dot{x}^2 \qquad (8\text{-}90)$$

Equating these energies results in defining the effective mass as

$$m_e = m_v + \frac{m_s}{3} + \frac{I}{a^2} + m_t\left(\frac{b}{a}\right)^2 \qquad (8\text{-}91)$$

In many cases it will be noted that $m_s/3$ for the spring is small compared to the other parts and may be neglected.

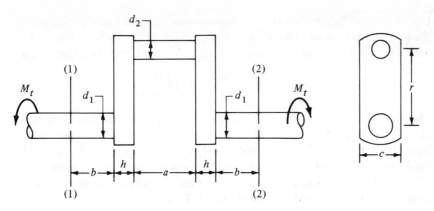

Figure 8-15

Reciprocating System

A multicylinder engine, composed of a crankshaft, connecting rods, pistons, a flywheel, and auxiliaries such as the cam system, generator, and transmission, is rather involved for an exact analysis. However, it can be reduced to an equivalent torsional system consisting of disks and shafts. Each piston, rod, and crank throw may be replaced by an equivalent disk having an effective moment of inertia defined by the actual arrangement, and the crank throw can be supplanted by a straight portion of shaft having a torsional stiffness equivalent to that of the crank. The auxiliaries may be similarly replaced by disks and shafts. This latter also involves consideration of the connecting gears or belts.

Consider one crank throw of a crankshaft. This is shown in simplified form in Fig. 8-15. The crank throw can be replaced by an equivalent length of straight shaft having the proper torsional rigidity. The method employed by Timoshenko† will be followed initially here. Assume first that the main bearings at sections 1 and 2 have clearances so that lateral displacements of these sections can occur. The angle of twist between sections 1 and 2 can then be expressed. This angle is due to (a) twist of the journals at 1 and 2, (b) twist of the crankpin, and (c) bending of the crank webs. Let

$$C_1 = \frac{\pi d_1^4 G}{32} = \text{torsional rigidity of the journal}$$

$$C_2 = \frac{\pi d_2^4 G}{32} = \text{torsional rigidity of the crankpin}$$

$$B = \frac{hc^3 E}{12} = \text{bending rigidity of the web}$$

† See S. Timoshenko, *Vibration Problems in Engineering*, 3rd ed. (New York: Van Nostrand, 1955), pp. 261–263.

Due to local deformations at their connections with the webs, the lengths of the journal and pin are taken, respectively, as $2b_1 = 2b + 0.9h$ and $a_1 = a + 0.9h$. The angle of twist θ, caused by torque M_t, may then be expressed as

$$\theta = \frac{M_t 2b_1}{C_1} + \frac{M_t a_1}{C_2} + 2\left(\frac{M_t}{B} r\right) \tag{8-92}$$

The first two parts are obtained from the angle-of-twist relation for torsion. In the last part, M_t/B represents the deflection at the crankpin due to bending of the web by the moment M_t, and multiplying by r gives the resulting angle of turning. This is doubled, since there are two webs.

The equivalent shaft is considered to have a uniform section, with the torsional rigidity defined by $C = \pi d^4 G/32$. Then

$$\theta = \frac{M_t l_e}{C} \tag{8-93}$$

where l_e is the length of the equivalent shaft. Equating the expressions defines l_e as

$$l_e = C\left(\frac{2b_1}{C_1} + \frac{a_1}{C_2} + \frac{2r}{B}\right) \tag{8-94}$$

If it is assumed that there are no clearances at the bearings, the analysis is much more complicated.† In reality, most cases will lie between these extremes, tending to approach the condition of complete constraint. This is because most bearings exhibit very little clearance. The condition of small clearance not only inhibits sidewise movement of the journals but increases their torsional stiffness.

Error also results from the beam-deflection theories used here, which are correct for long slender members, whereas, in general the webs and crankpins are short and thick.

It has been reported by Den Hartog‡ that experiments on crankshafts for large slow-speed engines show the variation in the equivalent length l_e to be given by

$$0.95l < l_e < 1.10l \tag{8-95}$$

where l is the length, between the centers of the main bearings, of the actual shaft. The lower limit applies for cranks having small throws and stiff webs, and the upper limit for those having large throws and flexible webs. In the tests the crankpin diameter was the same as that of the main journal. This is also used for the diameter of the equivalent shaft. If the crankpin diameter is different from that of the journal, the crank throw is replaced by a stepped

---

† See Timoshenko, *Vibration Problems in Engineering*, p. 263.
‡ See J. P. Den Hartog, *Mechanical Vibrations*, 4th ed. (New York: McGraw-Hill, 1956), pp. 185–186.

shaft. The diameters used are those of the journal and pin. The change in diameter is made at the section located at the center of the web. For high-speed, lightweight engines, the equivalent stiffness may be much smaller than that given above. The equivalent length is thereby increased beyond that given by Eq. 8-95, in some instances being as much as $2l$. In determining the stiffness and length of an equivalent shaft, the results of tests for similar-sized shafts and conditions are helpful.

By using the foregoing methods, each crank throw can be replaced by an equivalent length of straight shaft.

A method† of replacing the mass of the piston, connecting rod, and crank throw by a disk having an effective moment of inertia of the proper magnitude will now be considered. The moment of inertia I_c of the mass of the crankshaft (including the journal, webs of the throw, and crankpin) can be calculated by separating it into fundamentally shaped parts and using appropriate relations for determining the moments of inertia of these parts. The parallel-axis theorem is employed to transfer the moments of inertia to the common axis of the shaft.

The mass of the connecting rod and piston must also be considered. Referring to Fig. 8-16, note that the connecting rod exhibits plane motion. End 1 rotates about the shaft center O, and end 2 has a motion of translation the same as that of the piston P. The moment of inertia I_2 for the actual rod with respect to axis 2 at the piston pin can be calculated. It is then assumed that the mass of the rod is concentrated at its ends, being composed of the effective mass m_1 at the crankpin end and effective mass m_2 at the piston-pin end. Then

$$I_2 \approx m_1 l^2 \qquad \text{whence } m_1 \approx \frac{I_2}{l^2} \tag{8-96}$$

Since $m_1 + m_2 = M_R$ is the mass of the actual rod, then

$$m_2 = M_R - m_1 = M_R - \frac{I_2}{l^2} \tag{8-97}$$

The moment of inertia I_o' of m_1 with respect to the crank axis at O is expressed by

$$I_o' = m_1 r^2 \tag{8-98}$$

This may be added to the moment of inertia of the crank throw.

The remaining masses of the arrangement are located at the piston and consist of the mass of the piston, rings, and pin assembly and the connecting-rod mass m_2. Since these all move together, their masses may be added. Let this total mass be M_2. The kinetic energy of M_2 can be determined. Refer-ring to Fig. 8-17, ω is the angular velocity of the crank, B is the uppermost

† See Timoshenko, *Vibration Problems in Engineering*, pp. 263–264.

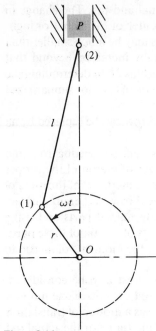

Figure 8-16

position of the piston assembly, and x represents the linear displacement of the mass M_2. Then x is defined by

$$x = (l + r) - (l \cos \beta + r \cos \omega t)$$
$$= l(1 - \cos \beta) + r(1 - \cos \omega t) \qquad (8\text{-}99)$$

Also,

$$r \sin \omega t = l \sin \beta$$

so that

$$\sin \beta = \frac{r}{l} \sin \omega t \qquad (8\text{-}100)$$

and

$$\cos \beta = \sqrt{1 - \sin^2 \beta} = \sqrt{1 - \frac{r^2}{l^2} \sin^2 \omega t}$$

$$\approx 1 - \frac{r^2}{2l^2} \sin^2 \omega t \qquad (8\text{-}101)$$

(The final form here is obtained by binomial expansion of the radical and then neglecting the terms beyond the second.) Substituting into Eq. 8-99 and rearranging gives

$$x = r(1 - \cos \omega t) + \frac{r^2}{2l} \sin^2 \omega t \qquad (8\text{-}102)$$

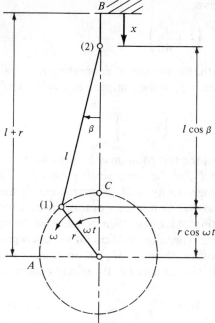

Figure 8-17

Taking the time derivative results in

$$\dot{x} = r\omega \sin \omega t + \frac{r^2\omega}{l}\sin \omega t \cos \omega t \qquad (8\text{-}103)$$

For M_2, the kinetic energy is then

$$T_2 = \tfrac{1}{2}M_2\dot{x}^2 = \tfrac{1}{2}M_2 r^2\omega^2\left(\sin \omega t + \frac{r}{l}\sin \omega t \cos \omega t\right)^2$$

The average kinetic energy for one revolution is

$$T_{2(\text{avg})} = \frac{1}{2\pi}\int_0^{2\pi} T_2\,d(\omega t)$$

$$= \frac{M_2 r^2\omega^2}{4\pi}\int_0^{2\pi}\left(\sin \omega t + \frac{r}{l}\sin \omega t \cdot \cos \omega t\right)^2 d(\omega t)$$

$$= \frac{M_2 r^2\omega^2}{4\pi}\left[\frac{\omega t}{2} - \frac{1}{4}\sin 2\omega t + \frac{2r}{3l}\sin^3 \omega t - \frac{r^2}{8l^2}\left(\frac{\sin 4\omega t}{4} - \omega t\right)\right]_0^{2\pi}$$

$$= \frac{1}{2}\left[\frac{M_2 r^2}{2}\left(1 + \frac{r^2}{4l^2}\right)\right]\omega^2 \qquad (8\text{-}104)$$

For an equivalent disk having moment of inertia I_{2e}, the kinetic energy is

$$T_2 = \tfrac{1}{2}I_{2e}\omega^2 \qquad (8\text{-}105)$$

Equating these latter two relations gives

$$I_{2e} = \frac{M_2 r^2}{2} \left(1 + \frac{r^2}{4l^2}\right) \qquad (8\text{-}106)$$

Finally, adding the moments of inertia for the crank, connecting-rod, and piston assembly gives the effective inertia I_e of the complete equivalent disk as

$$I_e = I_c + m_1 r^2 + \frac{M_2 r^2}{2} \left(1 + \frac{r^2}{4l^2}\right) \qquad (8\text{-}107)$$

A method† which is simpler than the foregoing may be used to determine the effective moment of inertia I_{2e} of the reciprocating mass M_2. In Fig. 8-17 when the crankpin is at A, the velocity of M_2 is approximately the same as that of the crankpin, provided that l/r is large. For this position, the moment of inertia of the reciprocating mass may then be assumed to be $M_2 r^2$. This situation also occurs at the other side position for the crankpin. When the crankpin is at the top or bottom, the mass M_2 has no velocity and hence a zero kinetic energy. Based on a linear average, the effective moment of inertia for M_2 is then

$$I_{2e} = \tfrac{1}{2} M_2 r^2 \qquad (8\text{-}108)$$

Note that, for large values of l/r, Eq. 8-106 reduces approximately to Eq. 8-108.

The foregoing methods involve rather broad approximations in certain steps. An equivalent system so determined therefore cannot be accepted as an exact or accurate representation of the actual arrangement. Accordingly, the vibration characteristics subsequently obtained for the equivalent system should be used with discretion. Experience and experimental results for similar systems are of considerable help in the accurate determination of equivalent systems.

Geared System

A torsional system that is connected by gears is shown in Fig. 8-18. This can be replaced by an equivalent system of two disks on a single shaft. Let n be the gear ratio of the connecting gears, that is, the ratio of the number of teeth in gear 1 to the number of teeth in gear 2. This is the same as the ratio d_1/d_2 of the pitch diameter d_1 for gear 1 to the pitch diameter d_2 for gear 2. Turning disk 1 through angle θ_1 will rotate disk 2 by the angle $\theta_2 = n\theta_1$. This will also define the ratio of the angular displacements for a vibration of the system. The time derivative $\dot{\theta}_2 = n\dot{\theta}_1$ similarly relates the angular velocities. The kinetic energy of disk 2 is expressed as

$$T_2 = \tfrac{1}{2} I_2 \dot{\theta}_2^2 = \tfrac{1}{2} n^2 I_2 \dot{\theta}_1^2$$

† See Den Hartog, *Mechanical Vibrations*, pp. 184–185.

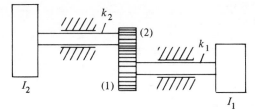

Figure 8-18

Then the effective inertia I_e for disk 2 referred to disk 1 is given by

$$I_e = n^2 I_2 \tag{8-109}$$

The two shafts will now be combined into an equivalent single shaft of proper stiffness, referred to shaft 1. Let the pitch diameters of gears 1 and 2 be $2r_1$ and $2r_2$, respectively. Fix disk 2 so that it cannot turn, and apply a torque of such magnitude to disk 1 that gear 2 rotates through an angle of 1 rad. The angle of twist for shaft 2 is also 1 rad, and the torque in shaft 2 is then $1 \times k_2$. The thrust (the force tangent to the pitch circle of the gear, and acting on the gear tooth) on gear 2 is $1 \times k_2/r_2$, so that the torque for gear 1 is $(k_2/r_2)r_1 = k_2 n$. Gear 1 will rotate through the angle $1 \cdot r_2/r_1 = 1/n$. Then, dividing the torque for gear 1 by the gear angle will give the equivalent stiffness for shaft 2 referred to shaft 1 as

$$\frac{k_2 n}{1/n} = n^2 k_2 \tag{8-110}$$

This can also be obtained by consideration of the potential energy. As before, let disk 2 be restrained from rotating. If a twisting moment is applied to disk 1 such that gear 1 rotates through angle θ_1, then gear 2 must turn through the angle $\theta_2 = n\theta_1$. The angle θ_2 also represents the angle of twist for shaft 2. Expressing the potential or strain energy for shaft 2 gives

$$V = \tfrac{1}{2}k_2\theta_2^2 = \tfrac{1}{2}k_2(n\theta_1)^2 = \tfrac{1}{2}n^2 k_2 \theta_1^2$$

Thus the stiffness for shaft 2 referred to shaft 1 is $n^2 k_2$, which checks Eq. 8-110.

Since the shafts are arranged end to end, they are in series, and the equivalent stiffness k_e for the single connecting shaft is given by

$$\frac{1}{k_e} = \frac{1}{k_1} + \frac{1}{n^2 k_2} \tag{8-111}$$

From the foregoing the equivalent system would be as shown in Fig. 8-19. In the analysis it was assumed that the gears had no mass and that they did not deform.

The analysis for the geared system applies also to a similar system in which the shafts are connected by a belt. The factor n would then represent

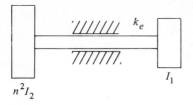

Figure 8-19

the ratio of the diameter of pulley 1 on shaft 1 to that of pulley 2 on shaft 2. In this case the flexibility of the belt would introduce a torsional spring between the two shafts. The stiffness constant k_3 for this part may be determined experimentally. The equivalent stiffness k_e would then be obtained from

$$\frac{1}{k_e} = \frac{1}{k_1} + \frac{1}{k_3} + \frac{1}{n^2 k_2} \tag{8-112}$$

8-12. FORCED MOTION AND STABILITY

Forcing conditions as well as damping can occur for systems of several degrees of freedom, such as that represented in Fig. 8-20. For harmonic forcing functions and viscous damping, the system can be analyzed by extending the methods of Sections 7-11, 7-12, and 7-14, although this becomes more difficult as the number of degrees of freedom increases.

An important aspect of the problem is the determination of dynamic stability. This is because self-excitation is typical for certain multidegree-of-freedom systems. Determination of the criteria for stability by the method outlined in Section 7-15 becomes very cumbersome for 3 and more degrees of freedom. A better method, based on the work of Routh[†] and Hurwitz,[‡] expresses these criteria as a determinant. The expansion of certain defined minors of the determinant then yield the conditions for stability.

Consider the system of Fig. 8-20, with self-exciting forcing conditions defined by

$$P_1 = P_{01}\dot{x}_1, \; P_2 = P_{02}\dot{x}_2, \ldots \tag{8-113}$$

which are typical. In writing the differential equations of motion, the forcing and damping terms may be combined, since both are linear functions of the velocities. The solutions may be assumed as

$$x_1 = C_1 e^{st}, \; x_2 = C_2 e^{st}, \ldots \tag{8-114}$$

Substituting these, and expanding the determinant for the resulting auxiliary equations, produces a characteristic equation of the general form

$$b_0 s^n + b_1 s^{n-1} + b_2 s^{n-2} + \cdots + b_{n-1} s + b_n = 0 \quad (b_0 > 0) \tag{8-115}$$

[†] See R. A. Frazer, W. J. Duncan, and A. R. Collar, *Elementary Matrices* (New York: Cambridge University Press, 1963), pp. 154–155.

[‡] See R. E. Gaskell, *Engineering Mathematics* (New York: Dryden Press, 1958), pp. 420–421.

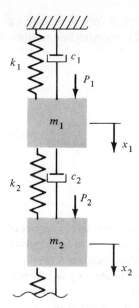

Figure 8-20

All orders, n, $n - 1$, $n - 2$, ... may be present. Note that the equation has been written so that the coefficient b_0 is positive.

First of all, to ensure stability it is necessary that no positive real roots of Eq. 8-115 exist. This requires that all b coefficients be positive.

The criteria for stability based on the s roots being complex will be considered next. The n roots for s may be taken as

$$p_1 + iq_1, p_1 - iq_1, p_2 + iq_2, p_2 - iq_2, \ldots \qquad (8\text{-}116)$$

For the system to be dynamically stable, p_1, p_2, ... must be negative. The criteria for this can be expressed as the following nth-order determinant:

$$H_n = \begin{vmatrix} b_1 & b_3 & b_5 & b_7 & \cdots & b_{2n-1} \\ b_0 & b_2 & b_4 & b_6 & \cdots & b_{2n-2} \\ 0 & b_1 & b_3 & b_5 & \cdots & b_{2n-3} \\ 0 & b_0 & b_2 & b_4 & \cdots & b_{2n-4} \\ \cdot & \cdot & \cdot & \cdot & \cdots & \cdot \\ \cdot & \cdot & \cdot & \cdot & \cdots & \cdot \\ \cdot & \cdot & \cdot & \cdot & \cdots & b_n \end{vmatrix} \qquad (8\text{-}117)$$

The coefficient b_j is replaced by zero if its subscript is less than zero or greater than n. All minors, indicated by the dashed lines, are then written. Thus

$$H_1 = b_1$$

$$H_2 = \begin{vmatrix} b_1 & b_3 \\ b_0 & b_2 \end{vmatrix}$$

$$H_3 = \begin{vmatrix} b_1 & b_3 & b_5 \\ b_0 & b_2 & b_4 \\ 0 & b_1 & b_3 \end{vmatrix}$$

(8-118)

$$\vdots$$

Then, if b_0 and all the determinants $H_1, H_2, \ldots, H_n$ are positive, every root of Eq. 8-115 will have a negative real part; that is, $p_1, p_2, \ldots$ will be negative, and the system will be dynamically stable. Proof of the foregoing statement, which is based on the original work of Routh[†] and Hurwitz,[‡] is not presented here, because it is quite lengthy and involved.

For a cubic equation, $n = 3$, and the determinant is written as

$$H = \begin{vmatrix} b_1 & b_3 & 0 \\ b_0 & b_2 & 0 \\ 0 & b_1 & b_3 \end{vmatrix}$$

(8-119)

which leads to

$$H_1 = b_1$$
$$H_2 = b_1 b_2 - b_0 b_3$$
$$H_3 = b_1 b_2 b_3 - b_0 b_3^2$$

(8-120)

The conditions for dynamic stability may then be written as

a. $$b_0 > 0$$
b. $$b_1 > 0$$
c. $$b_1 b_2 > b_0 b_3$$
d. $$b_1 b_2 b_3 > b_0 b_3^2$$

(8-121)

† See E. J. Routh, *On the Stability of a Given Motion* (London, 1877); also, E. J. Routh, *Rigid Dynamics*, Vol. 2 (London, 1892).

‡ See A. Hurwitz, *Math. Ann.*, 46, p. 211, 1895.

It should be noted that requirement (d) reduces to requirement (c), by dividing through by b_3, *only* if b_3 is positive. In this case b_3 being positive becomes a new condition replacing that originally expressed by requirement (d).

For a quartic equation, $n = 4$, and the governing determinant is

$$H = \begin{vmatrix} b_1 & b_3 & 0 & 0 \\ b_0 & b_2 & b_4 & 0 \\ 0 & b_1 & b_3 & 0 \\ 0 & b_0 & b_2 & b_4 \end{vmatrix} \tag{8-122}$$

Expanding the proper minors results in

$$H_1 = b_1$$
$$H_2 = b_1 b_2 - b_0 b_3$$
$$H_3 = b_1 b_2 b_3 - b_0 b_3^2 - b_1^2 b_4$$
$$H_4 = b_1 b_2 b_3 b_4 - b_1^2 b_4^2 - b_0 b_3^2 b_4 \tag{8-123}$$

The criteria for dynamic stability then are

a. $$b_0 > 0$$

b. $$b_1 > 0$$

c. $$b_1 b_2 > b_0 b_3$$

d. $$b_1 b_2 b_3 > b_0 b_3^2 + b_1^2 b_4$$

e. $$b_1 b_2 b_3 b_4 > b_1^2 b_4^2 + b_0 b_3^2 b_4 \tag{8-124}$$

The requirements represented by Eqs. 8-121 and 8-124 differ from those expressed in Section 7-15 for cubic and quartic equations, respectively. This is of no consequence, since either the conditions expressed here or those of Section 7-15 are correct and sufficient, and will lead to specifying identical physical conditions for ensuring stability.

For a fifth-degree equation, the controlling determinant is

$$H = \begin{vmatrix} b_1 & b_3 & b_5 & 0 & 0 \\ b_0 & b_2 & b_4 & 0 & 0 \\ 0 & b_1 & b_3 & b_5 & 0 \\ 0 & b_0 & b_2 & b_4 & 0 \\ 0 & 0 & b_1 & b_3 & b_5 \end{vmatrix} \tag{8-125}$$

This yields the following requirements:

a. $$b_0 > 0$$

b. $$H_1 = b_1 > 0$$

c. $$H_2 = b_1 b_2 - b_0 b_3 > 0$$

d. $$H_3 = b_1 b_2 b_3 + b_0 b_1 b_5 - b_0 b_3^2 - b_1^2 b_4 > 0$$

e. $$H_4 = b_1 b_2 b_3 b_4 + 2 b_0 b_1 b_4 b_5 + b_0 b_2 b_3 b_5$$
$$- b_1^2 b_4^2 - b_0 b_3^2 b_4 - b_1 b_2^2 b_5 - b_0^2 b_5^2 > 0$$

f. $$H_5 = b_1 b_2 b_3 b_4 b_5 + 2 b_0 b_1 b_4 b_5^2 + b_0 b_2 b_3 b_5^2$$
$$- b_1 b_2^2 b_5^2 - b_1^2 b_4^2 b_5 - b_0 b_3^2 b_4 b_5 - b_0^2 b_5^3 > 0 \qquad (8\text{-}126)$$

The dynamic-stability criteria for sixth and higher degree characteristic equations can be similarly obtained. However, the expansion of the higher-order determinants becomes increasingly more cumbersome. Since a sixth-degree equation can result for a system of only three degrees of freedom, the magnitude of the problem encountered in expanding the determinant becomes apparent. However, this difficulty can be largely overcome by setting up the determinant expansion on a digital computer.

PROBLEMS

8-1. For the system shown, (a) construct the free-body diagram for each mass element. (b) Using the method of force analysis, establish the differential equations of motion. (c) Express these in matrix form.

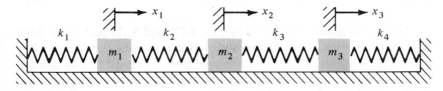

Problem 8-1

8-2. Determine the principal modes and natural frequencies for Prob. 8-1, taking $k_1 = k_2 = k_3 = k_4 = k$ and $m_1 = m_2 = m_3 = m$.

8-3. Determine the principal modes and natural frequencies for Prob. 8-1, taking $k_1 = k_2 = k_3 = k_4 = k$ and $m_1 = m$, $m_2 = 2m$, $m_3 = 3m$.

8-4. For the system shown, (a) construct the free-body diagram for each mass element. (b) Using moment analysis, establish the differential equations of motion. (c) Express these in matrix form.

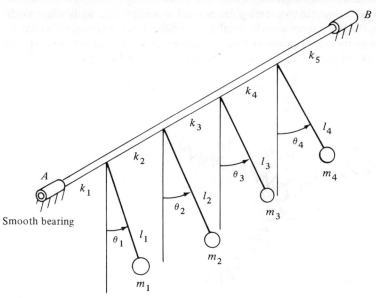

Smooth bearing

Problem 8-4

8-5. Determine the principal modes and natural frequencies for Prob. 8-4, taking
$k_1 = k_2 = k_3 = k_4 = k_5 = k$, $l_1 = l_2 = l_3 = l_4 = l$, $m_1 = m_2 = m_3 = m_4 = m$,
and $wl = k$.

8-6. Determine the principal modes and natural frequencies for Prob. 8-4, taking
$k_1 = k_2 = k_3 = k_4 = k$, $l_1 = l_2 = l_3 = l_4 = l$, $m_1 = 2m$, $m_2 = m$, $m_3 = 2m$,
$m_4 = 3m$, and $wl = k$.

8-7. For the torsional system shown, (a) construct the free-body diagram for each
inertial element (that is, each disk). (b) Using moment analysis, establish the
differential equations for angular motion. (c) Express these in matrix form.

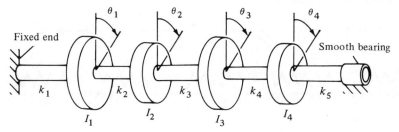

Problem 8-7

8-8. Determine the principal modes and natural frequencies for Prob. 8-7, taking all
torsional spring constants equal to k and all moments of inertia equal to I.

8-9. For the three-dimensional spring and mass arrangement, (a) construct the free-body diagram considering that m is at the origin for the equilibrium condition. (b) Using force analysis, establish the differential equations of motion. (c) Express these in matrix form. The direction angles with x, y, z axes, respectively, are as follows: (1) α_1, β_1, γ_1 for k_1; (2) α_2, β_2, γ_2 for k_2; (3) α_3, β_3, γ_3 for k_3.

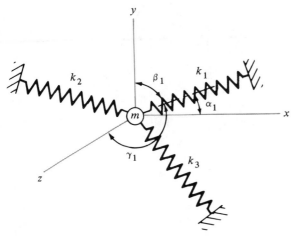

Problem 8-9

8-10. Referring to Prob. 8-9, take the fixed ends of k_1, k_2, and k_3, respectively, as located at $(\sqrt{2}/2, \sqrt{2}/2, 1)$, $(\sqrt{2}/2, \sqrt{2}/2, -1)$, and $(-1, 1, 0)$. Also $k_1 = k_2 = k_3 = k$. Determine the principal modes of vibration.

8-11. Refer to Prob. 8-9 and take the fixed ends of k_1, k_2, and k_3, respectively, as located at $(\sqrt{2}/2, \sqrt{2}/2, 1)$, $(\sqrt{2}/2, \sqrt{2}/2, -1)$, and $(-1, 1, 0)$. Also $k_1 = k$, $k_2 = 2k$, and $k_3 = 3k$. Determine the principal modes of vibration.

8-12. For the system shown, (a) determine the stiffness coefficients directly by consideration of displacements (see Section 8-3) and (b) establish the stiffness matrix. (c) Determine the mass matrix. (d) Write the resulting matrix form for the equations of motion.

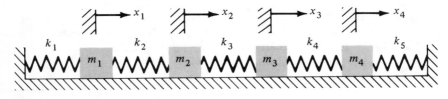

Problem 8-12

8-13. Determine the principal modes and natural frequencies for Prob. 8-12, taking all spring constants equal to k and all masses equal to m.

8-14. For the system shown, (a) determine the stiffness coefficients directly by considering displacements (see Section 8-3) and (b) establish the stiffness matrix.

(c) Determine the mass matrix. (d) Write the resulting matrix form of the equations of motion.

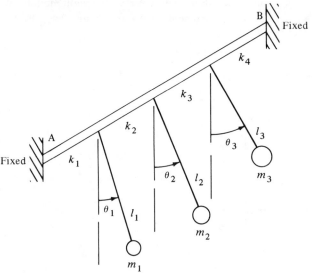

Problem 8-14

8-15. Determine the principal modes and natural frequencies for Prob. 8-14, taking $k_1 = k_2 = k_3 = k_4 = k$, $l_1 = l_2 = l_3 = l$, $m_1 = m_3 = 2m$, $m_2 = m$, and $wl = k$.

8-16. Determine the principal modes and natural frequencies for Prob. 8-14, taking $k_1 = k_2 = k_3 = k_4 = k$, $l_1 = l_2 = l_3 = l$, $m_1 = 2m$, $m_2 = m$, $m_3 = 3m$, and $wl = k$.

8-17. For the torsional system shown, (a) determine the stiffness matrix directly (see Section 8-3). (b) Obtain the mass or inertial matrix. (c) Write the resulting matrix form of the equations of motion.

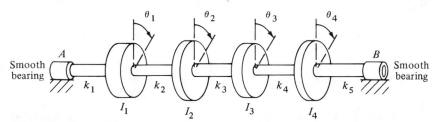

Problem 8-17

8-18. Determine the principal modes and natural frequencies for Prob. 8-17, taking $k_1 = k_2 = k_3 = k_4 = k_5 = k$, and $I_1 = I_4 = 2I$, $I_2 = I_3 = I$.

8-19. A vertical plane segment of a building frame is composed of columns and rigid girders, as shown. The columns are constrained against rotation at the upper and lower ends. The weights of the girders (including floor or roof, and so on) are shown, and the section and material properties for the columns are indicated. The columns are to be considered as massless. For small horizontal motion of the girders, (a) establish the stiffness and mass matrices, and (b) express the resulting matrix equation that governs the motion. (c) For $W_1 = W_2 = W_3 = W$, $I_1 = 3I$, $I_2 = 2I$, and $I_3 = I$ determine the principal modes and natural frequencies.

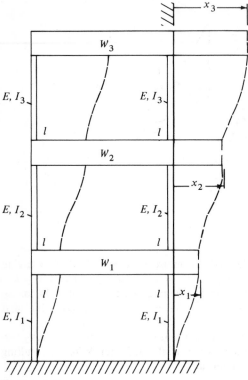

Problem 8-19

8-20. For a certain system, the stiffness and mass matrices are

$$[k] = k \begin{bmatrix} 3 & -1 & 0 \\ -1 & 4 & -2 \\ 0 & -2 & 3 \end{bmatrix} \quad [m] = m \begin{bmatrix} 2 & 0 & 0 \\ 0 & 1 & 0 \\ 0 & 0 & 1 \end{bmatrix}$$

(a) Determine the eigenvalues, and (b) obtain the eigenvectors from the adjoint matrix.

8-21. Using the stiffness and mass matrices of Prob. 8-20, determine the physical system represented.

8-22. Determine the flexibility matrix by the inverse of the stiffness matrix for Prob. 8-20. Then using this flexibility matrix, establish the matrix form of the displacement relations for the system.

8-23. Using the matrix equation from Prob. 8-22, determine the eigenvalues and eigenvectors.

8-24. For the simple beam and lumped-mass system shown, (a) determine the flexibility matrix, and (b) express the matrix form of the displacement equations for the system.

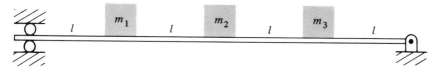

Problem 8-24

8-25. Determine the principal modes of vibration for Prob. 8-24, taking $m_1 = m_2 = m_3 = m$.

8-26. (a) Obtain the stiffness matrix by the inverse of the flexibility matrix for Prob. 8-24. (b) From this establish the matrix equation that governs motion of the system.

8-27. (a) Determine the stiffness matrix by the inverse of the flexibility matrix for Example 8-7. (b) From this establish the matrix equation that governs small vertical motion of the system.

8-28. For the system of Prob. 8-1, (a) express the kinetic-energy and potential-energy functions. (b) Obtain the Lagrange equations, and (c) write these in matrix form.

8-29. For the system shown, (a) establish the kinetic-energy and potential-energy functions. (b) Obtain the Lagrange equations, and (c) express these in matrix form.

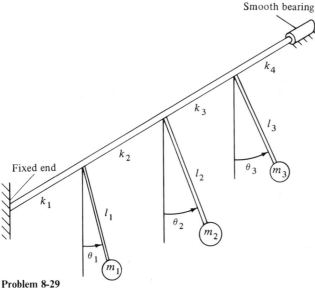

Problem 8-29

8-30. Determine the principal-mode solution for Prob. 8-29, taking $k_1 = k_2 = k_3 = k_4 = k$ and $m_1 = m_2 = m_3 = m$.

8-31. The arrangement shown consists of a weightless rigid member AB pivoted at O and containing small mass m_1 at A. Small mass m_2 is fastened to the bar at B by the weightless cord BC. (a) Determine the Lagrange equations and express these in matrix form. (b) For the case of $a = l = b$ and $m_1 = m = m_2$, obtain the principal-mode solution and interpret this for the type of motion involved. (c) For the case of $a = l$, $b = 2l$, and $m_1 = m = m_2$, obtain the principal modes.

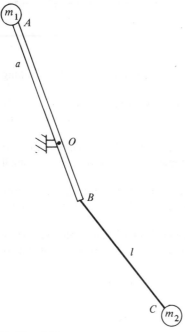

Problem 8-31

8-32. For the system shown, (a) establish the kinetic-energy and potential-energy functions. (b) Obtain the Lagrange equations and (c) express these in matrix form.

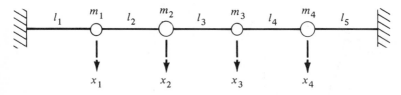

Problem 8-32   Tension in cord $= Q$

8-33. Determine the principal-mode solution for Prob. 8-32, taking $m_1 = m_2 = m_3 = m_4 = m$.

8-34. For the system of Prob. 8-9, express the kinetic-energy and potential-energy functions. (b) Determine the Lagrange equations, and (c) write these in matrix form.

8-35. Write the orthogonality relation between the principal modes obtained in Prob. 8-3.

8-36. Write the orthogonality relation between the principal modes determined in Prob. 8-8.

8-37. Establish the orthogonality relation between the principal modes determined in Prob. 8-15.

8-38. Express the orthogonality relation between the principal modes obtained in Prob. 8-19.

8-39. Establish the orthogonality relation between the eigenvectors calculated in Prob. 8-20.

8-40. A small electric generator is driven by a two-cylinder in-line internal-combustion engine. The dimensions and specifications for the crankshaft, generator, and flywheel are shown. Determine the equivalent torsional-vibration system, composed of disks and shafts.

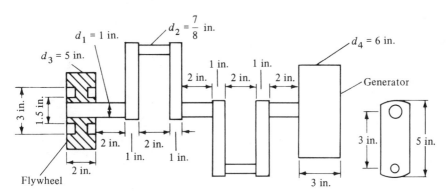

All steel: $\delta = 0.284$ lb/in.^{3}

$$G = 12 \times 10^6 \text{ lb/in.}^2$$
$$E = 30 \times 10^6 \text{ lb/in.}^2$$

Problem 8-40

8-41. Force $P = P_0 \dot{x}_2$ acts on mass element m_2 of the system shown. For $m_1 = m_2 = 1$ kg, $k = 2$ N/m and $P_0 = 1$ N·s/m, determine whether the system is dynamically stable for the following damping conditions: (a) Damping constant $c = 0.5$ N · s/m. (b) Damping constant $c = 1.5$ N · s/m.

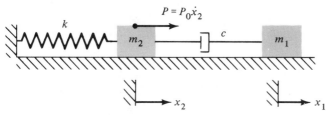

Problem 8-41

8-42. By a fundamental analysis of the system for Prob. 8-41, determine whether it is possible to make the system stable by each of the following: (a) Considering m_1, m_2, k, and P_0 to have the values listed, and permitting c to be increased without any limitation. (b) Considering m_1, m_2, and P_0 to have the values listed and $c = 1.5$ N $\cdot$ s/m, and permitting k to be increased without any limitation.

chapter
9

Numerical Analysis
for Lumped
Parameter Systems

9-1. INTRODUCTION

The classical analysis of multidegree-of-freedom systems was presented in the preceding chapter. This approach involves writing the differential equations which govern the motion of the system. Substitution of harmonic solutions results in a characteristic matrix for which the nth-order determinant must vanish. If n is large, the expansion of this determinant is cumbersome. Furthermore, solving the resulting nth-degree polynomial characteristic equation to obtain the natural frequencies is not simple. These limitations lead to the use of numerical analyses to determine the principal modes, or eigenvectors, and natural frequencies, or eigenvalues. Several methods of numerical analysis are presented in this chapter.

Certain numerical analyses make use of the amplitude equations, which have been determined by the classical approach, and involve a definite procedure for numerically satisfying these equations, thereby yielding the principal modes and frequencies. Such procedures include the Holzer and iteration methods. Other numerical methods employ concepts and analyses that are independent of the amplitude equations. One such procedure is the Rayleigh method.

Also presented in this chapter are transfer matrices and analyses related to them. Numerical methods of solution associated with the use of transfer matrices are shown, including the Myklestad method.

For the sake of simplicity, the numerical methods are illustrated here mainly for three degrees of freedom, but it should be recognized that they may be employed for systems having a large number of degrees of freedom.

9-2. RAYLEIGH'S METHOD

The general application of the Rayleigh method is to determine an approximate value for the lowest natural frequency of a conservative system based on an assumed configuration of the first mode. The error of the assumed modal pattern has a reduced effect on the error of the frequency determined by this method. The procedure and principles involved are presented below.

For small oscillations of the system, it can be stated that generally

$$T = T(\dot{q}_k)$$
= a second-degree function of the generalized velocities only
$$= a_{11}\dot{q}_1^2 + a_{22}\dot{q}_2^2 + \cdots + a_{12}\dot{q}_1\dot{q}_2 + \cdots \qquad (9\text{-}1)$$
$$V = V(q_k)$$
= a second-degree function of the generalized coordinates
$$= b_{11}q_1^2 + b_{22}q_2^2 + \cdots + b_{12}q_1q_2 + \cdots \qquad (9\text{-}2)$$

Also, for a principal mode

$$q_k = A_k \sin(\omega t + \phi)$$

$$\dot{q}_k = \omega A_k \cos(\omega t + \phi) \qquad (9\text{-}3)$$

Then

$$T = \omega^2 \cos^2(\omega t + \phi)(a_{11}A_1^2 + a_{22}A_2^2 + \cdots + a_{12}A_1A_2 + \cdots)$$
$$= \omega^2 \cos^2(\omega t + \phi)\bar{T} \qquad (9\text{-}4)$$

where $\bar{T}$ is the same function of A_k as T is of $\dot{q}_k$. Similarly,

$$V = \sin^2(\omega t + \phi)(b_{11}A_1^2 + b_{22}A_2^2 + \cdots + b_{12}A_1A_2 + \cdots)$$
$$= \sin^2(\omega t + \phi)\bar{V} \qquad (9\text{-}5)$$

where $\bar{V}$ is the same function of A_k as V is of q_k. Furthermore, since the sine and cosine have maximum values of unity, it can be noted that

$$T_{\text{max}} = \omega^2\bar{T}$$

$$V_{\text{max}} = \bar{V} \qquad (9\text{-}6)$$

Now, for a conservative system, $(T + V) = $ constant, and

$$\frac{d}{dt}(T + V) = 0$$

$$\frac{d}{dt}[\omega^2 \cos^2(\omega t + \phi)\bar{T} + \sin^2(\omega t + \phi)\bar{V}] = 0$$

$$-\omega^2 2\cos(\omega t + \phi) \cdot [\omega \sin(\omega t + \phi)] \cdot \bar{T} + 2\sin(\omega t + \phi)$$
$$\cdot [\omega \cos(\omega t + \phi)]\bar{V} = 0$$

$$\bar{V} - \omega^2\bar{T} = 0 \qquad (9\text{-}7)$$

whence

$$\omega_R^2 = \frac{\bar{V}}{\bar{T}}$$

and

$$\omega_R^2 \bar{T} = \bar{V} \tag{9-8}$$

or

$$T_{\max} = V_{\max} \tag{9-9}$$

where ω_R is called the Rayleigh frequency, and $\bar{V}/\bar{T}$ is the Rayleigh quotient. Since both $\bar{V}$ and $\bar{T}$ are functions of the amplitudes $(A_1, A_2, \ldots, A_k, \ldots, A_n)$ only, then if the mode configuration can be estimated reasonably accurately, the frequency can be determined quite simply from the Rayleigh quotient.

It is desirable to consider the error in frequency determined by this method. For $T = T(\dot{q}_k)$, then $\partial T/\partial q_k = 0$, and for the conservative system, Lagrange's equations are

$$\frac{d}{dt}\left(\frac{\partial T}{\partial \dot{q}_k}\right) + \frac{\partial V}{\partial q_k} = 0 \tag{9-10}$$

A study of the energy functions and of the derivatives reveals that

$$\frac{\partial V}{\partial q_k} = \text{a linear function of } q_k$$

$$= \sin\left(\omega t + \phi\right)\frac{\partial \bar{V}}{\partial A_k} \tag{9-11}$$

and

$$\frac{\partial T}{\partial \dot{q}_k} = \text{a linear function of } \dot{q}_k$$

$$= \omega \cos\left(\omega t + \phi\right)\frac{\partial \bar{T}}{\partial A_k} \tag{9-12}$$

whence

$$\frac{d}{dt}\left(\frac{\partial T}{\partial \dot{q}_k}\right) = -\omega^2 \sin\left(\omega t + \phi\right)\frac{\partial \bar{T}}{\partial A_k} \tag{9-13}$$

Substitution of these into Eq. 9-10 gives

$$\frac{\partial \bar{V}}{\partial A_k} - \omega^2 \frac{\partial \bar{T}}{\partial A_k} = 0 \tag{9-14}$$

and substituting the Rayleigh quotient (Eq. 9-8) results in

$$\frac{\partial \bar{V}}{\partial A_k} - \frac{\bar{V}}{\bar{T}} \frac{\partial \bar{T}}{\partial A_k} = 0$$

$$\frac{1}{\bar{T}} \frac{\partial \bar{V}}{\partial A_k} - \frac{\bar{V}}{\bar{T}^2} \frac{\partial \bar{T}}{\partial A_k} = 0$$

which can be written as

$$\frac{\partial}{\partial A_k} \left(\frac{\bar{V}}{\bar{T}} \right) = 0 \tag{9-15}$$

Thus in the region of the correct natural frequency the rate of change of the Rayleigh quotient relative to A_k is very small. Now, for a change ΔA_k in the amplitude A_k, the change in the Rayleigh quotient is defined by

$$\Delta \left(\frac{\bar{V}}{\bar{T}} \right) = \frac{\partial (\bar{V}/\bar{T})}{\partial A_k} \Delta A_k \tag{9-16}$$

and as ΔA_k becomes small,

$$\Delta \left(\frac{\bar{V}}{\bar{T}} \right) \to 0 \tag{9-17}$$

That is, in the region of the correct frequency, a difference or error in the amplitude A_k produces a discrepancy of smaller order in the Rayleigh quotient and hence in the Rayleigh frequency.

Since any error in amplitude represents a restraint that would increase the stiffness of the system, this means that the Rayleigh frequency would be greater than the correct frequency value. Thus Eq. 9-17 may be interpreted as defining a minimum value for the Rayleigh quotient.

Some means is needed for estimating the configuration of the vibration. The most common practice is to use the static equilibrium shape as the first-mode configuration. Accordingly, the resulting frequency determined is that of the first or lowest mode. This is usually the most important frequency that must be obtained and, in some cases, is the only information required.

It is also possible to merely estimate or assume a reasonable shape for the modal configuration. Although this procedure may be quite satisfactory, caution should be used in such an estimate so that some form of restraint on the system is not unintentionally included. This constraining condition may be quite subtle and not readily apparent. Such restraint can have a considerable effect in increasing the resulting Rayleigh frequency calculated compared to the correct value.

Generally it is not possible to estimate the configuration for the second and higher modes; consequently, the Rayleigh method is not feasible for such modes, but it can be used as a correction process or as the means for improving the calculated frequency for any mode if the modal shape has

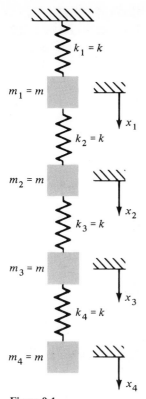

Figure 9-1

been determined by some other procedure. Thus, for example, if there appears to be some error in the roots of the frequency equation, and hence in the amplitude ratios obtained by the classical solution, the Rayleigh method can be used to correct the frequency value, and this can be used as an improved root of the frequency equation. This process can then be repeated, if necessary, to give further improvement. Such correction procedures may be applied to any mode.

EXAMPLE 9-1   Determine the Rayleigh frequency for the first principal mode for the system shown in Fig. 9-1.

SOLUTION   The first-mode configuration will be assumed to be the static equilibrium shape. For this,

$$\delta_1 = \frac{4W}{k} \qquad\qquad \delta_2 = \frac{3W}{k} + \delta_1 = \frac{7W}{k}$$

$$\delta_3 = \frac{2W}{k} + \delta_2 = \frac{9W}{k} \qquad \delta_4 = \frac{W}{k} + \delta_3 = \frac{10W}{k}$$

Only the relative value of displacements is needed. Hence the displacements can be divided by any common value, such as δ_4. Then, $\delta_1 = 0.4$, $\delta_2 = 0.7$, $\delta_3 = 0.9$, and $\delta_4 = 1$. These can be used as the amplitudes for expressing the energies. Thus

$$
\begin{aligned}
T_{max} &= \tfrac{1}{2}m(\omega A_1)^2 + \tfrac{1}{2}m(\omega A_2)^2 + \tfrac{1}{2}m(\omega A_3)^2 + \tfrac{1}{2}m(\omega A_4)^2 \\
&= \tfrac{1}{2}m\omega^2[(0.4)^2 + (0.7)^2 + (0.9)^2 + 1^2] = \tfrac{1}{2}m\omega^2(2.46) \\
V_{max} &= \tfrac{1}{2}k(A_1)^2 + \tfrac{1}{2}k(A_2 - A_1)^2 + \tfrac{1}{2}k(A_3 - A_2)^2 + \tfrac{1}{2}k(A_4 - A_3)^2 \\
&= \tfrac{1}{2}k[(0.4)^2 + (0.7 - 0.4)^2 + (0.9 - 0.7)^2 + (1 - 0.9)^2 \\
&= \tfrac{1}{2}k(0.30)
\end{aligned}
$$

Equating T_{max} and V_{max} and solving for the frequency gives

$$
\omega_R^2 = \frac{0.30}{2.46}\frac{k}{m}
$$

whence

$$
\omega_R = 0.3492\sqrt{\frac{k}{m}}
$$

Comparing this to the correct value of $0.3473\sqrt{k/m}$ shows that the Rayleigh frequency ω_R is 0.5471% higher.

9-3. HOLZER'S METHOD

Holzer's method is a well-known procedure which has had its widest application in the solution of torsional systems. The analysis is generally presented and justified on a physical basis. Actually, the method does not depend on the system being torsional. This can be shown, as can the basis for the method, by consideration of the amplitude equations for a vibrational system.

In the direct analysis of Section 8-2, the matrix form of the amplitude equations was written as follows (see Eq. 8-14):

$$
([k] - \xi[m])\{A\} = \{0\} \tag{9-18}
$$

For convenience, let it be assumed that no dynamic coupling exists so that $m_{ij} = 0$ for $i \neq j$, and the amplitude equations then exhibit only static coupling. If all the k_{ij} coefficients are present in the resulting amplitude equations, this would be designated as complete static coupling. Certain of these coefficients may be missing so that the coupling is incomplete. For example, consider the case for which the algebraic form of Eq. 8-14 is

$$(k_{11} - m_1 \xi)A_1 + k_{12} A_2 = 0$$

$$k_{21} A_1 + (k_{22} - m_2 \xi)A_2 + k_{23} A_3 = 0$$

$$k_{32} A_2 + (k_{33} - m_3 \xi)A_3 + k_{34} A_4 = 0$$

$$\vdots$$

$$(k_{n,n-1})A_{n-1} + (k_{nn} - m_n \xi)A_n = 0 \qquad (9\text{-}19)$$

so that any equation is coupled only to those on either side. That is, the first equation contains only A_1 and A_2; the second equation contains only A_1, A_2, and A_3; .... The equations thus exhibit *adjacent* coupling only. Such a set of n amplitude equations may be solved by the numerical procedure known as Holzer's method.

All the k and m values would be available. Referring to Eq. 9-19, if A_1 is set equal to 1, which is permissible and not restrictive, and a value of ξ is assumed, then A_2 can be determined from the first-amplitude equation. Then the substitution of these into the second equation will determine A_3. Similarly, A_4 can be calculated from the third equation, and so on, and A_n will be obtained from the next-to-last equation. The available calculated values can then be inserted in the last equation. Now, if the assumed ξ is correct, the last equation will come out to zero; that is, it will check. However, if ξ is incorrect, the value of the last equation will represent the error E. A new value of ξ can then be assumed and the entire process repeated. Eventually, the correct value of ξ can be found in this way, although several repetitions are usually required. It is helpful to plot the error E against ξ (see Fig. 9-2) as the calculations progress. The crossing points aid in the interpolation for the correct ξ values. The method can be used to determine all the natural frequencies for the system. The corresponding modes are defined by the amplitudes for the correct sets of calculations.

EXAMPLE 9-2   Using the Holzer method, determine the natural frequencies and corresponding principal modes for the mass-spring vibration-model system shown in Fig. 9-3, wherein $m_1 = 3m$, $m_2 = 2m$, $m_3 = m$; $k_1 = k_2 = k_3 = k$.

SOLUTION   The elements of the stiffness matrix are $k_{11} = 2k$, $k_{12} = -k$, $k_{13} = 0$; $k_{21} = -k$, $k_{22} = 2k$, $k_{23} = -k$; $k_{31} = 0$, $k_{32} = -k$, $k_{33} = k$. For the mass matrix, $m_{11} = 3m$, $m_{22} = 2m$, $m_{33} = m$, and all other elements are zero. Thus

$$[k] = \begin{bmatrix} 2k & -k & 0 \\ -k & 2k & -k \\ 0 & -k & k \end{bmatrix} \qquad [m] = \begin{bmatrix} 3m & 0 & 0 \\ 0 & 2m & 0 \\ 0 & 0 & m \end{bmatrix}$$

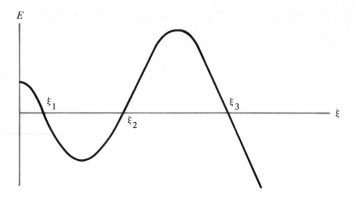

Figure 9-2

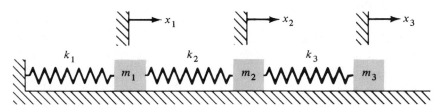

Figure 9-3

and the matrix equation of motion is

$$[m]\{\ddot{x}\} + [k]\{x\} = \{0\}$$

which results in the amplitude equation as

$$([k] - \xi[m])\{A\} = \{0\}$$

whence

$$\begin{bmatrix} (3\xi - 2) & 1 & 0 \\ 1 & (2\xi - 2) & 1 \\ 0 & 1 & (\xi - 1) \end{bmatrix} \begin{Bmatrix} A_1 \\ A_2 \\ A_3 \end{Bmatrix} = \begin{Bmatrix} 0 \\ 0 \\ 0 \end{Bmatrix}$$

where $\xi = (m/k)\omega^2$.

Setting $A_1 = 1$, rearranging, and writing this in algebraic form gives

$$A_2 = (2 - 3\xi)$$

$$A_3 = (2 - 2\xi)A_2 - 1$$

$$E = (1 - \xi)A_3 - A_2$$

which are now convenient for the Holzer procedure, as shown in Table 9-1. (These calculations were carried out on a small electronic calculator.) The nth calculation in each set represents the determinations for an acceptable small error and thus identifies a principal mode.

Table 9-1

Trial numbers	Assumed ξ	A_2	A_3	E
0–1	0	2	3	1
0–2	0.5	0.5	−0.5	−0.75
0–3	1	−1	−1	1
0–4	2	−4	7	−3
1–1	0.1	1.7	2.06	0.154
1–2	0.2	1.4	1.24	−0.408
1–3	0.12	1.64	1.8864	0.020 032
1–4	0.121	1.637	1.887 78	0.013 626
1–5	0.124	1.628	1.852 26	−0.000 542
1–6	0.123	1.631	1.860 77	0.000 899
1–7	0.1231	1.6307	1.859 92	0.000 265
1–8	0.1232	1.6304	1.859 07	−0.000 368
1–9	0.123 14	1.630 58	1.859 58	0.000 120
1–10	0.123 141	1.630 577	1.859 575	0.000 103 7
1–11	0.123 141 4	1.630 575 8	1.859 568 826	3.1176×10^{-6}
$\vdots$				
1–n	0.123 141 856 2	1.630 574 431	1.859 564 938	2.288×10^{-7}
$\vdots$				
2–n	0.758 024	−0.274 072	−1.132 637 693	8.617×10^{-7}
$\vdots$				
3–n	1.785 501	−3.356 503	4.273 072 926	$−5.64 \times 10^{-8}$

9-4. HOLZER'S METHOD FOR TORSIONAL SYSTEM

Consider the torsional system of Fig. 9-4, consisting of n disks on n continuous shafts. The moment of inertia of the jth disk is I_j, and the torsional stiffness of the jth shaft is k_j. A direct analysis of this system would lead to the following amplitude equations:

$$(k_1 - I_1\omega^2)A_1 - k_1 A_2 = 0$$
$$-k_1 A_1 + (k_1 + k_2 - I_2\omega^2)A_2 - k_2 A_3 = 0$$
$$-k_2 A_2 + (k_2 + k_3 - I_3\omega^2)A_3 - k_3 A_4 = 0$$
$$\vdots$$
$$-k_{n-1} A_{n-1} + (k_{n-1} + k_n - I_n\omega^2)A_n = 0 \qquad (9\text{-}20)$$

Using the procedures previously outlined, the Holzer method could be applied directly to the equations in the form shown. However, it is desirable to rearrange the equations so they will be convenient for this purpose. Such a rearrangement can be used to develop the method into its most common form.

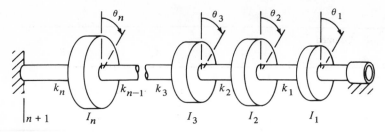

Figure 9-4

Rearranging the first equation for A_2 results in

$$A_2 = A_1 - \frac{I_1\omega^2 A_1}{k_1} \tag{9-21}$$

Similarly, solving the second equation for A_3 gives

$$A_3 = A_2 + \frac{k_1(A_2 - A_1)}{k_2} - \frac{I_2\omega^2 A_2}{k_2} \tag{9-22}$$

and substituting Eq. 9-21 into Eq. 9-22 yields

$$A_3 = A_2 - \left(\frac{I_1\omega^2 A_1 + I_2\omega^2 A_2}{k_2}\right) \tag{9-23}$$

From the third equation,

$$A_4 = A_3 + \frac{k_2(A_3 - A_2)}{k_3} - \frac{I_3\omega^2 A_3}{k_3} \tag{9-24}$$

whence substituting Eq. 9-23 gives

$$A_4 = A_3 - \left(\frac{I_1\omega^2 A_1 + I_2\omega^2 A_2 + I_3\omega^2 A_3}{k_3}\right) \tag{9-25}$$

These will lead to expressing the general form for the amplitude of disk j as

$$A_j = A_{j-1} - \frac{1}{k_{j-1}} \sum_{i=1}^{j-1} I_i\omega^2 A_i \tag{9-26}$$

This can be written for the fixed station $(n + 1)$, resulting in

$$A_{n+1} = A_n - \frac{1}{k_n} \sum_{i=1}^{n} I_i\omega^2 A_i \tag{9-27}$$

Since the $(n + 1)$ station is fixed, then A_{n+1} should be zero, and it will be zero if ω^2 is a correct value. If ω^2 is incorrect, then the value of A_{n+1} represents the error. The process can be repeated again and again, and values of A_{n+1} plotted against ω^2 to aid in interpolation, until all the natural frequencies and corresponding values of the principal modes have been obtained.

A physical interpretation can readily be made for Eqs. 9-21 through

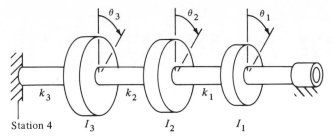

Figure 9-5

9-27. The maximum angular acceleration of disk number 1 is $\omega^2 A_1$, and the corresponding torque is equal to $I_1 \omega^2 A_1$. This is the same as the torque T_1 in shaft number 1. Similarly, the torque which acts on disk number 2 is $I_2 \omega^2 A_2$, but the torque T_2, acting on shaft number 2, is the sum of the preceding disk torques. Thus in Eq. 9-26 the summation term is the torque T_{j-1} acting in shaft $(j - 1)$. Dividing by the torsional stiffness k_{j-1} gives the angle of twist for the shaft $(j - 1)$. Subtracting this from A_{j-1} (the angular position of the disk $j - 1$) will then result in the angle A_j for the disk j.

The arrangement of Eq. 9-26 is convenient for setting up Holzer's method in tabular form, as shown in Example 9-3, below.

Holzer's method can also be used for a free-free torsional system (that is, a system with both ends free). In this case there is no station $(n + 1)$, and the end station occurs at the last disk n. Equation 9-26 then applies until A_n has been obtained. The torque T_n in the free shaft beyond the disk n is defined by

$$T_n = \sum_{i=1}^{n} I_i \omega^2 A_i \tag{9-28}$$

The torque T_n must be zero, and this becomes the condition that must be satisfied by the Holzer calculations. (In a stage of calculations for which $T_n \neq 0$, the value of T_n represents the harmonic torque amplitude, applied at the frequency assumed, which would maintain the harmonic motion of a pattern defined by the displacement amplitudes of the calculations.)

EXAMPLE 9-3   Using Holzer's method, determine the natural frequencies and corresponding principal modes for the system of Fig. 9-5. The data include $I_1 = 100$, $I_2 = 200$, $I_3 = 300$ in. lb sec²; $k_1 = 2 \times 10^6$, $k_2 = 3 \times 10^6$, $k_3 = 4 \times 10^6$ lb in./rad.

SOLUTION   The procedure discussed above, and specified by Eq. 9-26, can be conveniently carried out in tabular form, as shown in Table 9-2. This table is more or less self-explanatory.

Several stages of the calculations for the first mode are shown. The final set of calculations is also shown for each mode. These define the principal modes and natural frequencies. Such calculations can be continued until the desired refinement has been attained.

Table 9-2

(1)	(2)	(3)	(4)	(5)$_j$ = (5)$_{j-1}$ − (9)$_{j-1}$	(6) = (4) × (5)	(7) = $\sum_1^j$ (6)	(8)	(9) = (7)/(8)
ω^2 (rad/sec)^2 10^4	Station number	I (lb in. sec^2)	$I\omega^2 \times 10^{-6}$	A (rad)	$I\omega^2 A \times 10^{-6}$ (lb in.)	$\sum I\omega^2 A \times 10^{-6}$ (lb in.)	$k \times 10^{-6}$ (lb in./rad)	$(A_j - A_{j+1}) = (\sum I\omega^2 A)/k$ (rad)
	1	100	1	1	1	1	2	0.5000
	2	200	2	0.5000	1	2	3	0.6667
	3	300	3	−0.1667	−0.5000	1.5000	4	0.3750
	4			−0.5417				
5 × 10^3	1	100	0.5	1	0.5	0.5	2	0.2500
	2	200	1	0.7500	0.7500	1.2500	3	0.4167
	3	300	1.5	0.3333	0.5000	1.7500	4	0.4375
	4			−0.1042				
4 × 10^3	1	100	0.4	1	0.4	0.4	2	0.2000
	2	200	0.8	0.8000	0.6400	1.0400	3	0.3467
	3	300	1.2	0.4533	0.5440	1.5840	4	0.3960
	4			0.0573				
4336	1	100	0.4336	1	0.4336	0.4336	2	0.2168
	2	200	0.8672	0.7832	0.6792	1.1128	3	0.3709
	3	300	1.3008	0.4123	0.5363	1.6491	4	0.4123
	4			0.0000				
21 930	1	100	2.193	1	2.193	2.193	2	1.0965
	2	200	4.386	−0.0965	−0.4232	1.7698	3	0.5899
	3	300	6.579	−0.6864	−4.5158	−2.7460	4	−0.6865
	4			+0.0001				
42 070	1	100	4.207	1	4.207	4.207	2	2.1035
	2	200	8.414	−1.1035	−9.2848	−5.0778	3	−1.6926
	3	300	12.621	0.5891	7.4350	2.3572	4	0.5893
	4			−0.0002				

9-5. THE ITERATION METHOD

Consider a set of amplitude equations having the following matrix form:

$$\chi \begin{Bmatrix} A_1 \\ A_2 \\ A_3 \\ \cdot \\ A_n \end{Bmatrix} = \begin{bmatrix} b_{11} & b_{12} & b_{13} & \cdots & b_{1n} \\ b_{21} & b_{22} & b_{23} & \cdots & b_{2n} \\ b_{31} & b_{32} & b_{33} & \cdots & b_{3n} \\ \cdot & \cdot & \cdot & \cdots & \cdot \\ b_{n1} & b_{n2} & b_{n3} & \cdots & b_{nn} \end{bmatrix} \begin{Bmatrix} A_1 \\ A_2 \\ A_3 \\ \cdot \\ A_n \end{Bmatrix} \tag{9-29}$$

where χ is a factor related to the frequency, and b_{ij} are coefficients which are dependent on the physical constants of the system. Whether all the b coefficients are present depends on the system and the approach used in obtaining the differential equations. In certain cases, such as that of a massless beam with n lumped masses, the inverse approach is preferable and results in the presence of all b coefficients, or complete dynamic coupling of the equations. Because of this, Holzer's method is not practical for the solution of the problem. However, such a system of equations can be solved by the Stodola iteration method, as explained below.

For the iteration method, first a set of arbitrary values, including $A_1 = 1$, is assigned to the amplitudes. These values are substituted into the right side only of Eq. (9-29), and the right side is evaluated. This results in

$$\chi \begin{Bmatrix} A_1 \\ A_2 \\ A_3 \\ \cdot \\ A_n \end{Bmatrix} = \begin{Bmatrix} \gamma_1 \\ \gamma_2 \\ \gamma_3 \\ \cdot \\ \gamma_n \end{Bmatrix} \tag{9-30}$$

where $\gamma_1, \gamma_2, \gamma_3, \ldots, \gamma_n$ are numerical quantities. Dividing the right-hand column by γ_1 so as to normalize the values to the first element results in

$$\chi \begin{Bmatrix} A_1 \\ A_2 \\ A_3 \\ \cdot \\ A_n \end{Bmatrix} = \gamma_1 \begin{Bmatrix} 1 \\ \gamma_2/\gamma_1 \\ \gamma_3/\gamma_1 \\ \cdot \\ \gamma_n/\gamma_1 \end{Bmatrix} \tag{9-31}$$

Comparison of the two sides reveals that $\chi = \gamma_1$, $A_1 = 1$, $A_2 = \gamma_2/\gamma_1$, $A_3 = \gamma_3/\gamma_1, \ldots, A_n = \gamma_n/\gamma_1$. Thus the set of arbitrary (and probably incorrect) amplitudes initially assumed has led to a new set of values for the amplitudes. This new set may be substituted in the same way as the first set, thereby giving a third set of amplitude values. The process may be repeated again and again, until convergence has been reached; that is, until the values obtained no longer change. The final amplitudes and frequency factor so obtained will be correct to the number of significant figures carried in the calculations. (The number of significant figures used can be increased at any stage of the calculations.)

The process converges to the mode for which the eigenvalue is largest. (This will be proven in the next section.) If χ is the inverse frequency factor λ, the mode obtained will be the lowest or first (for which λ is largest and ω is the smallest frequency).

EXAMPLE 9-4   By the method of iteration, determine the first or lowest principal mode for the 3-mass elastic-beam system of Example 8-7.

SOLUTION   From the governing equation of motion for Example 8-7, the amplitude equation would be

a.
$$\lambda \begin{Bmatrix} A_1 \\ A_2 \\ A_3 \end{Bmatrix} = \begin{bmatrix} 9 & 28/3 & 4 \\ 14/3 & 16/3 & 5/2 \\ 4/3 & 5/3 & 1 \end{bmatrix} \begin{Bmatrix} A_1 \\ A_2 \\ A_3 \end{Bmatrix}$$

where $\lambda = EI/ml^3\omega^2$.

Now set $A_1 = 1$. In assuming arbitrary values for A_2 and A_3, it is desirable to select values which correspond to a first-mode configuration, since the solution process will converge to this. Thus $A_2 = 1 = A_3$ are conveniently chosen. These are substituted on the right-hand side, and the matrix product is obtained, as follows:

$$\lambda \begin{Bmatrix} A_1 \\ A_2 \\ A_3 \end{Bmatrix} = \begin{bmatrix} 9 & 28/3 & 4 \\ 14/3 & 16/3 & 5/2 \\ 4/3 & 5/3 & 1 \end{bmatrix} \begin{Bmatrix} 1 \\ 1 \\ 1 \end{Bmatrix}$$

$$= \begin{Bmatrix} 9 \times 1 + (28/3) \times 1 + 4 \times 1 \\ (14/3) \times 1 + (16/3) \times 1 + (5/2) \times 1 \\ (4/3) \times 1 + (5/3) \times 1 + 1 \times 1 \end{Bmatrix}$$

$$= \begin{Bmatrix} 22.\dot{3} \\ 12.5\dot{0} \\ 4 \end{Bmatrix}$$

Dividing by $\gamma_1 = 22.\dot{3}$ gives

$$\lambda \begin{Bmatrix} A_1 \\ A_2 \\ A_3 \end{Bmatrix} = 22.\dot{3} \begin{Bmatrix} 1 \\ 0.5597 \\ 0.1791 \end{Bmatrix}$$

These amplitude values are now used in the second stage of calculations. Thus

$$\lambda \begin{Bmatrix} A_1 \\ A_2 \\ A_3 \end{Bmatrix} = \begin{bmatrix} 9 & 28/3 & 4 \\ 14/3 & 16/3 & 5/2 \\ 4/3 & 5/3 & 1 \end{bmatrix} \begin{Bmatrix} 1 \\ 0.5597 \\ 0.1791 \end{Bmatrix}$$

$$= \begin{Bmatrix} 9 \times 1 + (28/3) \times 0.5597 + 4 \times 0.1791 \\ (14/3) \times 1 + (16/3) \times 0.5597 + (5/2) \times 0.1791 \\ (4/3) \times 1 + (5/3) \times 0.5597 + 1 \times 0.1791 \end{Bmatrix}$$

$$= \begin{Bmatrix} 14.9403 \\ 8.099\,48 \\ 2.445\,27 \end{Bmatrix} = 14.9403 \begin{Bmatrix} 1 \\ 0.542\,12 \\ 0.163\,67 \end{Bmatrix}$$

Repeating the process several more times finally results in

$$\lambda \begin{Bmatrix} A_1 \\ A_2 \\ A_3 \end{Bmatrix} = 14.7058 \begin{Bmatrix} 1 \\ 0.541\,425 \\ 0.163\,122 \end{Bmatrix}$$

The values will not change if this is used in an additional stage. From this

$$\omega = \sqrt{\frac{EI}{14.7058ml^3}} = 0.2608 \sqrt{\frac{EI}{ml^3}}$$

(The foregoing calculations were normalized to A_1; however, either A_2 or A_3 could have been used instead.)

9-6. CONVERGENCE OF ITERATION METHOD

The convergence of the method will now be shown. The amplitude Eq. 9-29 can be expressed as

$$\chi\{A\} = [b]\{A\} \qquad (9\text{-}32)$$

where the scalar matrix $[b]$ and the frequency factor χ depend on the method of formulation. This is satisfied by *all* principal modes. Thus using r to designate the mode, it can be stated that Eq. 9-32 contains every eigenvector $\{A^{(r)}\}$, with its associated eigenvalue χ_r. For any single eigenvector substituted into the right side of Eq. 9-32, the resulting vector obtained on the left side is proportional to $\{A^{(r)}\}$, the proportionality factor being the scalar matrix $[b]$.

The first *trial* configuration $\{A\}_1$ used in the iteration procedure may be expressed as the sum of the principal modes, as follows:

$$\{A\}_1 = C_1\{A^{(1)}\} + C_2\{A^{(2)}\} + C_3\{A^{(3)}\} + \cdots + C_n\{A^{(n)}\}$$

$$= \sum_{r=1}^{n} C_r\{A^{(r)}\} \qquad (9\text{-}33)$$

where the C coefficients are unspecified.

Based on reasoning of the preceding paragraph, then any eigenvector *component* $\{A^{(r)}\}$ of $\{A\}_1$ which is premultiplied by $[b]$ is proportionately reproduced. Accordingly, premultiplying $\{A\}_1$ by $[b]$ gives the new trial vector as

$$\{A\}_2 = [b]\{A\}_1 = \sum_{r=1}^{n} C_r \chi_r \{A^{(r)}\}$$

Next, premultiplying $\{A\}_2$ by $[b]$ yields

$$\{A\}_3 = [b]\{A\}_2 = \sum_{r=1}^{n} C_r \chi_r^2 \{A^{(r)}\}$$

$$\vdots$$

and calculation q gives

$$\{A\}_q = [b]\{A\}_{q-1} = \sum_{r=1}^{n} C_r \chi_r^{(q-1)} \{A^{(r)}\} \tag{9-34}$$

Now consider that $\chi_1 > \chi_2 > \chi_3 > \cdots > \chi_n$. Then as q increases, the first term in the series of Eq. 9-34 becomes larger and predominates over the other terms. As q becomes large, the vector $\{A\}_q$ will closely resemble the form of the eigenvector $\{A^{(1)}\}$ and

$$\lim_{q \to \infty} \{A\}_q = \{A^{(1)}\} \tag{9-35}$$

Thus for a sufficient number of iterations, the method converges to the mode for which the eigenvalue χ is largest. Accordingly, for a set of amplitude equations obtained by the displacement analysis using the flexibility matrix, the solution will converge to the fundamental or first mode and corresponding lowest frequency, as then $\chi_1 = \lambda_1 = C'/\omega_1^2$, where C' is a constant related to physical constants of the system. For the formulation based on the stiffness coefficients, the iteration will converge to the highest mode corresponding to the largest eigenvalue and greatest natural frequency, since $\chi_1 = \xi_n = C''\omega_n^2$, where C'' is a constant related to the physical constants of the system and ω_n represents the highest natural frequency.

In actual practice a finite number of iterations is sufficient to produce convergence to a configuration of acceptable accuracy.

9-7. ITERATION SOLUTION FOR THE HIGHER MODES

Since the iteration converges to the mode for which the eigenvalue χ is greatest, it is not possible to iterate Eq. 9-29 as it stands to obtain other modes. The mode for which the frequency factor χ is largest will be designated here as the first mode. In order to obtain the second mode (for which χ is second largest), the first mode must be eliminated from Eq. 9-29.

This is accomplished by using the orthogonality relation to sweep out the first mode.

For the iteration procedure, the configuration assumed can be considered to be composed of the sum of the principal modes or eigenvectors, and, accordingly, may be written as

$$\{A\} = C_1\{A^{(1)}\} + C_2\{A^{(2)}\} + C_3\{A^{(3)}\} + \cdots + C_n\{A^{(n)}\} \tag{9-36}$$

If this configuration were used in the iteration of Eq. 9-29, as outlined in Sections 9-5 and 9-6, it would converge to mode 1. However, if C_1 could be made equal to zero, then

$$\{A\} = C_2\{A^{(2)}\} + C_3\{A^{(3)}\} + \cdots + C_n\{A^{(n)}\} \tag{9-37}$$

and this form would converge to mode 2.

To determine the condition for which $C_1 = 0$, premultiply Eq. 9-36 by $\{A^{(1)}\}^T[m]$, resulting in

$$\{A^{(1)}\}^T[m]\{A\} = C_1\{A^{(1)}\}^T[m]\{A^{(1)}\} \tag{9-38}$$

since all other parts of the right side would be zero by virtue of the orthogonality for unlike modes. Now for this expression, if

$$\{A^{(1)}\}^T[m]\{A\} = 0 \tag{9-39}$$

then $C_1 = 0$ since $\{A^{(1)}\}^T[m]\{A^{(1)}\} \neq 0$. Equation 9-39 represents the orthogonality relation between mode 1 and the other modes. Thus to eliminate mode 1 from the assumed configuration $\{A\}$, the orthogonality relation is imposed on $\{A\}$, as defined by Eq. 9-36, resulting in the configuration specified by Eq. 9-37, for which the iteration procedure will converge to mode 2.

EXAMPLE 9-5 Obtain the second and third principal modes for the system of Example 8-7.

SOLUTION Using the first-mode results from Example 9-4, the orthogonality relation is written as

a.
$$\{A^{(1)}\}^T[m]\{A\} = \{0\}$$

$$\lfloor 1 \quad 0.541\,425 \quad 0.163\,122 \rfloor \begin{bmatrix} m & 0 & 0 \\ 0 & 2m & 0 \\ 0 & 0 & 3m \end{bmatrix} \begin{Bmatrix} A_1 \\ A_2 \\ A_3 \end{Bmatrix} = \{0\}$$

whence

$$1 \times m \times A_1 + 0.541\,425 \times 2m \times A_2 + 0.163\,122 \times 3m \times A_3 = 0$$

or

b.
$$A_1 + 1.082\,85A_2 + 0.489\,366A_3 = 0$$

(This relation would be satisfied by the amplitudes for modes 2 or 3 but not those for mode 1, as can be verified by substitution.) Simplifying and rearranging this relation, and introducing certain identities gives

c.
$$A_1 \equiv A_1$$

$$A_2 \equiv A_2$$

$$A_3 = -2.043\,46A_1 - 2.212\,76A_2$$

which may be written as the matrix relation

d.
$$\begin{Bmatrix} A_1 \\ A_2 \\ A_3 \end{Bmatrix} = \begin{bmatrix} 1 & 0 & 0 \\ 0 & 1 & 0 \\ -2.043\,46 & -2.212\,76 & 0 \end{bmatrix} \begin{Bmatrix} A_1 \\ A_2 \\ A_3 \end{Bmatrix}$$

This orthogonality relation is known as the *sweeping matrix*. It is imposed on amplitude equation (a) of Example 9-4 as follows:

e.
$$\lambda \begin{Bmatrix} A_1 \\ A_2 \\ A_3 \end{Bmatrix} = \begin{bmatrix} 9 & 28/3 & 4 \\ 14/3 & 16/3 & 5/2 \\ 4/3 & 5/3 & 1 \end{bmatrix} \begin{bmatrix} 1 & 0 & 0 \\ 0 & 1 & 0 \\ -2.043\,46 & -2.212\,76 & 0 \end{bmatrix} \begin{Bmatrix} A_1 \\ A_2 \\ A_3 \end{Bmatrix}$$

$$= \begin{bmatrix} 0.826\,160 & 0.482\,293 & 0 \\ -0.441\,983 & -0.198\,567 & 0 \\ -0.710\,127 & -0.546\,093 & 0 \end{bmatrix} \begin{Bmatrix} A_1 \\ A_2 \\ A_3 \end{Bmatrix}$$

In effect the amplitude column of equation (a) of Example 9-4 has been replaced by the orthogonality relation which contains only modes 2 and 3. Hence the resulting matrix equation contains these modes only.

The equations are now solved by iteration. It is desirable to start with assumed values for the A's that are of second mode form. Selecting $A_1 = 1$, $A_2 = 0$, and $A_3 = -1$, the solution then proceeds in the same manner as for the first mode. Thus

$$\lambda \begin{Bmatrix} A_1 \\ A_2 \\ A_3 \end{Bmatrix} = \begin{bmatrix} 0.826\,160 & 0.482\,293 & 0 \\ -0.441\,983 & -0.198\,567 & 0 \\ -0.710\,127 & -0.546\,093 & 0 \end{bmatrix} \begin{Bmatrix} 1 \\ 0 \\ -1 \end{Bmatrix}$$

$$= \begin{Bmatrix} 0.826\,160 \\ -0.441\,983 \\ -0.710\,127 \end{Bmatrix} = 0.826\,160 \begin{Bmatrix} 1 \\ -0.534\,985 \\ -0.859\,551 \end{Bmatrix}$$

These are then used in the next stage of the calculations. After several iterations, the results converge to

$$\lambda \begin{Bmatrix} A_1 \\ A_2 \\ A_3 \end{Bmatrix} = 0.535\,947 \begin{Bmatrix} 1 \\ -0.601\,735 \\ -0.711\,868 \end{Bmatrix}$$

from which

$$\omega = \sqrt{\frac{EI}{0.535\,947ml^3}} = 1.3660\sqrt{\frac{EI}{ml^3}}$$

The remaining mode may now be obtained. From the second-mode results, the orthogonality relation is

$$\lfloor 1 \quad -0.601\,735 \quad -0.711\,868 \rfloor \begin{bmatrix} m & 0 & 0 \\ 0 & 2m & 0 \\ 0 & 0 & 3m \end{bmatrix} \begin{Bmatrix} A_1 \\ A_2 \\ A_3 \end{Bmatrix} = 0$$

$$1 \times m \times A_1 - 0.601\,735 \times 2m \times A_2 - 0.711\,868 \times 3m \times A_3 = 0$$

or

f.
$$A_1 - 1.203\,47A_2 - 2.135\,60A_3 = 0$$

Solving this simultaneously with the other orthogonality relation (b) gives

g.
$$A_2 = -1.522\,958A_1$$

$$A_3 = \quad 1.326\,482A_1$$

which is satisfied only by the mode which is common to equations (b) and (f), and hence contains only mode 3. By including the identity $A_1 \equiv A_1$, these can be expressed as the matrix equation

h.
$$\begin{Bmatrix} A_1 \\ A_2 \\ A_3 \end{Bmatrix} = \begin{bmatrix} 1 & 0 & 0 \\ -1.522\,96 & 0 & 0 \\ 1.326\,48 & 0 & 0 \end{bmatrix} \begin{Bmatrix} A_1 \\ A_2 \\ A_3 \end{Bmatrix}$$

This is introduced into amplitude equation (a) of Example 9-4 as follows:

$$\lambda \begin{Bmatrix} A_1 \\ A_2 \\ A_3 \end{Bmatrix} = \begin{bmatrix} 9 & 28/3 & 4 \\ 14/3 & 16/3 & 5/2 \\ 4/3 & 5/3 & 1 \end{bmatrix} \begin{bmatrix} 1 & 0 & 0 \\ -1.522\,96 & 0 & 0 \\ 1.326\,48 & 0 & 0 \end{bmatrix} \begin{Bmatrix} A_1 \\ A_2 \\ A_3 \end{Bmatrix}$$

$$= \begin{bmatrix} 0.091\,626\,7 & 0 & 0 \\ -1.395\,87 & 0 & 0 \\ 0.121\,547 & 0 & 0 \end{bmatrix} \begin{Bmatrix} A_1 \\ A_2 \\ A_3 \end{Bmatrix}$$

Assuming as third mode values, $A_1 = 1$, $A_2 = -1$, and $A_3 = 1$, and substituting gives immediately

$$\lambda \begin{Bmatrix} A_1 \\ A_2 \\ A_3 \end{Bmatrix} = \begin{Bmatrix} 0.091\,626\,7 \\ -1.395\,87 \\ 0.121\,547 \end{Bmatrix} = 0.091\,626\,7 \begin{Bmatrix} 1 \\ -1.523\,43 \\ 1.326\,55 \end{Bmatrix}$$

which defines the third mode. From this,

$$\omega = \sqrt{\frac{EI}{0.091\,626\,7ml^3}} = 3.3036\sqrt{\frac{EI}{ml^3}}$$

The orthogonality relation (Eq. h) actually defines the third mode. However, the iteration following this is used to obtain the frequency. It also results in a minor change in the mode. This is due to the cumulative effect of rounding-off errors in the calculations.

Each mode is dependent on the results of the previous mode. This is one of the disadvantages of the iteration method, particularly when the system has a great many degrees of freedom. In order to minimize this disadvantage it is desirable to carry as many significant figures as possible in the calculations. A desk calculator is helpful, but programming the problem on a digital computer is even more desirable. Fortunately, the problem is not difficult to program.

9-8. FREQUENCY-ADJUSTMENT ITERATION METHOD

Referring to Section 9-3, the criterion for employing Holzer's method for determining the principal modes and frequencies is that the amplitude equations exhibit adjacent coupling *only*, resulting in a chain type or progressive form of calculations. That is, for any selected frequency value, the amplitude A_2 is determined directly from the first amplitude equation, A_3 is then obtained from the second amplitude equation, and so on, until the last amplitude equation yields the error value E.

If the amplitude equations show complete coupling, wherein each and every amplitude term occurs in all amplitude equations, Holzer's method would be impractical as it would require, for any assumed frequency, the simultaneous solution of $(n - 1)$ amplitude equations to obtain the amplitude values. Such a procedure would not be feasible for several degrees of freedom. In this case the iteration method, as explained in Sections 9-5 through 9-7, may be used. However, each higher mode determination is dependent on the amplitude values calculated in all lower modes previously obtained, using the orthogonality relations as a sweeping matrix for each of these lower modes. There is an accumulative error effect from each of the lower mode values which may be significant if many modes are involved. This objection in the iteration method for a system exhibiting complete coupling can be overcome by combining the Holzer method with the iteration procedure.

Referring to Eq. 9-29, the amplitude equations may be written in algebraic form as

$$\chi A_1 = b_{11} A_1 + b_{12} A_2 + b_{13} A_3 + \cdots + b_{1n} A_n$$

$$\chi A_2 = b_{21} A_1 + b_{22} A_2 + b_{23} A_3 + \cdots + b_{2n} A_n$$

$$\chi A_3 = b_{31} A_1 + b_{32} A_2 + b_{33} A_3 + \cdots + b_{3n} A_n$$

$$\vdots$$

$$\chi A_n = b_{n1} A_1 + b_{n2} A_2 + b_{n3} A_3 + \cdots + b_{nn} A_n \tag{9-40}$$

there being n such equations. All terms and coefficients are considered to be present, and the value of all b_{ij} coefficients are known. The frequency factor χ may now be assumed and A_1 may be set to unity. Then in the first $(n-1)$ amplitude equations there will be $(n-1)$ unknowns, namely, $A_2, A_3, \ldots, A_n$. These equations are independent and may be solved by iteration to yield the value of the amplitudes listed. Substituting these amplitude values into the last equation (the nth equation which has not been used), will yield the error value. For this purpose the last equation may be written in the form

$$-[b_{n1} A_1 + b_{n2} A_2 + b_{n3} A_3 + \cdots] + (\chi - b_{nn})A_n = E \qquad (9\text{-}41)$$

where E represents the error. In most instances the error calculated will not be zero. A new value for χ may then be assumed and the iteration process would be repeated, yielding new amplitude values, and a new error value. The entire procedure may be repeated again and again, until the error is reduced to an acceptable small value. For the final set of calculations, the frequency factor used and the amplitudes obtained define a principal mode.

In the procedure outlined the sign of the error is significant, as the correct frequency factor lies between a positive and negative value of the error. In this way a correct principal-mode frequency factor may be bracketed and eventually obtained. The error values obtained may be used in interpolating for the frequency to be assumed in the region of the correct value. Plotting the error against the frequency factor may help in this interpolation.

The foregoing procedure can be used to obtain all of the principal modes and frequencies above the first mode and each mode determination is independent of all other modes. The first mode would be obtained by the straight iteration process outlined in Section 9-5 without assuming a frequency value.

The procedure outlined above may appear to be lengthy and onerous, but such is not the case. After some experience, it can be carried out readily and easily. Significantly, whenever an amplitude value is calculated it should be used in any subsequent determinations of other amplitudes as this improves the rate of convergence of the calculations.

The entire process is convenient for digital computer programming, thereby easing the burden and improving the accuracy of the results.

EXAMPLE 9-6   Using the amplitude equations listed as (a) for Example 9-4, for the 3-mass elastic-beam system of Example 8-7, determine the second and third principal modes by the frequency-adjustment iteration method.

SOLUTION   The amplitude equations can be written in algebraic form as

a.
$$\lambda A_1 = 9A_1 + \tfrac{28}{3}A_2 + 4A_3$$
$$\lambda A_2 = \tfrac{14}{3}A_1 + \tfrac{16}{3}A_2 + \tfrac{5}{2}A_3$$
$$\lambda A_3 = \tfrac{4}{3}A_1 + \tfrac{5}{3}A_2 + 1A_3$$

Setting $A_1 = 1$, these can be rearranged in a form convenient for the frequency-adjustment iteration method, as follows:

b. $\qquad\qquad A_2 = [(\lambda - 9) - 4A_3] \times \frac{3}{28}$

c. $\qquad\qquad A_3 = [-\frac{14}{3} + (\lambda - \frac{16}{3})A_2] \times \frac{2}{5}$

d. $\qquad\qquad E = -\frac{4}{3} - \frac{5}{3}A_2 + (\lambda - 1)A_3$

Equations (b) and (c) will be used for amplitude determinations by iteration, for an assumed λ value. Equation (d) will be used for the error calculation.

Assuming the value $\lambda = 0.6$, as a possible second-mode value, Eqs. (b), (c), and (d) become

$$A_2 = [-8.4 - 4A_3] \times \frac{3}{28}$$

$$A_3 = [-\frac{14}{3} - 4.733\,33A_2] \times \frac{2}{5}$$

$$E = -\frac{4}{3} - \frac{5}{3}A_2 - 0.4A_3$$

Then, estimating the second-mode amplitudes as $A_1 = 1$, $A_2 = -1$, and $A_3 = -1$, and substituting gives first-stage iteration values as

$$A_2 = [-8.4 - 4 \times (-1)] \times \frac{3}{28} = -0.4714$$

$$A_3 = [-\frac{14}{3} - 4.733\,33 \times (-0.4714)] \times \frac{2}{5} = -0.9741$$

Stage-two values would then be

$$A_2 = [-8.4 - 4 \times (-0.9741)] \times \frac{3}{28} = -0.4825$$

$$A_3 = [-\frac{14}{3} - 4.733\,33 \times (-0.4825)] \times \frac{2}{5} = -0.9531$$

Stage-three values would be

$$A_2 = [-8.4 - 4 \times (-0.9531)] \times \frac{3}{28} = -0.4915$$

$$A_3 = [-\frac{14}{3} - 4.733\,33 \times (-0.4915)] \times \frac{2}{5} = -0.9361$$

This process is continued through several additional stages until the following are obtained:

$$A_2 = -0.530\,28$$

$$A_3 = -0.862\,67$$

Using these in the next stage results in the same values, so that the iteration has converged. The error is then calculated as

$$E = -\frac{4}{3} - \frac{5}{3}(-0.530\,28) - 0.4(-0.862\,67) = -0.104\,47$$

A new value of λ is now assumed and the foregoing process is repeated. For $\lambda = 0.5$, relations (b), (c), and (d) become

$$A_2 = [-8.5 - 4A_3] \times \frac{3}{28}$$

$$A_3 = [-\frac{14}{3} - 4.833\,33A_2] \times \frac{2}{5}$$

$$E = -\frac{4}{3} - \frac{5}{3}A_2 - 0.5A_3$$

Table 9-3

Assumed λ value	Final calculated values		
	A_2	A_3	E
0.6	$-0.530\,28$	$-0.862\,67$	$-0.104\,47$
0.535\,94	$-0.601\,75$	$-0.711\,94$	$-0.000\,040\,27$
0.535\,91*	$-0.601\,78$	$-0.711\,87$	$0.000\,004\,62$
0.535\,90	$-0.601\,79$	$-0.711\,85$	$0.000\,017\,18$
0.5333	$-0.604\,88$	$-0.705\,29$	$0.003\,958\,8$
0.525\,90	$-0.613\,75$	$-0.686\,45$	$0.015\,023$
0.5	$-0.645\,79$	$-0.618\,14$	$0.052\,053$
0.10	$-1.493\,06$	$1.258\,80$	$0.022\,176$
0.091\,62*	$-1.522\,82$	$1.326\,22$	$-0.000\,001\,42$
0.0916	$-1.523\,13$	$1.326\,87$	$-0.001\,120$
0.090	$-1.528\,95$	$1.340\,06$	$-0.004\,531$

* Accepted results for the second and third modes.

Setting $A_1 = 1$, $A_2 = -0.6$, and $A_3 = -0.9$ and iterating through several stages results in

$$A_2 = -0.645\,79$$

$$A_3 = -0.618\,14$$

$$E = 0.052\,053$$

Interpolating suggests using $\lambda = 0.5333$. The subsequent iteration eventually results in

$$A_2 = -0.604\,88$$

$$A_3 = -0.705\,29$$

$$E = 0.003\,958\,8$$

Continuing the foregoing process finally gives

$$\lambda = 0.535\,91$$

$$A_2 = -0.601\,780$$

$$A_3 = -0.711\,869$$

$$E = 0.000\,004\,617\,5$$

Considering the number of significant places carried in the calculations, this error is acceptable and the results represent a principal mode, which is recognized as the second mode.

In a similar manner the third-mode values are calculated to be

$$\lambda = 0.091\,62$$

$$A_2 = -1.522\,82$$

$$A_3 = 1.326\,22$$

$$E = -0.000\,001\,42$$

The results of the various calculations involved, including those for the second mode, are summarized in Table 9-3. The accepted results for the second and third modes are indicated by an asterisk.

9-9. TRANSFER MATRICES FOR MASS-SPRING SYSTEMS

An alternative analysis or formulation using transfer matrices can be employed for obtaining relations that govern the motion of a discrete system composed of lumped masses connected by massless elastic parts. The properties and conditions are expressed by state vectors at sections or points immediately adjacent to the sides of a discrete mass. Specifically, a *state vector* is a column matrix which contains the components of the displacements, forces, and moments at a point or station adjacent to a mass. Such a state vector can then be transferred to another location by a *transfer matrix*, there being two types. A *point transfer* matrix transfers a state vector from a location on one side of a mass to the other side, at the same designated station—and thus is a transfer at a point. A *field transfer* matrix transfers a state vector across a spatial distance or field of the system from a station at one mass to a station at another mass.

The relations resulting from the use of transfer matrices lead to a solution in which the natural frequencies and mode shapes may be determined from the characteristic equation if the number of degrees of freedom is small or otherwise by a numerical procedure such as Holzer's method.

For the sake of simplicity the construction of transfer matrices will be presented first for the rectilinear mass-spring and similar systems. Consider the general mass-spring system of Fig. 9-6(a) which is restricted to move in the horizontal direction only. The corresponding free-body diagrams are shown in Fig. 9-6(b). The letters L and R are used to denote the left-hand and right-hand sides, respectively, of a mass station.

For spring k_j the following relations can be written

$$x_j^L = x_{j-1}^R + \frac{F_{j-1}^R}{k_j} \tag{9-42}$$

$$F_j^L = F_{j-1}^R \tag{9-43}$$

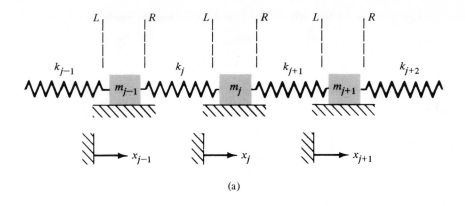

(a)

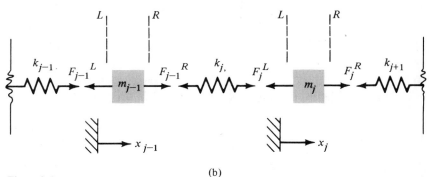

(b)

Figure 9-6

In matrix form these would be written as

$$\begin{Bmatrix} x \\ F \end{Bmatrix}_j^L = \begin{bmatrix} 1 & 1/k \\ 0 & 1 \end{bmatrix}_j \begin{Bmatrix} x \\ F \end{Bmatrix}_{j-1}^R \tag{9-44}$$

Or, more concisely,

$$\{v\}_j^L = [F]_j\{v\}_{j-1}^R \tag{9-45}$$

where

$$\{v\}_j^L = \begin{Bmatrix} x \\ F \end{Bmatrix}_j^L \qquad \{v\}_{j-1}^R = \begin{Bmatrix} x \\ F \end{Bmatrix}_{j-1}^R \tag{9-46}$$

are state vectors for the displacements and internal forces at stations j and $j - 1$, respectively. The scalar matrix

$$[F]_j = \begin{bmatrix} 1 & 1/k \\ 0 & 1 \end{bmatrix}_j \tag{9-47}$$

is the field transfer matrix which relates $\{v\}_j^L$ to $\{v\}_{j-1}^R$.

For mass m_j in Fig. 9-6(b), the following equations apply:

$$x_j^R = x_j^L \tag{9-48}$$

$$F_j^R - F_j^L = m_j \ddot{x}_j \tag{9-49}$$

Now for harmonic motion of m_j, $x_j = A_j \sin(\omega t + \phi)$ and $\ddot{x}_j = -\omega^2 A_j \sin(\omega t + \phi)$, so that

$$\ddot{x}_j = -\omega^2 x_j \tag{9-50}$$

Equation 9-49 then becomes

$$F_j^R = -m_j \omega^2 x_j^L + F_j^L \tag{9-51}$$

In matrix form Eqs. 9-48 and 9-51 would be expressed as

$$\begin{Bmatrix} x \\ F \end{Bmatrix}_j^R = \begin{bmatrix} 1 & 0 \\ -m\omega^2 & 1 \end{bmatrix}_j \begin{Bmatrix} x \\ F \end{Bmatrix}_j^L \tag{9-52}$$

or

$$\{v\}_j^R = [P]_j \{v\}_j^L \tag{9-53}$$

Here, $\{v\}_j^R$ and $\{v\}_j^L$ are state vectors for the displacements and internal forces to the right and left of mass m_j, respectively, and

$$[P]_j = \begin{bmatrix} 1 & 0 \\ -m\omega^2 & 1 \end{bmatrix}_j \tag{9-54}$$

is the scalar point transfer matrix, which relates $\{v\}_j^R$ and $\{v\}_j^L$. Substituting Eq. 9-45 into Eq. 9-53 gives

$$\{v\}_j^R = [P]_j [F]_j \{v\}_{j-1}^R$$

$$= [Q]_j \{v\}_{j-1}^R \tag{9-55}$$

where (see Eqs. 9-47 and 9-54)

$$[Q]_j = [P]_j [F]_j \tag{9-56}$$

$$= \begin{bmatrix} 1 & 0 \\ -m\omega^2 & 1 \end{bmatrix}_j \begin{bmatrix} 1 & \dfrac{1}{k} \\ 0 & 1 \end{bmatrix}_j \tag{9-57}$$

$$= \begin{bmatrix} (1 \times 1 + 0 \times 0) & \left(1 \times \dfrac{1}{k} + 0 \times 1\right) \\ (-m\omega^2 \times 1 + 1 \times 0) & \left(-m\omega^2 \times \dfrac{1}{k} + 1 \times 1\right) \end{bmatrix}_j$$

$$= \begin{bmatrix} 1 & \dfrac{1}{k} \\ -m\omega^2 & \left(1 - \dfrac{m\omega^2}{k}\right) \end{bmatrix}_j \tag{9-58}$$

This procedure can be continued. Thus

$$\{v\}_n^R = [Q]_{n!}\{v\}_0^R \tag{9-59}$$

where

$$[Q]_{n!} = [P]_n[F]_n[P]_{n-1}[F]_{n-1} \cdots [P]_1[F]_1 \tag{9-60}$$

Equation 9-59 expresses the state vector $\{v\}_n^R$ at the nth station in terms of the state vector $\{v\}_0^R$ at the initial station.

Noting that Eq. 9-59 represents two algebraic equations, and that usually a boundary condition, such as $x = 0$ or $F = 0$, would be known at each end of the system, then the equations can be solved to yield the natural frequencies and principal modes of vibration. The precise manner in which this would be carried out is not readily apparent, and will be illustrated below. Two methods are suitable. The first yields the nth-order characteristic equation, which can then be solved for the n natural frequencies. Each natural frequency can then be substituted into individual stages of the determinations to give the amplitude ratios and thus the modal pattern. This procedure is feasible only if the number of degrees of freedom is small.

The second method is to follow a numerical procedure such as Holzer's method. By assuming a value for ω^2 and assigning a value to the unknown boundary condition at station 0, then the unknown parameters of x and F can be determined at Station 1 by the related matrix multiplication. This process can be continued from one station to the next until the final station n is reached and the boundary condition there is checked, yielding the error. A new value of ω^2 would then be assumed and the entire procedure would be repeated, resulting in a new error value. The process can be continued until the error is brought to zero—or rather, to an acceptable small value.

EXAMPLE 9-7   For the mass-spring system shown in Fig. 9-7, use transfer matrices to determine the principal modes and natural frequencies by the first method described above. Assume $k_1 = k_2 = k_3 = k$ and $m_1 = m$, $m_2 = 2m$.

SOLUTION   The boundary conditions for Sta. 0^R are $x = 0$, $F = F_0$ and for Sta. 3^R are $x = 0$, $F = F_3$, where F_3 is unknown. The first transfer matrix would be written for Sta. 1^R in terms of Sta. 0^R as

a.

$$\begin{Bmatrix} x \\ F \end{Bmatrix}_1^R = \begin{bmatrix} 1 & \dfrac{1}{k_1} \\ -m_1\omega^2 & \left(1 - \dfrac{m_1\omega^2}{k_1}\right) \end{bmatrix}_1 \begin{Bmatrix} x \\ F \end{Bmatrix}_0^R$$

$$= \begin{bmatrix} 1 & \dfrac{1}{k} \\ -m\omega^2 & \left(1 - \dfrac{m\omega^2}{k}\right) \end{bmatrix}_1 \begin{Bmatrix} 0 \\ F \end{Bmatrix}_0^R$$

Next, the transfer matrix for Sta. 2 is written as follows:

b.
$$
\begin{Bmatrix} x \\ F \end{Bmatrix}_2^R =
\begin{bmatrix}
1 & \dfrac{1}{k_2} \\[2ex]
-m_2\omega^2 & \left(1 - \dfrac{m_2\omega^2}{k_2}\right)
\end{bmatrix}_2
\begin{Bmatrix} x \\ F \end{Bmatrix}_1^R
$$

$$
=
\begin{bmatrix}
1 & \dfrac{1}{k} \\[2ex]
-2m\omega^2 & \left(1 - \dfrac{2m\omega^2}{k}\right)
\end{bmatrix}_2
\begin{bmatrix}
1 & \dfrac{1}{k} \\[2ex]
-m\omega^2 & \left(1 - \dfrac{m\omega^2}{k}\right)
\end{bmatrix}_1
\begin{Bmatrix} 0 \\ F \end{Bmatrix}_0^R
$$

$$
=
\begin{bmatrix}
\Big| & 1 \times \dfrac{1}{k} + \dfrac{1}{k}\left(1 - \dfrac{m\omega^2}{k}\right) \\[2ex]
\Big| & -2m\omega^2 \times \dfrac{1}{k} + \left(1 - \dfrac{2m\omega^2}{k}\right)\left(1 - \dfrac{m\omega^2}{k}\right)
\end{bmatrix}_2
\begin{Bmatrix} 0 \\ F \end{Bmatrix}_0^R
$$

$$
=
\begin{bmatrix}
\Big| & \left(\dfrac{2}{k} - \dfrac{m\omega^2}{k^2}\right) \\[2ex]
\Big| & \left(1 - \dfrac{5m\omega^2}{k} + \dfrac{2m^2\omega^4}{k^2}\right)
\end{bmatrix}_2
\begin{Bmatrix} 0 \\ F \end{Bmatrix}_0^R
$$

The computations in the first column of certain of the transfer matrices shown by the vertical line are omitted as they are not needed, due to the zero in the first row of the state vector for Sta. 0^R. Continuing,

c.
$$
\begin{Bmatrix} x \\ F \end{Bmatrix}_3^R =
\begin{bmatrix}
1 & \dfrac{1}{k_3} \\[2ex]
-m_3\omega^2 & \left(1 - \dfrac{m_3\omega^2}{k_3}\right)
\end{bmatrix}_3
\begin{Bmatrix} x \\ F \end{Bmatrix}_2^R
$$

$$
=
\begin{bmatrix}
1 & \dfrac{1}{k} \\[2ex]
0 & 1
\end{bmatrix}_3
\begin{bmatrix}
\Big| & \left(\dfrac{2}{k} - \dfrac{m\omega^2}{k^2}\right) \\[2ex]
\Big| & \left(1 - \dfrac{5m\omega^2}{k} + \dfrac{2m^2\omega^4}{k^2}\right)
\end{bmatrix}_2
\begin{Bmatrix} 0 \\ F \end{Bmatrix}_0^R
$$

$$
=
\begin{bmatrix}
\Big| & 1 \times \left(\dfrac{2}{k} - \dfrac{m\omega^2}{k^2}\right) + \dfrac{1}{k} \times \left(1 - \dfrac{5m\omega^2}{k} + \dfrac{2m^2\omega^4}{k^2}\right) \\[2ex]
\Big| & 0 \times \left(\dfrac{2}{k} - \dfrac{m\omega^2}{k^2}\right) + 1 \times \left(1 - \dfrac{5m\omega^2}{k} + \dfrac{2m^2\omega^4}{k^2}\right)
\end{bmatrix}_3
\begin{Bmatrix} 0 \\ F \end{Bmatrix}_0^R
$$

$$
=
\begin{bmatrix}
\Big| & \left(\dfrac{3}{k} - \dfrac{6m\omega^2}{k^2} + \dfrac{2m^2\omega^4}{k^3}\right) \\[2ex]
\Big| & \left(1 - \dfrac{5m\omega^2}{k} + \dfrac{2m^2\omega^4}{k^2}\right)
\end{bmatrix}_3
\begin{Bmatrix} 0 \\ F_0 \end{Bmatrix}_0^R
$$

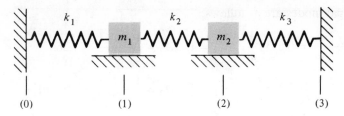

Figure 9-7

From the first algebraic equation of this final matrix,

$$0 = \left(\frac{3}{k} - \frac{6m\omega^2}{k^2} + \frac{2m^2\omega^4}{k^3} \right) F_0$$

whence

$$\frac{m^2\omega^4}{k^2} - \frac{3m\omega^2}{k} + \tfrac{3}{2} = 0$$

which represents the characteristic equation and has the roots

$$\frac{m\omega^2}{k} = \frac{3 - \sqrt{3}}{2}, \qquad \frac{3 + \sqrt{3}}{2}$$

This defines the natural frequencies as

$$\omega_1^2 = \frac{3 - \sqrt{3}}{2} \frac{k}{m}, \qquad \omega_2^2 = \frac{3 + \sqrt{3}}{2} \frac{k}{m}$$

Setting $x_1 = 1$ and also substituting ω_1^2 into Eq. (a) gives

$$F_0 = k, \qquad F_1 = \frac{\sqrt{3} - 1}{2} k$$

Then substituting these into Eq. (b) gives

$$x_2 = \frac{1 + \sqrt{3}}{2}$$

Similarly, setting $x_1 = 1$ and substituting ω_2^2 into Eq. (a) gives

$$F_0 = k, \qquad F_1 = \frac{-\sqrt{3} - 1}{2} k$$

and substituting these into Eq. (b) yields

$$x_2 = \frac{1 - \sqrt{3}}{2}$$

Thus the principal modes are as follows:

$$\omega_1^2 = \frac{3 - \sqrt{3}}{2}\frac{k}{m} = 0.6334\frac{k}{m}, \quad A_1 = 1, \quad A_2 = \frac{1 + \sqrt{3}}{2} = 1.366$$

$$\omega_2^2 = \frac{3 + \sqrt{3}}{2}\frac{k}{m} = 2.366\frac{k}{m}, \quad A_1 = 1, \quad A_2 = \frac{1 - \sqrt{3}}{2} = -0.361$$

9-10. TRANSFER MATRICES FOR TORSIONAL SYSTEMS

For a torsional system composed of disks on a massless shaft, the analysis and formulation of the transfer matrices are identical to that of the rectilinear mass-spring system. Thus considering the torsional systems and corresponding free-body diagrams shown in Fig. 9-8, the transfer-matrix relations for the torsional shaft (spring) K_j are

$$\left\{ \begin{matrix} \theta \\ M_t \end{matrix} \right\}_j^L = \begin{bmatrix} 1 & \frac{1}{K} \\ 0 & 1 \end{bmatrix}_j \left\{ \begin{matrix} \theta \\ M_t \end{matrix} \right\}_{j-1}^R \tag{9-61}$$

or in general form

$$\{v\}_j^L = [F]_j\{v\}_{j-1}^R \tag{9-62}$$

For disk I_j, the matrix relations for harmonic motion are

$$\left\{ \begin{matrix} \theta \\ M_t \end{matrix} \right\}_j^R = \begin{bmatrix} 1 & 0 \\ -I\omega^2 & 1 \end{bmatrix}_j \left\{ \begin{matrix} \theta \\ M_t \end{matrix} \right\}_j^L \tag{9-63}$$

having the general form

$$\{v\}_j^R = [P]_j\{v\}_j^L \tag{9-64}$$

Substituting Eq. 9-62 into Eq. 9-64 gives

$$\{v\}_j^R = [P]_j[F]_j\{v\}_{j-1}^R$$
$$= [Q]_j\{v\}_{j-1}^R \tag{9-65}$$

where

$$[Q]_j = \begin{bmatrix} 1 & \frac{1}{K} \\ -I\omega^2 & \left(1 - \frac{I\omega^2}{K}\right) \end{bmatrix}_j \tag{9-66}$$

If Eq. 9-65 is applied to successive stations of the torsional system, then

$$\{v\}_n^R = [Q]_{n!}\{v\}_0^R \tag{9-67}$$

where

$$[Q]_{n!} = [P]_n[F]_n[P]_{n-1}[F]_{n-1} \cdots [P]_1[F]_1 \tag{9-68}$$

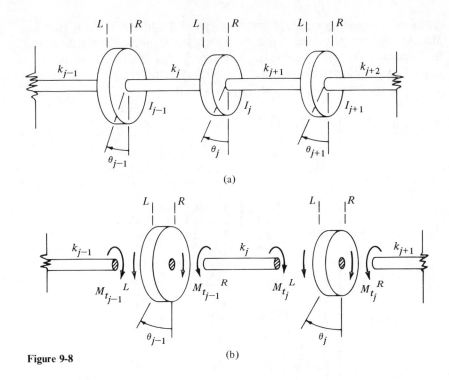

Figure 9-8

(a)

(b)

Equations 9-61 through 9-68 are the same as those for the mass-spring system except that m has been replaced by I, x has been replaced by θ, and the K now represents a torsional spring constant.

EXAMPLE 9-8 For the torsional system shown in Fig. 9-9, employ transfer matrices to determine the principal modes and natural frequencies by the second method described above (the Holzer procedure).

SOLUTION The boundary conditions for Sta. 0^R are $\theta = 0$, $M_t = M_{t_0} = 1$ and for Sta. 3^R are $\theta = \theta_3$, $M_t = 0$, where θ_3 is unknown. Assuming the first trial frequency value as $\omega^2 = 1.0(K/I)$, and referring to Eq. 9-65, the first transfer matrix for Sta. 1^R in terms of Sta. 0^R would be written as

a.
$$
\begin{Bmatrix} \theta \\ M_t \end{Bmatrix}_1^R = \begin{bmatrix} 1 & \dfrac{1}{2K} \\ -3I \times \dfrac{K}{I} & \left(1 - \dfrac{3I}{2K} \times \dfrac{K}{I}\right) \end{bmatrix}_1 \begin{Bmatrix} 0 \\ M_{t_0} \end{Bmatrix}_0^R
$$

$$
= \begin{bmatrix} & \dfrac{1}{2K} \\ & -\dfrac{1}{2} \end{bmatrix}_1 \begin{Bmatrix} 0 \\ 1 \end{Bmatrix}_0^R
$$

As in the preceeding example, the elements of the first row of the transfer matrix shown by the vertical line are not included as they are not needed, due to the zero in the first row of the state vector for Sta. 0^R. This condition also applies to subsequent transfer matrices in this example.

Next, the transfer matrix for Sta. 2^R is written,

b.
$$\begin{Bmatrix} \theta \\ M_t \end{Bmatrix}_2^R = \begin{bmatrix} 1 & \dfrac{1}{1.5K} \\ -2I\dfrac{K}{I} & \left(1 - \dfrac{2I}{1.5K} \times \dfrac{K}{I}\right) \end{bmatrix}_2 \begin{bmatrix} & \dfrac{1}{2K} \\ & -\dfrac{1}{2} \end{bmatrix}_1 \begin{Bmatrix} 0 \\ 1 \end{Bmatrix}_0^R$$

$$= \begin{bmatrix} 1 & \dfrac{1}{1.5K} \\ -2K & -\dfrac{1}{3} \end{bmatrix}_2 \begin{bmatrix} & \dfrac{1}{2K} \\ & -\dfrac{1}{2} \end{bmatrix}_1 \begin{Bmatrix} 0 \\ 1 \end{Bmatrix}_0^R$$

$$= \begin{bmatrix} & 1 \times \dfrac{1}{2K} + \dfrac{1}{1.5K} \times \left(-\dfrac{1}{2}\right) \\ & -2K \times \dfrac{1}{2K} - \dfrac{1}{3} \times \left(-\dfrac{1}{2}\right) \end{bmatrix}_2 \begin{Bmatrix} 0 \\ 1 \end{Bmatrix}_0^R$$

$$= \begin{bmatrix} & \dfrac{1}{6K} \\ & -\dfrac{5}{6} \end{bmatrix}_2 \begin{Bmatrix} 0 \\ 1 \end{Bmatrix}_0^R$$

Next,

c.
$$\begin{Bmatrix} \theta \\ M_t \end{Bmatrix}_3^R = \begin{bmatrix} 1 & \dfrac{1}{K} \\ -I \times \dfrac{K}{I} & \left(1 - \dfrac{I}{K} \times \dfrac{K}{I}\right) \end{bmatrix}_3 \begin{bmatrix} & \dfrac{1}{6K} \\ & -\dfrac{5}{6} \end{bmatrix}_2 \begin{Bmatrix} 0 \\ 1 \end{Bmatrix}_0^R$$

$$= \begin{bmatrix} 1 & \dfrac{1}{K} \\ -K & 0 \end{bmatrix}_3 \begin{bmatrix} & \dfrac{1}{6K} \\ & -\dfrac{5}{6} \end{bmatrix}_2 \begin{Bmatrix} 0 \\ 1 \end{Bmatrix}_0^R$$

$$= \begin{bmatrix} & 1 \times \dfrac{1}{6K} + \dfrac{1}{K} \times \left(-\dfrac{5}{6}\right) \\ & -K \times \dfrac{1}{6K} + 0 \times \left(-\dfrac{5}{6}\right) \end{bmatrix}_3 \begin{Bmatrix} 0 \\ 1 \end{Bmatrix}_0^R$$

$$= \begin{bmatrix} & -\dfrac{2}{3K} \\ & -\dfrac{1}{6} \end{bmatrix}_3 \begin{Bmatrix} 0 \\ 1 \end{Bmatrix}_0^R$$

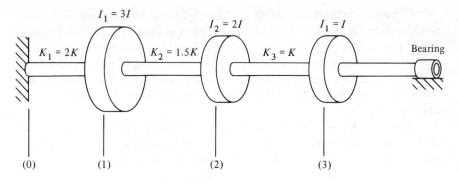

Figure 9-9

From the second algebraic equation of this matrix, $M_{t_3} = -\frac{1}{6}$ representing the error since M_{t_3} should be zero.

A new value of ω^2 is now chosen and the process is repeated, resulting in a new error. This is continued until the M_{t_3} error is reduced to an acceptable value, and the corresponding ω^2 is then a correct natural frequency value. As in Sections 9-3 and 9-4, plotting the error versus the ω^2 value aids in interpolating for the correct frequency values. Such continued calculations are not shown here, but the correct (acceptable) ω^2 values can be verified as

$$\omega_1^2 = 0.2168\,\frac{K}{I}, \quad \omega_2^2 = 1.0965\,\frac{K}{I}, \quad \omega_3^2 = 2.1035\,\frac{K}{I}$$

The corresponding errors are

$$M_{t_3}^{(1)} = 0.000\,00, \quad M_{t_3}^{(2)} = 0.000\,13, \quad M_{t_3}^{(3)} = 0.000\,23$$

The principal-mode amplitude ratios can be determined by substituting the natural frequency values into the transfer matrix equations for each station.

9-11. TRANSFER MATRICES FOR FLEXURAL SYSTEMS, AND THE MYKLESTAD METHOD

Field and point transfer matrices may also be used to obtain relations that govern the flexural motion and vibrations of lumped-mass massless elastic-beam systems. The elements of the state vector $\{v\}_j$ at station j are the displacement y_j, the slope θ_j, the internal moment M_j, and the internal shear force V_j.

Although this approach may be used to obtain the characteristic equation and then solve for the natural frequencies and principal modes in the classical manner, even with few degrees of flexural freedom this procedure is cumbersome and impractical. However, the transfer matrices obtained are suitable for solution by the Myklestad method, which is essentially a Holzer procedure applied to the beam problem.

Consider a continuous beam containing several lumped masses. Figure 9-10(b) shows a segment of the beam between station $(j-1)$ and station j, and the mass m_j at station j, with the shear force V and bending moment M appropriately indicated. Figure 9-10(c) also shows the same portion of the beam, arranged so that the deflection y and slope θ, as well as V and M are indicated. The letters L and R indicate the left and right side, respectively, of the mass station.

Referring to Fig. 9-10(b) and (c) and recognizing the continuity of deflection, slope, and moment, and applying Newton's second law to the vertical force condition, the following relations can be set down:

$$y_j^R = y_j^L \qquad V_j^L - V_j^R = m_j \ddot{y}_j$$

and since for harmonic motion

$$\ddot{y}_j = -m\omega^2 y_j$$

then

$$\theta_j^R = \theta_j^L$$

$$M_j^R = M_j^L$$

$$V_j^R = V_j^L - m_j \ddot{y}_j$$

$$= V_j^L + m_j \omega^2 y_j \tag{9-69}$$

These may be expressed in matrix form as

$$\begin{Bmatrix} y \\ \theta \\ M \\ V \end{Bmatrix}_j^R = \begin{bmatrix} 1 & 0 & 0 & 0 \\ 0 & 1 & 0 & 0 \\ 0 & 0 & 1 & 0 \\ m\omega^2 & 0 & 0 & 1 \end{bmatrix}_j \begin{Bmatrix} y \\ \theta \\ M \\ V \end{Bmatrix}_j^L \tag{9-70}$$

or

$$\{v\}_j^R = [P]_j \{v\}_j^L \tag{9-71}$$

$[P]_j$ is the scalar point transfer matrix which relates state vectors $\{v\}_j^R$ and $\{v\}_j^L$.

Influence coefficients here are defined as follows:

α_{Vj} = slope of the beam at station j relative to the tangent at station $(j-1)$, due to unit force at j

α_{Mj} = slope of the beam at station j relative to the tangent at station $(j-1)$, due to unit moment at j

d_{Vj} = tangential deviation or vertical distance from point on the beam at station j to the tangent at station $(j-1)$ due to unit force at j

d_{Mj} = tangential deviation or vertical distance from point on the beam at station j to the tangent at station $(j-1)$ due to unit moment at j

The beam segment l_j is considered to be a cantilever beam fixed at station $(j-1)$ and acted on by a unit condition (force or moment) at station j, and

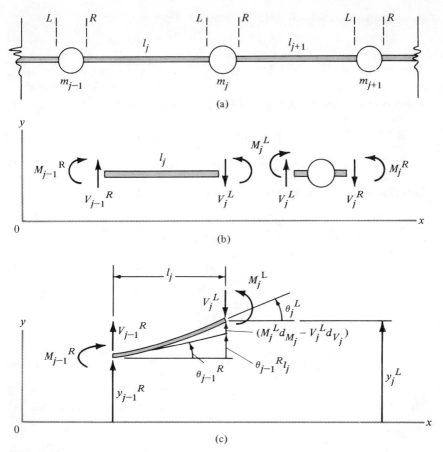

Figure 9-10

linear elastic-beam theory is applied. Then the influence coefficients can be derived† as

$$\alpha_{Vj} = \frac{l_j^2}{2B_j}$$

where $B_j = E_j I_j$ is the stiffness factor of beam segment l_j and

$$\alpha_{Mj} = \frac{l_j}{B_j}$$

$$d_{Vj} = \frac{l_j^3}{3B_j}$$

$$d_{Mj} = \frac{l_j^2}{2B_j} \tag{9-72}$$

† See deflection theory by moment-area method in strength of materials text.

Referring to Fig. 9-10(b) or (c), it is noted that

$$M_j^L = M_{j-1}^R + V_{j-1}^R l_j$$
$$V_j^L = V_{j-1}^R \tag{9-73}$$

Based on the geometry of Fig. 9-10(c), the following relations can be written:

$$y_j^L = y_{j-1}^R + \theta_{j-1}^R l_j + M_j^L d_{Mj} - V_j^L d_{Vj}$$
$$\theta_j^L = \theta_{j-1}^R + M_j^L \alpha_{Mj} - V_j \alpha_{Vj} \tag{9-74}$$

Substituting Eqs. 9-72 and 9-73 into Eq. 9-74 results in

$$y_j^L = y_{j-1}^R + \theta_{j-1}^R l_j + (M_{j-1}^R + V_{j-1}^R l_j)\frac{l_j^2}{2B_j} - V_{j-1}^R \frac{l_j^3}{3B_j}$$

$$= y_{j-1}^R + \theta_{j-1}^R l_j + M_{j-1}^R \frac{l_j^2}{2B_j} + V_{j-1}^R \frac{l_j^3}{6B_j} \tag{9-75}$$

$$\theta_j^L = \theta_{j-1}^R + (M_{j-1}^R + V_{j-1}^R l_j)\frac{l_j}{B_j} - V_{j-1}^R \frac{l_j^2}{2B_j}$$

$$= \theta_{j-1}^R + M_{j-1}^R \frac{l_j}{B_j} + V_{j-1}^R \frac{l_j^2}{2B_j} \tag{9-76}$$

Equations 9-75, 9-76, and 9-73 are now expressed in matrix form as follows:

$$\begin{Bmatrix} y \\ \theta \\ M \\ V \end{Bmatrix}_j^L = \begin{bmatrix} 1 & l & \dfrac{l^2}{2B} & \dfrac{l^3}{6B} \\ 0 & 1 & \dfrac{l}{B} & \dfrac{l^2}{2B} \\ 0 & 0 & 1 & l \\ 0 & 0 & 0 & 1 \end{bmatrix}_j \begin{Bmatrix} y \\ \theta \\ M \\ V \end{Bmatrix}_{j-1}^R \tag{9-77}$$

or

$$\{v\}_j^L = [F]_j \{v\}_{j-1}^R \tag{9-78}$$

where $[F]_j$ is the scalar field matrix which relates state vectors $\{v\}_j^L$ and $\{v\}_{j-1}^R$.

Substituting Eq. 9-78 into Eq. 9-71 yields the matrix $[Q]_j$ expressing $\{v\}_j^R$ in terms of $\{v\}_{j-1}^R$, as follows:

$$\{v\}_j^R = [P]_j [F]_j \{v\}_{j-1}^R$$

$$= [Q]_j \{v\}_{j-1}^R \tag{9-79}$$

where

$$[Q]_j = [P]_j[F]_j \tag{9-80}$$

$$= \begin{bmatrix} 1 & 0 & 0 & 0 \\ 0 & 1 & 0 & 0 \\ 0 & 0 & 1 & 0 \\ m\omega^2 & 0 & 0 & 1 \end{bmatrix}_j \begin{bmatrix} 1 & l & \dfrac{l^2}{2B} & \dfrac{l^3}{6B} \\ 0 & 1 & \dfrac{l}{B} & \dfrac{l^2}{2B} \\ 0 & 0 & 1 & l \\ 0 & 0 & 0 & 1 \end{bmatrix}_j \tag{9-81}$$

Carrying out the multiplication of these two matrices gives

$$[Q]_j = \begin{bmatrix} 1 & l & \dfrac{l^2}{2B} & \dfrac{l^3}{6B} \\ 0 & 1 & \dfrac{l}{B} & \dfrac{l^2}{2B} \\ 0 & 0 & 1 & l \\ m\omega^2 & m\omega^2 l & \dfrac{m\omega^2 l^2}{2B} & \left(1 + \dfrac{m\omega^2 l^3}{6B}\right) \end{bmatrix}_j \tag{9-82}$$

Thus Eq. 9-79 can be written as

$$\begin{Bmatrix} y \\ \theta \\ M \\ V \end{Bmatrix}_j = \begin{bmatrix} 1 & l & \dfrac{l^2}{2B} & \dfrac{l^3}{6B} \\ 0 & 1 & \dfrac{l}{B} & \dfrac{l^2}{2B} \\ 0 & 0 & 1 & l \\ m\omega^2 & m\omega^2 l & \dfrac{m\omega^2 l^2}{2B} & \left(1 + \dfrac{m\omega^2 l^3}{6B}\right) \end{bmatrix}_j \begin{Bmatrix} y \\ \theta \\ M \\ V \end{Bmatrix}_{j-1}^{R} \tag{9-83}$$

Equation 9-83, or its more general form Eq. 9-79, may be successively applied to transfer the state vector station by station, progressing from one end of the system to the other. This may be represented as follows:

$$\{v\}_1^R = [Q]_1 \{v\}_0^R$$

$$\{v\}_2^R = [Q]_2 \{v\}_1^R$$

$$\{v\}_3^R = [Q]_3 \{v\}_2^R$$

$$\vdots$$

$$\{v\}_n^R = [Q]_n \{v\}_{n-1}^R \tag{9-84}$$

which leads to

$$\{v\}_n^R = [Q]_{n:} \{v\}_0^R \tag{9-85}$$

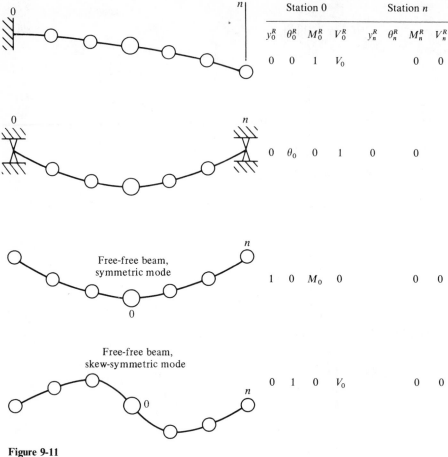

	Station 0				Station n		
y_0^R	θ_0^R	M_0^R	V_0^R	y_n^R	θ_n^R	M_n^R	V_n^R
0	0	1	V_0			0	0
0	θ_0	0	1			0	0
1	0	M_0	0			0	0
0	1	0	V_0			0	0

Free-free beam, symmetric mode

Free-free beam, skew-symmetric mode

Figure 9-11

where

$$[Q]_{n!} = [Q]_n[Q]_{n-1}[Q]_{n-2} \cdots [Q]_1 \qquad (9\text{-}86)$$

Equation 9-84 or Eq. 9-83 may be used for carrying out the Myklestad method of determining the principal-mode shapes and natural frequencies for flexural systems. The method is essentially a Holzer type of computation. That is, calculations are based on an assumed frequency and proceed successively through sets of relations to a final expression which determines an error value. A new frequency is then assumed, and the calculations are repeated. Eventually the correct frequency, corresponding to a zero or small error, can be obtained. The mode is defined by the amplitudes for the correct set of calculations. In this manner all the modes and natural frequencies can be determined independently of each other. At the initial station 0, two boundary conditions, such as y and θ or y and M, would be known. Another

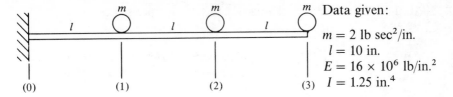

Data given:

$m = 2$ lb sec^2/in.
$l = 10$ in.
$E = 16 \times 10^6$ lb/in.^{2}
$I = 1.25$ in.^{4}

(0)   (1)   (2)   (3)

Figure 9-12

of the boundary values here may be set equal to unity, and the remaining boundary value may be designated by appropriate subscript, such as for V_0. Also, the physical constants l, m, and B ($= EI$) for station j would be available. Then Eq. 9-83 expresses the state vector (and thus y, θ, M, and V) at station 1. Reapplying Eq. 9-83 then gives the state vector at station 2, and so on, station by station, until the final station is reached, where again two boundary conditions would be known and, accordingly, an error value can be appropriately determined.

Boundary conditions for some common types of beams are shown in Fig. 9-11. The reader should study these sufficiently to justify the boundary conditions listed.

EXAMPLE 9-9   For the uniform cantilever beam shown in Fig. 9-12, write the necessary transfer matrices and determine the natural frequencies and corresponding principal modes of vibration, using the Myklestad method.

SOLUTION   The boundary conditions are $y_0^R = 0$, $\theta_0^R = 0$, and $M_3^R = 0$, $V_3^R = 0$. Also, $M_0^R = 1$ and $V_0^R = V_0$. Assume a first trial value of $\omega^2 = 30\,000$ (rad/sec)^2. The transfer matrix expressing $\{v\}_1^R$ in terms of $\{v\}_0^R$ can now be written.

a.

$$
\left\{ \begin{array}{c} y \\ \theta \\ M \\ V \end{array} \right\}_1^R
=
\begin{bmatrix}
1 & l & \dfrac{l^2}{2B} & \dfrac{l^3}{6B} \\
0 & 1 & \dfrac{l}{B} & \dfrac{l^2}{2B} \\
0 & 0 & 1 & l \\
\omega^2 m & \omega^2 ml & \dfrac{\omega^2 ml^2}{2B} & \left(1 + \dfrac{\omega^2 ml^3}{6B}\right)
\end{bmatrix}_1
\left\{ \begin{array}{c} y \\ \theta \\ M \\ V \end{array} \right\}_0^R
$$

$$
=
\begin{bmatrix}
1 & 10 & 2.5 \times 10^{-6} & 8.3 \times 10^{-6} \\
0 & 1 & 0.5 \times 10^{-6} & 2.5 \times 10^{-6} \\
0 & 0 & 1 & 10 \\
6 \times 10^4 & 6 \times 10^5 & 0.15 & 1.5
\end{bmatrix}_1
\left\{ \begin{array}{c} 0 \\ 0 \\ M \\ V \end{array} \right\}_0^R
$$

Next, the transfer matrix relating $\{v\}_2^R$ to $\{v\}_0^R$ is expressed as

b.
$$
\begin{Bmatrix} y \\ \theta \\ M \\ V \end{Bmatrix}_2^R =
\begin{bmatrix}
1 & 10 & 2.5 \times 10^{-6} & 8.3 \times 10^{-6} \\
0 & 1 & 0.5 \times 10^{-6} & 2.5 \times 10^{-6} \\
0 & 0 & 1 & 10 \\
6 \times 10^4 & 6 \times 10^5 & 0.15 & 1.5
\end{bmatrix}_2
$$

$$
\times
\begin{bmatrix}
& & 2.5 \times 10^{-6} & 8.3 \times 10^{-6} \\
& & 0.5 \times 10^{-6} & 2.5 \times 10^{-6} \\
& & 1 & 10 \\
& & 0.15 & 1.5
\end{bmatrix}_1
\begin{Bmatrix} 0 \\ 0 \\ M \\ V \end{Bmatrix}_0^R
$$

$$
=
\begin{bmatrix}
& & 11.25 \times 10^{-6} & 70.83 \times 10^{-6} \\
& & 1.375 \times 10^{-6} & 11.25 \times 10^{-6} \\
& & 2.5 & 25 \\
& & 0.825 & 5.75
\end{bmatrix}_2
\begin{Bmatrix} 0 \\ 0 \\ M \\ V \end{Bmatrix}_0^R
$$

The elements for columns one and two of the transfer matrix are shown by vertical lines because these are not needed due to the zeros in rows one and two of state vector $\{v\}_0^R$ for station zero.

Finally, the transfer matrix which expresses $\{v\}_3^R$ in terms of $\{v\}_0^R$ is written.

c.
$$
\begin{Bmatrix} y \\ \theta \\ M \\ V \end{Bmatrix}_3 =
\begin{bmatrix}
1 & 10 & 2.5 \times 10^{-6} & 8.3 \times 10^{-6} \\
0 & 1 & 0.5 \times 10^{-6} & 2.5 \times 10^{-6} \\
0 & 0 & 1 & 10 \\
6 \times 10^4 & 6 \times 10^5 & 0.15 & 1.5
\end{bmatrix}_3
$$

$$
\times
\begin{bmatrix}
& & 11.25 \times 10^{-6} & 70.83 \times 10^{-6} \\
& & 1.375 \times 10^{-6} & 11.25 \times 10^{-6} \\
& & 2.5 & 25 \\
& & 0.825 & 5.75
\end{bmatrix}_2
\begin{Bmatrix} 0 \\ 0 \\ M \\ V \end{Bmatrix}_0^R
$$

$$
\begin{Bmatrix} y \\ \theta \\ M \\ V \end{Bmatrix}_3^R =
\begin{bmatrix}
& & 38.125 \times 10^{-6} & 293.75 \times 10^{-6} \\
& & 4.6875 \times 10^{-6} & 38.125 \times 10^{-6} \\
& & 10.75 & 82.5 \\
& & 3.1125 & 23.375
\end{bmatrix}_3
\begin{Bmatrix} 0 \\ 0 \\ M \\ V \end{Bmatrix}_0^R
$$

Setting $M_0 = 1$ and $M_3 = 0$, and writing the third algebraic equation gives

$$
0 = 10.75 + 82.5 V_0
$$

which yields

$$
V_0 = \frac{-10.75}{82.5} = -0.130
$$

Then from the fourth equation

$$V_3 = 3.1125 \times 1 + 23.375 \times (-0.130)$$

$$= 3.1125 - 3.0458 = 0.6667 = E$$

representing the error E due to the incorrect ω^2 selected.

By continuing the procedure for newly assumed ω^2 values, the following are finally obtained: For $\omega^2 = 36\,678$,

$$V_0 = -0.138\,346, \quad V_3 = 0.000\,00 = E$$

This error value is acceptable for the number of significant figures used in the calculations. The mode shape can now be determined by substituting these final M_0 and V_0 values into Eqs. (a), (b), and (c), resulting in the following:

$$y_1 = 1.347\,12 \times 10^{-6}; \quad y_2 = 1.600\,43 \times 10^{-6}; \quad y_3 = -1.061\,58 \times 10^{-6}$$

Normalizing these to y_3 gives

$$y_1 = -1.268\,98; \quad y_2 = -1.507\,59; \quad y_3 = 1$$

This represents the second mode of vibration.

Similarly, the first and third modes may be found. In this manner the following results were obtained:

For $\omega^2 = 855.5$,

$$V_0 = -0.040\,005, \quad V_3 = 0.000\,001 = E$$

$$y_1 = 0.158\,423, \quad y_2 = 0.531\,649, \quad y_3 = 1$$

These are deemed to be acceptable, and represent the first mode.

For $\omega^2 = 264\,800$,

$$V_0 = -0.208\,483, \quad V_3 = 0.000\,278 = E$$

$$y_1 = 4.461\,531, \quad y_2 = -3.237\,614, \quad y_3 = 1$$

These are acceptable for the third mode.

PROBLEMS

9-1. Determine the Rayleigh frequency of the first mode for the stretched-cord system shown, assuming the configuration to be the same as for static equilibrium.

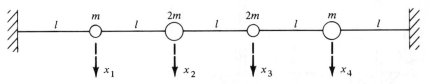

Problem 9-1   Tension in cord = Q

9-2. Determine the first-mode frequency by the Rayleigh method. Assume the configuration to be the same as the static equilibrium shape due to torsional loads proportional to the moments of inertia of the disks. (Suggestion: Check the result against the solution for Prob. 8-8.)

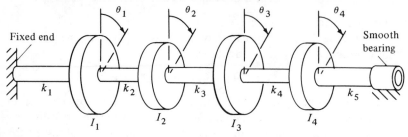

$$(k_1 = k_2 = k_3 = k_4 = k_5 = k, \; I_1 = I_2 = I_3 = I_4 = I)$$

Problem 9-2

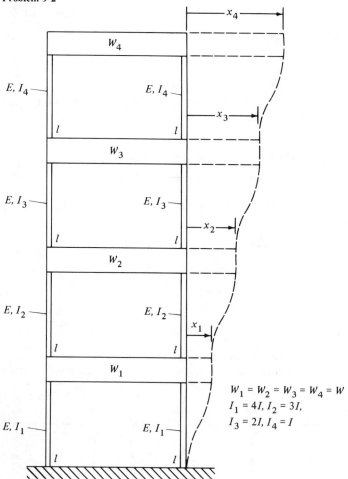

$W_1 = W_2 = W_3 = W_4 = W$
$I_1 = 4I, I_2 = 3I,$
$I_3 = 2I, I_4 = I$

Problem 9-3

9-3. Calculate the Rayleigh frequency for the building frame shown. Consider the column ends to be constrained against turning. Assume the first-mode configuration to be the same as the static-equilibrium shape due to loads proportional to the girder weights.

9-4. Determine the Rayleigh frequency for the uniform, weightless, cantilever beam and lumped-mass system shown.

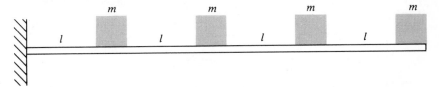

Problem 9-4

9-5. Calculate the Rayleigh frequency for the uniform, weightless, simply supported beam and lumped-mass system shown.

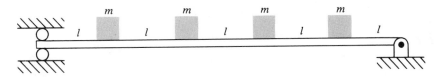

Problem 9-5

9-6. For a certain lumped system, the stiffness and mass matrices are given as

$$[k] = k \begin{bmatrix} 1 & -1 & 0 \\ -1 & 2 & -1 \\ 0 & -1 & 1 \end{bmatrix} \quad [m] = m \begin{bmatrix} 1 & 0 & 0 \\ 0 & 3 & 0 \\ 0 & 0 & 2 \end{bmatrix}$$

Using Holzer's method, determine a principal mode and frequency.

9-7. Using the Holzer method, calculate the principal modes and natural frequencies for the system of Prob. 9-6.

9-8. For the torsional system shown, determine a principal mode and frequency by the Holzer procedure. Take $k_1 = k_3 = k_5 = k$, $k_2 = 2k$, $k_4 = 3k$, and $I_1 = I_2 = I_3 = I_4 = I$.

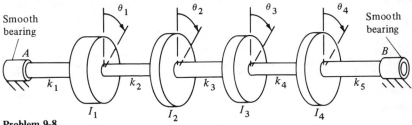

Problem 9-8

9-9. Using the Holzer method, calculate the principal modes and natural frequencies for the torsional system of Prob. 9-8.

9-10. Using the Holzer method, calculate a principal mode and natural frequency for the torsional system of Prob. 9-2.

9-11. Using the Holzer method, calculate the principal modes and corresponding natural frequencies for the torsional system of Prob. 9-2.

9-12. Use the Holzer method to calculate a principal mode and natural frequency for the building frame shown. The column ends are constrained against rotation.

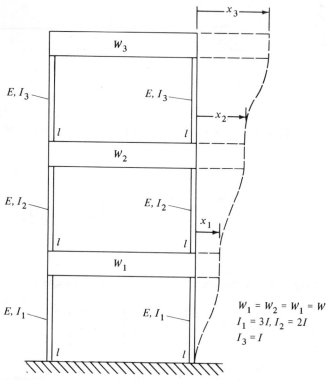

Problem 9-12

9-13. Use the Holzer method to calculate the principal modes and corresponding natural frequencies for the building frame of Prob. 9-12. The column ends are constrained against turning.

9-14. For the system shown, determine the first mode and frequency using the iteration procedure. (Refer to Prob. 8-24 for the influence coefficients, and so on.)

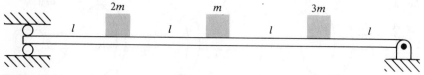

Problem 9-14

9-15. Using the iteration procedure, determine all principal modes and frequencies for the system of Prob. 9-14.

9-16. Determine the flexibility coefficients and write the matrix form of the displacement equations for the stretched-cord system shown.

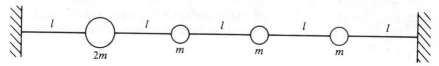

Problem 9-16   Tension in cord = Q

9-17. Obtain the fundamental mode and frequency by the iteration procedure for the
system of Prob. 9-16.

9-18. Determine the second mode and frequency by the iteration procedure for the
system of Prob. 9-16.

9-19. Determine the flexibility influence coefficients and write the matrix form of the
displacement equations for the system of Prob. 9-4.

9-20. For the system of Prob. 9-19, obtain the fundamental mode and frequency by
the iteration procedure.

9-21. Determine the second mode and frequency by the iteration procedure for the
system of Prob. 9-19.

Problems 9-22 through 9-25 are to be solved by the Myklestad method. Assume
that $m = 2$ lb sec²/in., $l = 10$ in., $E = 16 \times 10^6$ lb/in.², and $I = 1.25$ in.⁴

9-22. Obtain the natural frequencies and the corresponding principal modes for the
uniform, weightless, cantilever beam shown.

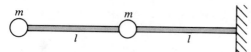

Problem 9-22

9-23. Determine the symmetric mode and the corresponding frequency for the uniform, weightless, free-free beam shown.

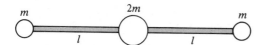

Problem 9-23

9-24. Determine the antisymmetric mode and the corresponding frequency for the
uniform, weightless, free-free beam of Prob. 9-23.

9-25. Solve for the first mode and frequency of vibration for the uniform, weightless,
simply supported beam shown.

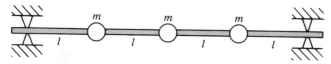

Problem 9-25

9-26. Using the Myklestad method, obtain the expression for the natural frequency of vibration for a weightless, simply supported beam with a concentrated mass at the center.

9-27. Using the Myklestad method, obtain the expression for the natural frequency of vibration for a weightless cantilever beam with a concentrated mass at the free end.

chapter
10

Vibration of
Continuous Systems

10-1. INTRODUCTION

So far the systems analyzed have consisted of separate mass and elastic members. Such discrete systems form a large part of the arrangements occurring in machines and structures which are investigated with respect to vibrations. However, elastic bodies also serve as machine and structural components and, accordingly, it is desirable to determine their vibration characteristics. In elastic bodies the mass and elasticity are continuously distributed. Members of this type include stretched cords or cables, shafts or rods with torsional freedom, bars with longitudinal freedom, and beams or bars which can oscillate laterally. In analyzing such members it is assumed that the material is homogeneous and isotropic and that it follows Hooke's law.

For discrete systems, the number of generalized coordinates needed, the number of degrees of freedom, and the number of principal vibration modes were finite, but in distributed systems these all become infinite. The location coordinate then becomes a variable, and the free-body diagram used is for a differential element. The governing relation obtained is a partial differential equation. This changes the solution procedure somewhat, although many of the techniques are, in reality, extensions of those for the multidegree problem.

The following section starts with the analysis of the stretched-cord elastic member. This is done because it is simpler to visualize conditions, displacements, motion, and so on, for this case, and these concepts carry over to other members that are more important from an engineering viewpoint.

10-2. THE STRETCHED ELASTIC CORD

Consider the case of a tightly stretched elastic cord or cable which is free to vibrate in the x-y plane, as shown in Fig. 10-1(a). Note that the coordinate y is a function of both x and t. Thus

$$y = y(x, t) \tag{10-1}$$

Accordingly, parameters such as slope, velocity, and acceleration will have to be designated by partial derivatives; that is, slope $= \partial y/\partial x$, velocity $= \partial y/\partial t$, and acceleration $= \partial^2 y/\partial t^2$. The displacement and slope of the cord are to be considered as small, so that the tension F does not change along the member, and the sine and tangent of the cord angle may be assumed to be equal. The dynamic free-body diagram of a differential element of the string is shown in Fig. 10-1(b). For the element length dx, the change in slope is defined by $(\partial^2 y/\partial x^2)\, dx$, as shown. The mass of the element is $dm = \gamma\, dx$, where γ is the mass of the cord per unit length. By considering the vertical components of the string tension, the differential equation of motion for the element is expressed as

$$F\left(\frac{\partial y}{\partial x} + \frac{\partial^2 y}{\partial x^2}\, dx\right) - F\frac{\partial y}{\partial x} = (\gamma\, dx)\frac{\partial^2 y}{\partial t^2} \tag{10-2}$$

whence

$$\frac{\partial^2 y}{\partial x^2} = \frac{1}{c^2}\frac{\partial^2 y}{\partial t^2}$$

or

$$\frac{\partial^2 y}{\partial x^2} - \frac{1}{c^2}\frac{\partial^2 y}{\partial t^2} = 0 \tag{10-3}$$

where $c = \sqrt{F/\gamma}$. The solution to this is taken to have the form

$$y = XT \tag{10-4}$$

where $X = X(x)$ and $T = T(t)$. This assumes that there are no mixed x, t terms in function y, or that the independent variables x and t are separable. The physical meaning of the product form XT for y is that X defines the configuration of the cord and T specifies its motion. The form of the cord is always the same, and all points move in step with each other. Substituting the assumed solution into the differential equation results in

$$X''T - \frac{1}{c^2}T''X = 0 \tag{10-5}$$

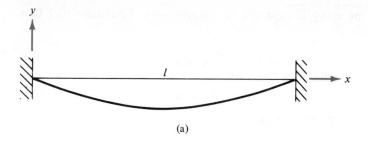

(a)

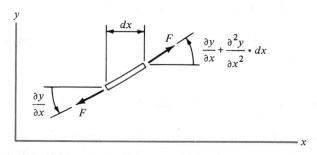

Figure 10-1 (b)

where $X'' = d^2 X/dx^2$ and $T'' = d^2 T/dt^2$. These are expressed as ordinary derivatives, since X is a function of x only, and T is a function of t only. Equation 10-5 can be rearranged as

$$c^2 \frac{X''}{X} = \frac{T''}{T} = \mu \qquad (10\text{-}6)$$

where μ is a constant. That is, the ratios X''/X and T''/T are proportional for all values of x and t. For this to hold, the ratios must be constant. Hence, from Eq. 10-6, the following can be obtained:

$$X'' - \frac{\mu}{c^2} X = 0$$

$$T'' - \mu T = 0 \qquad (10\text{-}7)$$

The solution to these depends on whether (a) $\mu > 0$, (b) $\mu = 0$, or (c) $\mu < 0$. If $\mu > 0$, the solution is hyperbolic, and if $\mu = 0$, the solution is a linear algebraic function. These will satisfy the differential equation (Eq. 10-3) but not the boundary conditions, as specified. Accordingly, they are inadmissible as solutions here.

On the other hand, case (c) does lead to a proper solution. Since μ is negative, it is convenient to set

$$\mu = -\omega^2 \qquad (10\text{-}8)$$

The reason for using ω here will be evident soon. Equations 10-7 then become

$$X'' + \left(\frac{\omega}{c}\right)^2 X = 0$$

$$T'' + \omega^2 T = 0 \tag{10-9}$$

and the solutions for these would be

$$X = A \sin \frac{\omega}{c} x + B \cos \frac{\omega}{c} x$$

$$T = C_1 \sin \omega t + C_2 \cos \omega t \tag{10-10}$$

so that

$$y = \left(A \sin \frac{\omega}{c} x + B \cos \frac{\omega}{c} x\right)(C_1 \sin \omega t + C_2 \cos \omega t) \tag{10-11}$$

where A, B, C_1, and C_2 are arbitrary constants, and ω is a circular frequency which is yet to be evaluated. The first part of the expression defines the form of the cord; the second part specifies its movement.

The constants of the solution can be evaluated from the available boundary conditions for the cord, which will be independent of time, and from the conditions of motion at a specified time. If the stretched cord is fastened to fixed points at the ends, then the boundary conditions are $y = 0$ at $x = 0$ and $y = 0$ at $x = l$, and these must hold for all values of t. Applying these to the solution leads to

$$0 = B(C_1 \sin \omega t + C_2 \cos \omega t) \tag{10-12}$$

and

$$0 = \left(A \sin \frac{\omega}{c} l + B \cos \frac{\omega}{c} l\right)(C_1 \sin \omega t + C_2 \cos \omega t) \tag{10-13}$$

Since these must hold for any value of t, then

$$B = 0 \quad \text{and} \quad A \sin \frac{\omega}{c} l = 0 \tag{10-14}$$

Because A cannot vanish,

$$\sin \frac{\omega}{c} l = 0 \tag{10-15}$$

from which

$$\frac{\omega}{c} l = \pi, \, 2\pi, \, 3\pi, \, \ldots, \, n\pi, \, \ldots \tag{10-16}$$

The frequency is then given by

$$\omega = \frac{n\pi}{l}c = \frac{n\pi}{l}\sqrt{\frac{F}{\gamma}} \tag{10-17}$$

where $n = 1, 2, 3, \dots$ represents the order of the mode. The solution can then be written as

$$y = \sin\frac{\omega}{c}x(C_1' \sin\omega t + C_2' \cos\omega t) \tag{10-18}$$

where $C_1' = AC_1$ and $C_2' = AC_2$.

The velocity is defined by the partial derivative with respect to time,

$$\frac{\partial y}{\partial t} = \sin\frac{\omega}{c}x(C_1' \cos\omega t - C_2' \sin\omega t)\omega \tag{10-19}$$

The initial conditions of motion may now be specified. For example, if no motion (that is, zero velocity everywhere) and a principal-mode configuration occur at $t = 0$, then setting $\partial y/\partial t = 0$ results in $C_1' = 0$ and

$$y = C_2' \sin\frac{\omega}{c}x \cos\omega t \tag{10-20}$$

Now setting $t = 0$ yields

$$y_{t=0} = C_2' \sin\frac{\omega}{c}x \tag{10-21}$$

which defines the principal-mode shape for the initial condition. The constant C_2' may be replaced by y_0, representing the maximum value of y. The solution then is

$$y = y_0 \sin\frac{\omega}{c}x \cos\omega t$$

$$= y_0 \sin\frac{n\pi}{l}x \cos\frac{n\pi}{l}\sqrt{\frac{F}{\gamma}}t \tag{10-22}$$

where $n = 1, 2, 3, \dots$. The configuration is defined by $y_0 \sin(n\pi/l)x$, and the manner in which the motion takes place is given by $\cos(n\pi/l)\sqrt{F/\gamma}\,t$. The modal form of the cord is defined by $t = 0$, and the motion alternates between this and the inverse configuration occurring one-half cycle later. The ratio of the amplitude of any two points is always the same. The entire cord goes through the equilibrium position simultaneously. There is no apparent progression; the cord simply appears to move up and down. Such motion is called a *standing wave*. The length of the wave is defined by the length of one sine wave for $\sin(n\pi/l)x$; that is, for

$$\frac{n\pi x_\lambda}{l} = 2\pi$$

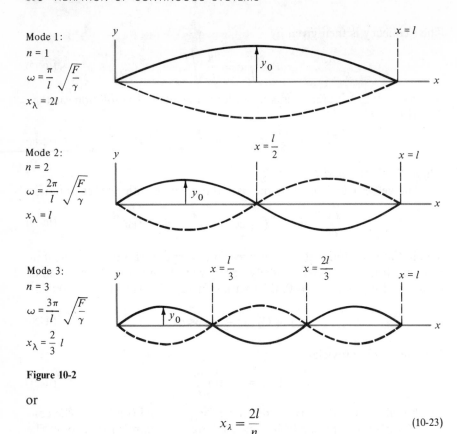

Mode 1:
$n = 1$
$\omega = \dfrac{\pi}{l}\sqrt{\dfrac{F}{\gamma}}$
$x_\lambda = 2l$

Mode 2:
$n = 2$
$\omega = \dfrac{2\pi}{l}\sqrt{\dfrac{F}{\gamma}}$
$x_\lambda = l$

Mode 3:
$n = 3$
$\omega = \dfrac{3\pi}{l}\sqrt{\dfrac{F}{\gamma}}$
$x_\lambda = \dfrac{2}{3}l$

Figure 10-2

or

$$x_\lambda = \frac{2l}{n} \qquad (10\text{-}23)$$

where x_λ is the wavelength. The first three principal modes of vibration are plotted in Fig. 10-2.

If the initial conditions had been a principal-mode velocity distribution and zero displacement, then the solution becomes

$$y = y_0 \sin \frac{\omega}{c} x \sin \omega t \qquad (10\text{-}24)$$

For initial conditions consisting of both principal-mode motion (velocity) and displacement, the solution is

$$y = y_0 \sin \frac{\omega}{c} x \sin (\omega t + \phi) \qquad (10\text{-}25)$$

In the solutions expressed by Eqs. 10-24 and 10-25, ω and c are defined by Eq. 10-17.

It is important to note the initial conditions required to produce a principal mode for the distributed system. The motion represented by Eq. 10-22, described above and shown in Fig. 10-2, would require that the elastic cord be given the exact sinusoidal form which is defined by $t = 0$. This would be difficult to invoke. Any other initial conditions (such as a sinusoidal velocity distribution and zero displacements) which would result in a principal mode would be equally difficult to arrange. In general, initial conditions are much simpler, and the resulting motion is a nonharmonic motion of nonharmonic form, composed of harmonic parts. (The manner of determining the solution for such general initial conditions is explained in Section 10-7.)

A comparison of the solution and the motion here with those for a lumped-mass system reveals much similarity, the main difference being in the number of degrees of freedom. For the discrete mass system, in any principal mode each mass oscillates harmonically. For the distributed mass-elastic case, any principal mode consists of harmonic motion of each point mass of the system.

10-3. TRAVELING-WAVE SOLUTION

The standing-wave motion may be represented as two traveling half-amplitude waves which move in opposite directions with equal velocities. In order to show this the solution expressed by Eq. 10-22 may be written as

$$y = \frac{y_0}{2} \left[\left(\sin \frac{\omega}{c} x \cos \omega t + \cos \frac{\omega}{c} x \sin \omega t \right) \right.$$

$$\left. + \left(\sin \frac{\omega}{c} x \cos \omega t - \cos \frac{\omega}{c} x \sin \omega t \right) \right] \qquad (10\text{-}26)$$

$$= \frac{y_0}{2} \left[\sin \frac{\omega}{c} (x + ct) + \sin \frac{\omega}{c} (x - ct) \right] \qquad (10\text{-}27)$$

From analytic geometry, in each sine function here, ct represents a shifting term. Thus for $\sin (\omega/c)(x + ct)$, the reference for x is shifted to the left by the amount ct. This shift is proportional to the time t. Hence, the $\sin (\omega/c) \times (x + ct)$ curve is moving to the left. The velocity of such movement is $d(-ct)/dt = -c$. In a similar manner the other sine function, $\sin (\omega/c) \times (x - ct)$, is moving to the right at a rate of c. The amplitude of both moving sine curves is $y_0/2$. Such moving curves are called *traveling waves*. The described traveling-wave motion can be readily visualized.

This is a particular case of the general wave solution

$$y = y_1(x + ct) + y_2(x - ct) \qquad (10\text{-}28)$$

for the differential equation (Eq. 10-3). The function y_1 is an undefined function of $(x + ct)$ and y_2 is an unspecified function of $(x - ct)$. These functions are arbitrary and not necessarily of the same form. If the forms differ in shape or magnitude, the two oppositely traveling waves which they represent do not add up to a standing wave but, instead, to a net wave which also travels.

To show that this function is a proper solution to Eq. 10-3, consider its partial derivatives

$$\frac{\partial^2 y}{\partial x^2} = y_1''(x + ct) + y_2''(x - ct)$$

$$\frac{\partial^2 y}{\partial t^2} = c^2 y_1''(x + ct) + c^2 y_2''(x - ct) \tag{10-29}$$

where the double prime represents the second derivative of the function. Substitution of these into the wave equation (Eq. 10-3) reveals that the differential equation is satisfied.

10-4. OTHER BOUNDARY CONDITIONS

A different boundary condition than that previously described can prevail at the cable end—the cable can be fastened to a smooth pin or bearing which is free to move in a smooth slot or rail arranged perpendicular to the cable, as shown in Fig. 10-3. (If both ends are arranged in this manner, then the figure should be considered as a top view, so that the cable is free to move only in a horizontal plane.) This type of end will not support a component of force at right angles to the cable and hence the boundary condition is defined by

$$F \frac{\partial y}{\partial x} = 0 \tag{10-30}$$

so that $\partial y / \partial x = 0$. Such a boundary condition leads to a different form for the particular solution to the differential equation.

EXAMPLE 10-1   Obtain expressions which define the principal modes and natural frequencies for a tightly stretched cable having ends that move freely in slots at 90 degrees to the direction of the cable. Figure 10-4 represents a top view of the arrangement.

SOLUTION   The general solution given by Eq. 10-11 is

a. $$y = \left(A \sin \frac{\omega}{c} x + B \cos \frac{\omega}{c} x \right)(C_1 \sin \omega t + C_2 \cos \omega t)$$

from which

b. $$\frac{\partial y}{\partial x} = \frac{\omega}{c} \left(A \cos \frac{\omega}{c} x - B \sin \frac{\omega}{c} x \right)(C_1 \sin \omega t + C_2 \cos \omega t)$$

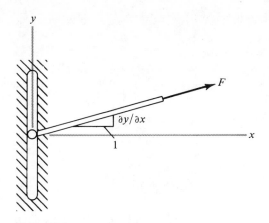

Figure 10-3

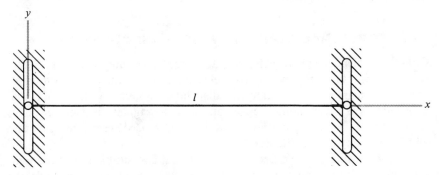

Figure 10-4

For a smooth slot, there can be no y component of force, hence at the ends the slope is zero. Thus

$$\frac{\partial y}{\partial x} = 0$$

at $x = 0$ and $x = l$, which for $x = 0$ leads to

$$0 = \frac{\omega}{c} A(C_1 \sin \omega t + C_2 \cos \omega t)$$

so that $A = 0$. For $x = l$, the result is

$$0 = \frac{\omega}{c}\left(-B \sin \frac{\omega}{c} l\right)(C_1 \sin \omega t + C_2 \cos \omega t)$$

which requires that

$$\sin \frac{\omega}{c} l = 0$$

whence

$$\frac{\omega}{c} l = \pi, 2\pi, 3\pi, \ldots, n\pi$$

where n is the mode number. The solution is then

$$y = \cos \frac{\omega}{c} x (C'_1 \sin \omega t + C'_2 \cos \omega t)$$

with

$$\omega = n\pi \frac{c}{l} \qquad n = 1, 2, 3, \ldots, n$$

The principal-mode configurations are defined by $\cos (\omega/c)x$ and thus have the form of a cosine curve.

10-5. TORSIONAL VIBRATION OF A CIRCULAR ROD

Consider a slender cylindrical rod that is free to vibrate in torsion. Since there are an infinite number of mass-element sections, there are an infinite number of degrees of freedom for the distributed system. In Fig. 10-5(a) the coordinate x is taken along the rod, and θ is the angular displacement of a section. Since $\theta = \theta(x, t)$, partial derivatives will be needed for expressing the angular velocity and acceleration.

Before setting up the differential equation, it is desirable to note certain relations for the torsion problem. For static elastic conditions, the angle-of-twist relation for a shaft of circular section is

$$\Delta\theta = \frac{M_t \, \Delta l}{GJ} \qquad \text{or} \qquad \frac{\Delta\theta}{\Delta l} = \frac{M_t}{GJ}$$

where $\Delta\theta$ is the angle of twist for two sections Δl distance apart, M_t is the torque or twisting moment acting on the shaft, G is the modulus of rigidity for the rod material, and J is the polar moment of inertia for the circular section. For the problem here, where θ is a function of t as well as x, the relation would be written as

$$\frac{\partial\theta}{\partial x} = \frac{M_t}{GJ} \tag{10-31}$$

The partial derivative of this gives

$$\frac{\partial M_t}{\partial x} = GJ \frac{\partial^2\theta}{\partial x^2} \tag{10-32}$$

Next, consider the member of uniform thickness b, shown in Fig. 10-6. The x axis is perpendicular to the surfaces of this member, and r is the variable

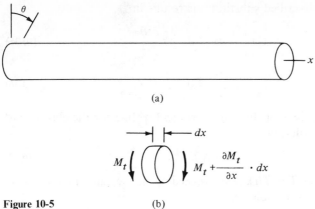

(a)

(b)

Figure 10-5

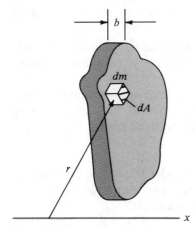

Figure 10-6

distance from the x axis to the differential mass element dm of the member. Then I, the mass moment of inertia with respect to the x axis for the member, is defined by

$$I = \int r^2 \, dm = \int r^2 \, dA \, b\rho = b\rho \int r^2 \, dA = b\rho J \qquad (10\text{-}33)$$

where dA is the surface area of the element in the plane of the section, ρ is the mass per unit volume, and J is the polar moment of the section area with respect to the x axis.

Returning now to the free-body diagram for the differential element shown in Fig. 10-5(b), the differential equation for rotation motion of the element would be

$$\left(M_t + \frac{\partial M_t}{\partial x} \, dx \right) - M_t = I \frac{\partial^2 \theta}{\partial t^2} \qquad (10\text{-}34)$$

Making the above-described substitutions results in

$$GJ \frac{\partial^2 \theta}{\partial x^2} dx = dx \, \rho J \frac{\partial^2 \theta}{\partial t^2} \tag{10-35}$$

$$\frac{\partial^2 \theta}{\partial x^2} - \frac{1}{c^2} \frac{\partial^2 \theta}{\partial t^2} = 0 \tag{10-36}$$

where $c = \sqrt{G/\rho}$. This is of the same form as Eq. 10-3 for the elastic-cord case. Assuming the solution

$$\theta = XT \tag{10-37}$$

where $X = X(x)$ and $T = T(t)$, and proceeding in the same manner as in Section 10-2 results in the solution

$$\theta = \left(A \sin \frac{\omega}{c} x + B \cos \frac{\omega}{c} x \right) (C_1 \sin \omega t + C_2 \cos \omega t) \tag{10-38}$$

In this case also, the hyperbolic and linear forms of the solutions are inadmissible, except as cases corresponding to boundary conditions for situations of little importance.

Either end of the member may be free or fixed against rotation. For a free end, the moment M_t must be zero so that from Eq. 10-31 this boundary condition is specified as $\partial\theta/\partial x = 0$. For an end fixed against rotation, the boundary condition is $\theta = 0$.

EXAMPLE 10-2   Determine the principal modes of torsional vibration for a free-free rod (both ends free), including expressions for the natural frequencies. Show figures for the first three principal modes.

SOLUTION   The boundary conditions will be

$$\frac{\partial \theta}{\partial x} = 0 \quad \text{at} \quad x = 0 \quad \text{and} \quad x = l$$

which holds for all values of t. Taking the partial derivative of Eq. 10-38 with respect to x gives

a. $$\frac{\partial \theta}{\partial x} = \frac{\omega}{c} \left(A \cos \frac{\omega}{c} x - B \sin \frac{\omega}{c} x \right) (C_1 \sin \omega t + C_2 \cos \omega t)$$

Then at $x = 0$,

b. $$\frac{\omega}{c} A = 0 \quad \text{and} \quad A = 0$$

For $x = l$,

c. $$\frac{\omega}{c} B \sin \frac{\omega}{c} l = 0 \quad \text{and} \quad \sin \frac{\omega}{c} l = 0$$

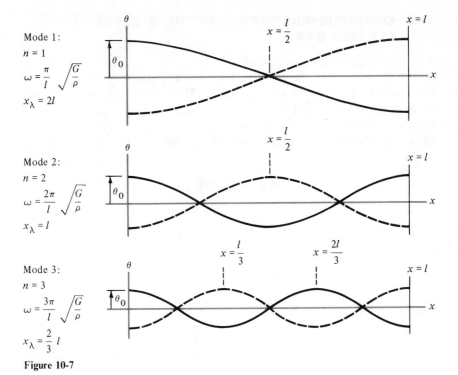

Mode 1:

$n = 1$

$\omega = \dfrac{\pi}{l} \sqrt{\dfrac{G}{\rho}}$

$x_\lambda = 2l$

Mode 2:

$n = 2$

$\omega = \dfrac{2\pi}{l} \sqrt{\dfrac{G}{\rho}}$

$x_\lambda = l$

Mode 3:

$n = 3$

$\omega = \dfrac{3\pi}{l} \sqrt{\dfrac{G}{\rho}}$

$x_\lambda = \dfrac{2}{3} l$

Figure 10-7

whence

d.
$$\frac{\omega}{c} l = \pi, \; 2\pi, \; 3\pi, \; \ldots, \; n\pi, \; \ldots$$

so that

e.
$$\omega = \frac{n\pi}{l} c = \frac{n\pi}{l} \sqrt{\frac{G}{\rho}}$$

where $n = 1, 2, 3, \ldots$ gives the order of the mode.

If the rod has a principal mode configuration but no motion at $t = 0$, the solution becomes

f.
$$\theta = \theta_0 \cos \frac{\omega}{c} x \cos \omega t$$

$$= \theta_0 \cos \frac{n\pi}{l} x \cos \frac{n\pi}{l} \sqrt{\frac{G}{\rho}} t$$

where θ_0 is the maximum angular displacement. The first part defines the configuration of the rod for a principal mode. The wavelength can be verified as $x_\lambda = 2l/n$.

The solution should be studied with a view to visualizing the configurations and motion for the principal modes. The first three principal modes are shown in Fig. 10-7.

10-6. LONGITUDINAL VIBRATION OF A SLENDER ELASTIC BAR

A bar is capable of oscillating longitudinally. For a uniform slender bar it can be assumed that the sections will remain plane in such vibrations, provided elastic conditions are maintained. The deformation and strain can be defined from Fig. 10-8, where u represents the displacement of a plane section having the equilibrium position defined by coordinate x, δ is total strain or deformation, and ε is unit strain. Thus for the element having length dx,

$$\delta = \left(u + \frac{\partial u}{\partial x} dx \right) - u$$

$$= \frac{\partial u}{\partial x} dx$$

and

$$\varepsilon = \frac{\delta}{dx} = \frac{\partial u}{\partial x} \tag{10-39}$$

Then from Hooke's law, the unit stress σ is

$$\sigma = E\varepsilon = E \frac{\partial u}{\partial x} \tag{10-40}$$

where E is the tension-compression modulus of elasticity for the material. The force P on the section is

$$P = A\sigma = AE \frac{\partial u}{\partial x} \tag{10-41}$$

where A is the cross-sectional area of the bar. Then

$$\frac{\partial P}{\partial x} = AE \frac{\partial^2 u}{\partial x^2} \tag{10-42}$$

The free-body diagram of the mass element dm is shown in Fig. 10-9. From this, the differential equation of motion is

$$\left(P + \frac{\partial P}{\partial x} dx \right) - P = \rho A \, dx \frac{\partial^2 u}{\partial t^2} \tag{10-43}$$

where ρ is the mass per unit volume for the material. By substituting in Eq. 10-42 and simplifying,

$$E \frac{\partial^2 u}{\partial x^2} = \rho \frac{\partial^2 u}{\partial t^2}$$

$$\frac{\partial^2 u}{\partial x^2} - \frac{1}{c^2} \frac{\partial^2 u}{\partial t^2} = 0 \tag{10-44}$$

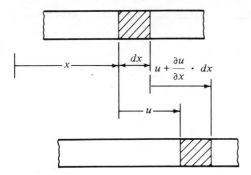

Figure 10-8

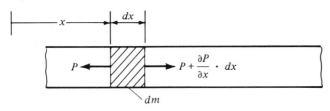

Figure 10-9

where $c = \sqrt{E/\rho}$. Since this is of the same form as Eq. 10-3, the differential equation can be solved by taking

$$u = XT \tag{10-45}$$

where $X = X(x)$ and $T = T(t)$. After discarding the inadmissible cases, the solution will be

$$u = \left(A \sin \frac{\omega}{c} x + B \cos \frac{\omega}{c} x \right)(C_1 \sin \omega t + C_2 \cos \omega t) \tag{10-46}$$

Either end may be free or fixed. For a free end, $P = 0$ and by Eq. 10-41 this boundary condition becomes $\partial u/\partial x = 0$. For a fixed end, the motion is restrained and the boundary condition is $u = 0$.

EXAMPLE 10-3   Obtain relations which define the principal modes and natural frequencies of vibration for a uniform slender bar having one end free and the other end fixed. Show the form of the first three principal modes.

SOLUTION   The boundary conditions can be set as

$$u = 0 \qquad \text{at } x = 0$$

for the fixed end;

$$\frac{\partial u}{\partial x} = 0 \qquad \text{at } x = l$$

for the free end. These will hold for all values of time t. The first condition leads to

a.
$$B = 0$$

From

$$\frac{\partial u}{\partial x} = \frac{\omega}{c}\left(A \cos \frac{\omega}{c}x - B \sin \frac{\omega}{c}x\right)(C_1 \sin \omega t + C_2 \cos \omega t)$$

the second condition results in

b.
$$\frac{\omega}{c}A \cos \frac{\omega}{c}l = 0 \qquad \text{and} \qquad \cos \frac{\omega}{c}l = 0$$

so that

c.
$$\frac{\omega}{c}l = \frac{\pi}{2}, \frac{3\pi}{2}, \frac{5\pi}{2}, \ldots, \left(n - \frac{1}{2}\right)\pi, \ldots$$

or

d.
$$\omega = \left(n - \frac{1}{2}\right)\pi\frac{c}{l} = \left(n - \frac{1}{2}\right)\frac{\pi}{l}\sqrt{\frac{E}{\rho}}$$

where $n = 1, 2, 3, \ldots$ gives the order of the mode.

Assuming for initial conditions that the velocity of all sections is zero at $t = 0$ and that a principal-mode configuration occurs then, the solution becomes

e.
$$u = u_0 \sin \frac{\omega}{c}x \cos \omega t$$

$$= u_0 \sin \left[\left(n - \frac{1}{2}\right)\pi\frac{x}{l}\right] \cos \left[\left(n - \frac{1}{2}\right)\frac{\pi}{l}\sqrt{\frac{E}{\rho}}t\right]$$

where u_0 represents the maximum displacement that occurs.

The sine part defines the configuration of the bar in a principal-mode vibration. One cycle of this function specifies the wavelength x_λ as

f.
$$\left(n - \frac{1}{2}\right)\pi\frac{x_\lambda}{l} = 2\pi$$

$$x_\lambda = \frac{2l}{n - 1/2} = \frac{4l}{2n - 1}$$

The first three principal modes are represented in Fig. 10-10. The configuration and motion for the principal-mode vibrations can be visualized without difficulty. It should be kept in mind that the displacements occur along the direction of the member.

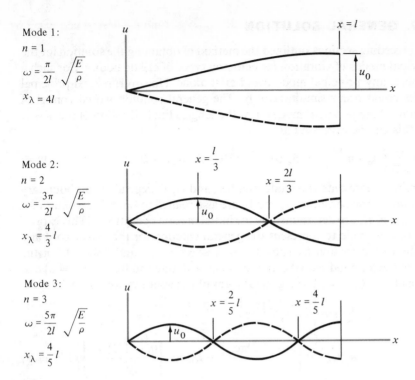

Mode 1:
$n = 1$
$$\omega = \frac{\pi}{2l}\sqrt{\frac{E}{\rho}}$$
$x_\lambda = 4l$

Mode 2:
$n = 2$
$$\omega = \frac{3\pi}{2l}\sqrt{\frac{E}{\rho}}$$
$x_\lambda = \frac{4}{3}l$

Mode 3:
$n = 3$
$$\omega = \frac{5\pi}{2l}\sqrt{\frac{E}{\rho}}$$
$x_\lambda = \frac{4}{5}l$

Figure 10-10

The order of magnitude of the frequency at which these vibrations take place is of interest here. This is indicated by the illustrative problem that follows.

EXAMPLE 10-4 Determine the natural frequency of the first three modes of longitudinal vibration of a slender aluminum-alloy bar 0.5 in. square and 100 in. long. The bar is fixed at the upper end and is free at the lower end. The material has a unit weight of 0.100 lb/in.³, and $E = 10^7$ lb/in.².

SOLUTION

$$f_1 = \frac{\omega_1}{2\pi} = \frac{1}{4l}\sqrt{\frac{E}{\rho}} = \frac{1}{4 \times 100}\sqrt{\frac{386}{0.1} \times 10^7} = \frac{19.647 \times 10^4}{400} = 491.2 \text{ Hz}$$

$$f_2 = \frac{\omega_2}{2\pi} = \frac{3}{4l}\sqrt{\frac{E}{\rho}} = 3f_1 = 1474 \text{ Hz}$$

$$f_3 = \frac{\omega_3}{2\pi} = \frac{5}{4l}\sqrt{\frac{E}{\rho}} = 5f_1 = 2456 \text{ Hz}$$

10-7. GENERAL SOLUTION

The preceding sections outlined the method of obtaining the solution for the principal modes of vibration for common types of elastic bodies. For such a member, any principal mode could exist alone. However, several principal modes could occur simultaneously. The general solution would consist of the sum of the principal modes. For the longitudinal vibration of the elastic bar, this can be expressed as

$$u = \sum_{n=1}^{\infty} \left(A_n \sin \frac{\omega_n}{c} x + B_n \cos \frac{\omega_n}{c} x \right) (C_n \sin \omega_n t + D_n \cos \omega_n t) \tag{10-47}$$

wherein n represents the mode number, and ω_n is dependent on boundary conditions. The constants A_n and B_n are also related to boundary conditions, and C_n and D_n are determined by the initial motion conditions. (This expression can be made to represent the general solution for the elastic cord and for the rod in torsion by substituting, respectively, y and θ for u.) For the case of the longitudinal vibration of a bar with one end fixed (at $x = 0$) and one end free (at $x = l$), the general form of the solution would be

$$u = \sum_{n=1}^{\infty} \sin \left[\frac{(n - 1/2)\pi}{l} x \right]$$

$$\times \left\{ C_n \sin \left[\frac{(n - 1/2)\pi c}{l} t \right] + D_n \cos \left[\frac{(n - 1/2)\pi c}{l} t \right] \right\} \tag{10-48}$$

The constants have been modified so that C_n and D_n now contain A_n.

Since the general form of the solution is expressed by trigonometric series, the evaluation of the constants is somewhat involved. The methods of Section 5-2 for determining Fourier coefficients pertain here. For $t = 0$, if the displacements are given by $u_{t=0} = U(x) = U$ and the velocities by $\dot{u}_{t=0} = U'(x) = U'$, then for the case of a bar with one end free and one end fixed, substituting $t = 0$ into Eq. 10-48 results in

$$U = \sum_{n=1}^{\infty} D_n \sin \left[\frac{(n - 1/2)\pi x}{l} \right] \tag{10-49}$$

and placing $t = 0$ into the time derivative of Eq. 10-48 gives

$$U' = \sum_{n=1}^{\infty} \frac{(n - 1/2)\pi c}{l} C_n \sin \left[\frac{(n - 1/2)\pi x}{l} \right] \tag{10-50}$$

Multiplying both sides of Eq. 10-49 by $\sin [(n - 1/2)\pi x/l] \, dx$ and integrating over the interval from $x = 0$ to $x = l$ gives

$$\int_0^l U \sin \left[\frac{(n - 1/2)\pi x}{l} \right] dx = \int_0^l D_n \sin^2 \left[\frac{(n - 1/2)\pi x}{l} \right] dx$$

$$= D_n \left\{ \frac{x}{2} - \frac{\sin \left[\frac{2(n - 1/2)\pi x}{l} \right]}{4(n - 1/2)\pi/l} \right\} \Bigg|_0^l = D_n \frac{l}{2} \tag{10-51}$$

$(n = 1, 2, 3, \ldots)$ whence

$$D_n = \frac{2}{l} \int_0^l U \sin \left[\frac{(n - 1/2)\pi x}{l} \right] dx \qquad (10\text{-}52)$$

Similar treatment of Eq. 10-50 results in

$$\int_0^l U' \sin \left[\frac{(n - 1/2)\pi x}{l} \right] dx = \int_0^l \frac{(n - 1/2)\pi c}{l} C_n \sin^2 \left[\frac{(n - 1/2)\pi x}{l} \right] dx$$

$$= \left[\frac{(n - 1/2)\pi c}{l} C_n \right] \frac{l}{2} \qquad (10\text{-}53)$$

$(n = 1, 2, 3, \ldots)$ from which

$$C_n = \frac{2}{(n - 1/2)\pi c} \int_0^l U' \sin \left[\frac{(n - 1/2)\pi}{l} x \right] dx \qquad (10\text{-}54)$$

The coefficients C_n and D_n can now be calculated from Eqs. 10-54 and 10-52, provided that U and U' are specified and the expressions are integrable.

The boundary conditions which result in the occurrence of a principal mode alone are difficult to impose and hence are unusual. Accordingly, the procedure outlined above is of more general application, in that it enables the particular solution to be written for boundary conditions which are likely to exist.

EXAMPLE 10-5   Consider the arrangement of Fig. 10-11, in which axial force P is applied to the free end of the bar and then suddenly released. Obtain the general solution for this case.

SOLUTION   The unit strain ε_0 due to P can be determined from the unit stress $\sigma_0 = P/A$ by $\varepsilon_0 = \sigma_0/E$. The initial conditions then are

a. $\qquad\qquad u = \varepsilon_0 x \qquad$ and $\qquad \dot{u} = 0 \qquad$ at $t = 0$

The second condition will result in the coefficients C_n vanishing when U' is set equal to zero in Eq. 10-54. The remaining coefficients D_n are determined by replacing U by $\varepsilon_0 x$ in Eq. 10-52 and integrating. Thus

b.
$$D_n = \frac{2}{l} \int_0^l \varepsilon_0 x \sin \left[\frac{(n - 1/2)\pi x}{l} \right] dx$$

$$= \frac{2\varepsilon_0}{l} \left\{ \frac{l^2}{(n - 1/2)^2 \pi^2} \sin \left[\frac{(n - 1/2)\pi x}{l} \right] \right.$$

$$\left. - \frac{lx}{(n - 1/2)\pi} \cos \left[\frac{(n - 1/2)\pi x}{l} \right] \right\}_0^l$$

$$= \frac{8\varepsilon_0 l}{(2n - 1)^2 \pi^2} (-1)^{n-1}$$

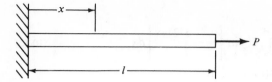

Figure 10-11

Equation 10-48 then becomes

c.
$$u = \frac{8\varepsilon_0 l}{\pi^2} \sum_{n=1}^{\infty} \left\{ \frac{(-1)^{n-1}}{(2n-1)^2} \sin\left[\frac{(n-1/2)\pi x}{l}\right] \cos\left[\frac{(n-1/2)\pi ct}{l}\right] \right\}$$

defining the free oscillatory motion for the specified conditions. The nature of the series represented by relation (c) should be carefully noted. It is a summation of the product of the terms rather than the product of the summation. In other words, expansion of the expression would be written as

d.
$$u = \frac{8\varepsilon_0 l}{\pi^2} \left[\sin\left(\frac{\pi}{2}\frac{x}{l}\right) \cos\left(\frac{\pi}{2}\frac{ct}{l}\right) - \frac{1}{9}\sin\left(\frac{3}{2}\pi\frac{x}{l}\right) \cos\left(\frac{3}{2}\pi\frac{ct}{l}\right) \right.$$
$$\left. + \frac{1}{25}\sin\left(\frac{5}{2}\pi\frac{x}{l}\right) \cos\left(\frac{5}{2}\pi\frac{ct}{l}\right) - \frac{1}{49}\sin\left(\frac{7}{2}\pi\frac{x}{l}\right) \cos\left(\frac{7}{2}\pi\frac{ct}{l}\right) + \cdots \right]$$

This should be studied sufficiently to gain an understanding of the motion for this case.

10-8. COMBINED LUMPED AND DISTRIBUTED MASS SYSTEM

Consider the system represented by Fig. 10-12, consisting of an elastic bar or a spring and a discrete mass M. The elastic member is free to oscillate longitudinally and is considered to contain a total uniformly distributed mass $m = \rho A l$, where ρ is the mass per unit volume. The term A is the cross-sectional area of the rod, but for the spring, A represents an effective area. The mass of the spring per unit length is designated by $\gamma = A\rho$. The equivalent spring constant for the bar can be obtained from the relation $\delta = Pl/AE$ for axial loading. From this,

$$k = \frac{P}{\delta} = \frac{AE}{l}$$

where k is the spring constant and E is the tension-compression modulus of elasticity. The differential equation governing the longitudinal displacement u of the rod has already been obtained, in Section 10-6, as

$$\frac{\partial^2 u}{\partial x^2} - \frac{1}{c^2}\frac{\partial^2 u}{\partial t^2} = 0 \qquad 0 \le x \le l \tag{10-55}$$

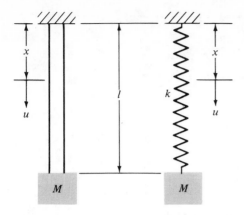

Figure 10-12

where

$$c = \sqrt{\frac{E}{\rho}} \qquad \text{for the rod}$$

and

$$c = \sqrt{\frac{E}{\rho}} = \sqrt{\frac{kl}{A\rho}} = \sqrt{\frac{kl}{\gamma}} \qquad \text{for the spring} \qquad (10\text{-}56)$$

For the fixed upper end of the elastic member, the following condition holds:

$$u = 0 \qquad \text{at } x = 0 \qquad (10\text{-}57)$$

The dynamic free-body diagram for the lumped-mass M at the lower end of the rod is shown in Fig. 10-13, where by Eq. 10-41, the tensile force in the rod is $P = AE(\partial u/\partial x)_{x=l}$. The acceleration of mass M is $(\partial^2 u/\partial t^2)_{x=l}$. Newton's second law is expressed here as

$$M \left(\frac{\partial^2 u}{\partial t^2} \right)_{x=l} = - AE \left(\frac{\partial u}{\partial x} \right)_{x=l} \qquad (10\text{-}58)$$

From Section 10-6, the solution to Eq. 10-55 is

$$u = \left(B_1 \sin \frac{\omega}{c} x + B_2 \cos \frac{\omega}{c} x \right) (C_1 \sin \omega t + C_2 \cos \omega t) \qquad (10\text{-}59)$$

The conditions listed must hold for all values of time t. Equation 10-57 requires that $B_2 = 0$, and Eq. 10-58 results in

$$M\omega^2 \sin \left(\frac{\omega}{c} l \right) = AE \frac{\omega}{c} \cos \left(\frac{\omega}{c} l \right) \qquad (10\text{-}60)$$

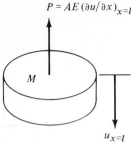

$$P = AE \left(\partial u / \partial x \right)_{x=l}$$

$$u_{x=l}$$

Figure 10-13

which may be rearranged as

$$\left(\frac{\omega l}{c} \right) \tan \left(\frac{\omega l}{c} \right) = \frac{Al}{M} \frac{E}{c^2}$$

$$= \frac{Al\rho}{M} \tag{10-61}$$

This can be expressed in the form

$$\alpha = \beta \tan \beta \tag{10-62}$$

where

$$\alpha = \frac{Al\rho}{M} = \frac{m}{M} = \frac{w}{W} \quad \text{and} \quad \beta = \frac{\omega l}{c} \tag{10-63}$$

The symbol α is the ratio of the mass or weight of the bar or spring to that of the suspended member. Equation 10-62 represents the frequency equation. For each value assigned to α, the corresponding values of β which satisfy Eq. 10-62 can be found. These values, β_1, β_2, β_3, ..., represent the frequency factors for the principal modes for the single α value. That is, for $\alpha = \alpha'$,

$$\omega_1 = \frac{c}{l} \beta'_1, \ \omega_2 = \frac{c}{l} \beta'_2, \ \omega_3 = \frac{c}{l} \beta'_3, \ \dots \tag{10-64}$$

Such a set of frequency values is determined for each value of α investigated. For a given value of α, the principal modes will then be defined by

$$u_n = \sin \left(\frac{\beta_n x}{l} \right) \left[C_{1n} \sin \left(\frac{\beta_n ct}{l} \right) + C_{2n} \cos \left(\frac{\beta_n ct}{l} \right) \right] \tag{10-65}$$

where n denotes the order of the principal mode. The general form of any free vibration can then be obtained by superposition of the principal modes and can thus be written as

$$u = \sum_{n=1}^{\infty} \left\{ \sin \left(\frac{\beta_n x}{l} \right) \left[C_{1n} \sin \left(\frac{\beta_n ct}{l} \right) + C_{2n} \cos \left(\frac{\beta_n ct}{l} \right) \right] \right\} \tag{10-66}$$

Table 10-1

$\alpha =$	0.010	0.100	0.500	1.00	2.00	5.00	10.00
$\beta_1 =$	0.100	0.311	0.653	0.860	1.077	1.314	1.429
$\alpha =$	20.0	100.0	∞				
$\beta_1 =$	1.496	1.5553	$\pi/2$				

The constants C_{1n} and C_{2n} are determined from the initial conditions of motion in a manner similar to that employed in Section 10-7.

For an assigned value of α, the first-mode frequency, given by β_1, is of special interest. Representative values are listed in Table 10-1. These values can be readily verified. The two limiting cases are significant. First, if $m \to 0$, then $\alpha \to 0$ and $\tan \beta \to \beta$, so that

$$\beta_1^2 = \alpha$$

$$\beta_1 = \sqrt{\alpha}$$

$$\frac{\omega_1 l}{c} = \sqrt{\frac{Al\rho}{M}}$$

$$\omega_1 = \sqrt{\frac{c^2 A\rho}{lM}} = \sqrt{\frac{AE}{lM}} = \sqrt{\frac{k}{M}} \tag{10-67}$$

which coincides, as it should, with the Eq. 2-6 for a lumped mass suspended by a massless spring.

Second, if $M \to 0$, then $\alpha \to \infty$, $\beta_1 \to \pi/2$, and $\omega_1 l/c = \pi/2$, so that

$$\omega_1 = \frac{\pi}{2}\frac{c}{l} = \frac{\pi}{2l}\sqrt{\frac{E}{\rho}} \tag{10-68}$$

which agrees, as would be expected, with Eq. (d) of Example 10-3 for the first mode of longitudinal vibration for a uniform bar, without a suspended mass, fixed at one end and free at the other.

EXAMPLE 10-6 Determine the first-mode frequency for the data given in Example 2-5 ($w = 2.5$ lb, $W = 5$ lb, $l = 100$ in., $A = 0.25$ in.^{2}, and $E = 10^7$ lb/in.^{2}).

SOLUTION

$$c = \sqrt{\frac{E}{\rho}} = \sqrt{\frac{EgAl}{w}} = \sqrt{\frac{10^7 \times 386 \times 0.25 \times 100}{2.5}}$$

$$= \sqrt{386 \times 10^8} = 19.647 \times 10^4 \text{ in./sec}$$

Then $w/W = 2.5/5 = 0.5$, and from Eq. 10-62 and Table 10-1, $\beta_1 = 0.65$. From Eq. 10-63,

$$\omega_1 = \frac{c}{l}\beta_1 = \frac{19.647 \times 10^4}{100} \times 0.653 = 1283 \text{ rad/sec}$$

$$f_1 = \frac{\omega_1}{2\pi} = 204 \text{ Hz}$$

EXAMPLE 10-7 Determine the frequency for the second and third modes of vibration for the system of Example 10-6.

SOLUTION By considering a plotting (not shown) of $\tan \beta$ against β, it is observed that the next β which satisfies Eq. 10-62 for $\alpha = 0.5$ will occur slightly above $\beta = 180°$. Eventually, this is found to be at $\beta = 188.64°$, for which

$$\beta_2 = \frac{188.64}{57.296} = 3.2924 \text{ rad}$$

$$\tan \beta_2 = 0.1519$$

(Thus $\beta_2 \tan \beta_2 = 3.2924 \times 0.1519 = 0.5001$.) Similarly, the next β value will occur very slightly above $\beta = 360°$. This is finally determined to occur at $\beta = 364.50°$. Then

$$\beta_3 = \frac{364.50}{57.296} = 6.3617 \text{ rad}$$

$$\tan \beta_3 = 0.0787$$

(Whence $\beta_3 \tan \beta_3 = 6.3617 \times 0.0787 = 0.5007$.) From the foregoing,

$$\omega_2 = \frac{c}{l}\beta_2 = \frac{19.647 \times 10^4}{100} \times 3.2924 = 6469 \text{ rad/sec}$$

$$\omega_3 = \frac{c}{l}\beta_3 = \frac{19.647 \times 10^4}{100} \times 6.3617 = 12\,500 \text{ rad/sec}$$

and

$$f_2 = \frac{\omega_2}{2\pi} = 1030 \text{ Hz}$$

$$f_3 = \frac{\omega_3}{2\pi} = 1989 \text{ Hz}$$

It is interesting to compare the results of this example and of Example 10-6 with those of the first three modes, as given in Example 10-4 for the same rod without the end mass. The suspended mass has the effect, as would be expected, of reducing the frequency of each corresponding principal mode.

10-9. FREE LATERAL VIBRATION OF ELASTIC BARS

Bars are subject to transverse or bending vibrations. Since such members are typical of many structures, investigation of this type of oscillation is rather important. For the bar of Fig. 10-14, x is the coordinate along the neutral axis of the member, and y measures the lateral displacement or deflection of the beam. Note that for the transverse motion, $y = y(x, t)$. Before the differential equation of motion for the member is developed, pertinent relations for beams will be introduced and adapted to the present dynamic conditions.

Figure 10-15 shows an incremental portion of an elastic beam, considered to be subjected to static loading. Here, $G - N$ is the neutral plane. A section of the beam will exhibit plane rotation. However, section CD has been reoriented to its original position so that all the rotation is shown by section AB, which has been turned to position $A'B'$. The *change* in slope for the neutral plane is $(d^2y/dx^2)\,dx$. All slopes and angles of rotation here are considered to be small. The total strain δ at the outer fibers is AA', occurring at distance $c = A'N$ from the neutral plane. (Point O is the center of the curvature.) Then from equal angles,

$$\frac{d^2y}{dx^2}dx = \frac{\delta}{c}$$

whence

$$\frac{d^2y}{dx^2} = \frac{\delta}{c\,dx} = \frac{\varepsilon}{c}$$

where ε is the unit strain at the outer fibers. By Hooke's law and the bending-stress relation, for elastic conditions

$$\frac{d^2y}{dx^2} = \frac{\sigma/E}{c}$$

$$= \frac{Mc/I}{Ec} = \frac{M}{EI}$$

where σ is the unit stress at the outer fibers. Then

$$M = EI\frac{d^2y}{dx^2} \tag{10-69}$$

Figure 10-16 is a free-body diagram for a differential element of a beam, for a static loading condition, where V is the shear force and M is the bending moment on the section. The intensity of loading is w (force per unit length), being taken positive upward, so that the load is then $w\,dx$, as shown. The change dV in the shear and the change dM in the moment are also positive, as shown. Then for equilibrium, the following moment relation can be written

$$M + dM = M + V\,dx + w\,dx\frac{dx}{2}$$

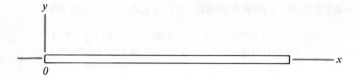

Figure 10-14

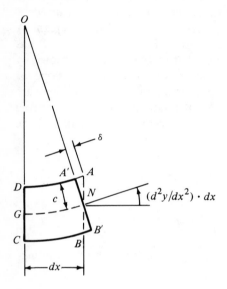

Figure 10-15

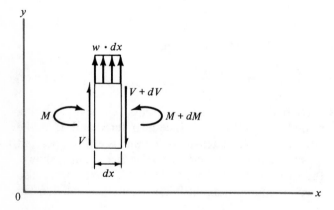

Figure 10-16

By neglecting the higher-order term $(dx)^2$,

$$dM = V\,dx \quad \text{and} \quad V = \frac{dM}{dx} \tag{10-70}$$

The summation of force relation for equilibrium gives

$$V + dV = V + w\,dx$$

whence

$$dV = w\,dx \quad \text{and} \quad w = \frac{dV}{dx} \tag{10-71}$$

Equations 10-69 and 10-70 may be substituted into Eq. 10-71, and if EI is constant, then

$$w = \frac{d}{dx}\left(\frac{dM}{dx}\right) = \frac{d^2M}{dx^2} = \frac{d^2}{dx^2}\left(EI\frac{d^2y}{dx^2}\right) = EI\frac{d^4y}{dx^4} \tag{10-72}$$

which is a useful relation for the analysis of elastic beams. For the condition of motion, Eq. 10-72 is written as

$$EI\frac{\partial^4 y}{\partial x^4} = w \tag{10-73}$$

Consider now an incremental mass element of length dx. The accelerating force acting on this mass is $\gamma\,dx\,\partial^2 y/\partial t^2$, where γ is the mass of the beam per unit length. Then the opposing force upon the beam element is $-\gamma\,dx\,\partial^2 y/\partial t^2$. Hence the intensity of force on the beam is

$$w = -\gamma\frac{\partial^2 y}{\partial t^2} \tag{10-74}$$

Placing this in Eq. 10-73 gives

$$EI\frac{\partial^4 y}{\partial x^4} = -\gamma\frac{\partial^2 y}{\partial t^2} \tag{10-75}$$

as the differential equation for the lateral motion of the bar. This may be written as

$$\frac{\partial^4 y}{\partial x^4} + a^4\frac{\partial^2 y}{\partial t^2} = 0 \tag{10-76}$$

where $a^4 = \gamma/EI$.

The reason for using the fourth power for the term a will be observed later. Note that a^4 is necessarily positive, which is in harmony with the fact that γ, E, and I are positive. The solution will be assumed as

$$y = XT \tag{10-77}$$

where $X = X(x)$ and $T = T(t)$. Then

$$\frac{\partial^4 y}{\partial x^4} = T\frac{d^4 X}{dx^4} \quad \text{and} \quad \frac{\partial^2 y}{\partial t^2} = X\frac{d^2 T}{dt^2} \tag{10-78}$$

Substituting these into the differential equation results in

$$T\frac{d^4 X}{dx^4} + a^4 X\frac{d^2 T}{dt^2} = 0 \tag{10-79}$$

from which

$$-\frac{1}{a^4}\frac{d^4 X/dx^4}{X} = \frac{d^2 T/dt^2}{T} = \mu \tag{10-80}$$

Since this must hold for all values of x and of t, the ratios here must be constant, and hence μ is a constant. For $\mu \geq 0$, inadmissible solutions result, as could be shown. Disregarding these, the case for $\mu < 0$ will be considered. The condition of μ being negative can be specified by

$$\mu = -b^4 \tag{10-81}$$

Then from Eq. 10-80, the following is obtained:

$$\frac{d^2 T}{dt^2} + b^4 T = 0 \tag{10-82}$$

The solution to this may be written as

$$T = C_1 \sin \omega t + C_2 \cos \omega t \tag{10-83}$$

where $\omega = b^2$.

The reason for using ω here is evident, but it should be noted that ω is undefined, since b has not yet been determined. It is reasonable that ω is not definable at this stage, since it must depend on boundary conditions for the beam, that is, on the type of support or restraint.

The function X is next considered. From Eqs. 10-80 and 10-81, the following is written:

$$-\frac{1}{a^4}\frac{d^4 X/dx^4}{X} = -b^4$$

$$\frac{d^4 X}{dx^4} - \lambda^4 X = 0 \tag{10-84}$$

where $\lambda^4 = (ab)^4$ is positive. For the solution, the exponential form

$$X = Ce^{sx} \tag{10-85}$$

is assumed, where s is to be determined by satisfying the differential equation. Substituting this into Eq. 10-84 gives the auxiliary relation

$$s^4 - \lambda^4 = 0 \tag{10-86}$$

which has the roots

$$s = \pm \lambda, \quad \pm i\lambda \tag{10-87}$$

so that the solution for Eq. 10-84 is

$$X = C_3 e^{\lambda x} + C_4 e^{-\lambda x} + C_5 e^{i\lambda x} + C_6 e^{-i\lambda x} \tag{10-88}$$

By letting $C_3 = (B + A)/2$, $C_4 = (B - A)/2$, $C_5 = (D - iC)/2$, and $C_6 = (D + iC)/2$, this can be written as

$$X = A \left(\frac{e^{\lambda x} - e^{-\lambda x}}{2} \right) + B \left(\frac{e^{\lambda x} + e^{-\lambda x}}{2} \right)$$

$$+ C \left[-i \left(\frac{e^{i\lambda x} - e^{-i\lambda x}}{2} \right) \right] + D \left(\frac{e^{i\lambda x} + e^{-i\lambda x}}{2} \right)$$

$$= A \sinh \lambda x + B \cosh \lambda x + C \sin \lambda x + D \cos \lambda x \tag{10-89}$$

where

$$\lambda^2 = \omega a^2 = \omega \sqrt{\frac{\gamma}{EI}} \tag{10-90}$$

The solution to partial differential Eq. 10-76 is thus

$$y = (A \sinh \lambda x + B \cosh \lambda x + C \sin \lambda x + D \cos \lambda x)$$

$$\times (C_1 \sin \omega t + C_2 \cos \omega t) \tag{10-91}$$

The constants A, B, C, and D can be determined from available boundary conditions for the beam, and C_1 and C_2 can be established from the initial conditions for the motion. Either end of the beam may be fixed or free. For example, if one end is taken to be fixed and the other end free, the boundary conditions may be expressed as

$$\left. \begin{array}{l} y = 0 \\[2mm] \dfrac{\partial y}{\partial x} = 0 \end{array} \right\} \quad \text{for } x = 0$$

$$\left. \begin{array}{l} M = 0 = \dfrac{\partial^2 y}{\partial x^2} \\[3mm] V = 0 = \dfrac{\partial^3 y}{\partial x^3} \end{array} \right\} \quad \text{for } x = l \tag{10-92}$$

These hold, irrespective of time. The first two conditions† result in

$$C = -A$$

$$D = -B \tag{10-93}$$

---

† It should be noted that $d(\sinh x)/dx = \cosh x$ and $d(\cosh x)/dx = \sinh x$, and that $\sinh (0) = 0$ and $\cosh (0) = 1$.

The second two conditions and the substitution of Eq. 10-93 result in

$$A(\cosh \lambda l + \cos \lambda l) + B(\sinh \lambda l - \sin \lambda l) = 0$$

$$A(\sinh \lambda l + \sin \lambda l) + B(\cosh \lambda l + \cos \lambda l) = 0 \qquad (10\text{-}94)$$

This represents an eigenvalue problem. The determinant of A and B must be zero, and this condition establishes the characteristic equation having an infinite number of roots which specify the characteristic number or frequency value. Setting the determinant to zero and expanding results in†

$$\cosh \lambda l \cos \lambda l + 1 = 0 \qquad (10\text{-}95)$$

Since $\cosh \lambda l$ is an increasing exponential and $\cos \lambda l$ is cyclic so that it has recurring values, as shown in Fig. 10-17, there will be an infinite number of values of λl that satisfy this relation. The first few consecutive roots of Eq. 10-95 are

$$\lambda_1 l = 1.875$$

$$\lambda_2 l = 4.694$$

$$\lambda_3 l = 7.855$$

$$\lambda_4 l = 10.996$$

$$\lambda_5 l = 14.137$$

$$\vdots$$

These values can be used in Eq. 10-90 to establish the natural frequencies of the principal modes. For example, using λ_1 gives

$$\left(\frac{1.875}{l}\right)^2 = \omega_1 \sqrt{\frac{\gamma}{EI}}$$

$$\omega_1 = \frac{3.515}{l^2} \sqrt{\frac{EI}{\gamma}} \qquad (10\text{-}96)$$

This is the lowest natural frequency, corresponding to the first principal mode. Other frequencies may be similarly determined.

Substituting the first-mode value of $\lambda_1 l = 1.875$ into either relation of Eq. 10-94 determines

$$A = -0.7341B \qquad (10\text{-}97)$$

The remaining constants can be determined from the initial conditions of motion. For example, if the principal-mode configuration but no veloc-

† Note that $\cosh^2 \beta - \sinh^2 \beta = 1$.

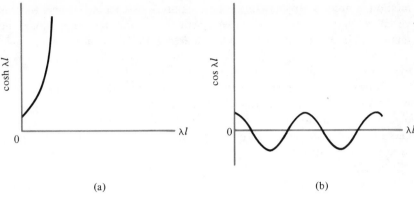

(a)   (b)

Figure 10-17

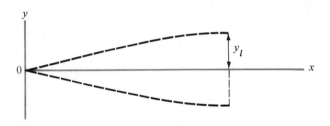

Figure 10-18

ity occurs at $t = 0$, then $C_1 = 0$; and if $y = y_l$ at $x = l$ for $t = 0$, the solution becomes

$$
y = 0.5001 y_l \left[\cosh \frac{1.875}{l} x - \cos \frac{1.875}{l} x \right.
$$

$$
\left. - 0.7341 \left(\sinh \frac{1.875}{l} x - \sin \frac{1.875}{l} \right) \right] \cos \left(\frac{3.515}{l^2} \sqrt{\frac{EI}{\gamma}} t \right) \qquad (10\text{-}98)
$$

The configuration of the member is shown in Fig. 10-18. It should be noted that the curve is not sinusoidal but, rather, a sine form of curve modified by the exponential components. The beam then oscillates harmonically, with all points along the curve remaining in step, or maintaining the same ratio of displacements relative to each other at all times.

10-10. RAYLEIGH METHOD FOR DISTRIBUTED SYSTEMS

The Rayleigh method can be applied to distributed mass-elastic systems to determine the frequency for the first mode. This lowest natural frequency is sometimes the only measure required for the vibration being investigated, and the Rayleigh value is often sufficiently accurate for such a purpose. This method is much simpler than the exact analysis, particularly where there is variation in the mass and in the elasticity or stiffness factor along the member. The method can be used for different types of elastic members, such as rods in torsion, the longitudinal vibration of bars, and so on, but it will be explained here only for the transverse vibration of beams.

Consider the beam of Fig. 10-19(a), containing distributed mass and with coordinates as shown. In order to apply the Rayleigh method, it is first necessary to develop expressions for the maximum kinetic and potential energies. Using mass element dm shown in Fig. 10-19(b), the kinetic energy at any time is expressed by

$$T = \int \frac{\dot{y}^2 \, dm}{2} \tag{10-99}$$

where it is understood that the integral and subsequent integrals here are performed over the length of the member. Let y maximum be designated by Y and $\dot{y}$ maximum by $\dot{Y}$. Then Y and $\dot{Y}$ vary along the member but are independent of time, so that $Y = Y(x)$ and $\dot{Y} = \dot{Y}(x)$. Also $\dot{Y} = \omega Y$, as the motion is harmonic. Then the maximum kinetic energy is given by

$$T_{\text{max}} = \int \frac{\dot{Y}^2 \, dm}{2} = \int \frac{(\omega Y)^2 \, dm}{2} = \frac{\omega^2}{2} \int Y^2 \gamma \, dx \tag{10-100}$$

where γ is the mass per unit length.

The potential energy can be derived from the work done by the elastic forces in returning to the neutral configuration. Consider the free-body diagram shown in Fig. 10-19(c) for an incremental portion of the beam. Neglecting the shear forces, the differential potential energy dV may be expressed as

$$dV = \tfrac{1}{2}(M + dM) \, d\theta$$

$$= \frac{M \, d\theta}{2} \tag{10-101}$$

if the higher-order value $dM \cdot d\theta$ is dropped. Then

$$V = \int \frac{M \, d\theta}{2} = \frac{1}{2} \int \left(EI \frac{d^2y}{dx^2} \right)\left(\frac{d^2y}{dx^2} dx \right)$$

$$= \frac{1}{2} \int EI \left(\frac{d^2y}{dx^2} \right)^2 dx \tag{10-102}$$

whence

$$V_{\text{max}} = \frac{1}{2} \int EI \left(\frac{d^2Y}{dx^2} \right)^2 dx \tag{10-103}$$

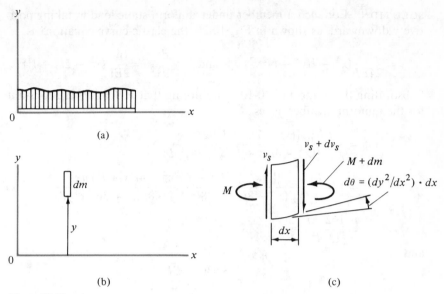

Figure 10-19

Note that in Eq. 10-100, the mass density γ may vary along the member, and in Eq. 10-103, E and I may be functions of x. Since Y and d^2Y/dx^2 are functions of x, in order to express the energies the system configuration must be available. In general, this is not known but must be assumed. Although it may be simple to assume a reasonable configuration for the beam or to estimate its shape, this may unintentionally introduce a restraint on the system which can cause a considerable error in the Rayleigh frequency determination. Most commonly, the static equilibrium shape is used as an estimate of the configuration of the first mode. Then the integrals can be performed and the Rayleigh frequency obtained from the quotient that results from equating T_{max} and V_{max}. Thus

$$\omega_R^2 = \frac{\int_0^l EI(d^2Y/dx^2)^2 \, dx}{\int_0^l \gamma Y^2 \, dx} \qquad (10\text{-}104)$$

The beam may be stepped due to the member being built of different parts or of a telescoping construction. Then E, I, and γ may be constant but different for each of several reaches of the beam. In such case the sum of the integrals over the separate portions are used. Thus

$$\omega_R^2 = \frac{E_1 I_1 \int_0^{x_1} (d^2Y/dx^2)^2 \, dx + E_2 I_2 \int_{x_1}^{x_2} (d^2Y/dx^2)^2 \, dx + \cdots}{\gamma_1 \int_0^{x_1} Y^2 \, dx + \gamma_2 \int_{x_1}^{x_2} Y^2 \, dx + \cdots}$$

$$(10\text{-}105)$$

EXAMPLE 10-8 Determine the Rayleigh frequency for the first vibration mode of a uniform cantilever beam.

SOLUTION   For such a member under uniform static load w, taking positive y downward, as shown in Fig. 10-20, the elastic-curve equation† is

$$y = \frac{w}{24EI}(x^4 - 4lx^3 + 6l^2x^2) \quad \text{and} \quad \frac{d^2y}{dx^2} = \frac{w}{2EI}(x^2 - 2lx + l^2)$$

Substituting these into Eq. 10-104, and noting that E, I, and γ are constant for the uniform member, gives

$$\omega_R^2 = \frac{EI \int_0^l [(x^2 - 2lx + l^2)/2]^2 \, dx}{\gamma \int_0^l [(x^4 - 4lx^3 + 6l^2x^2)/24]^2 \, dx}$$

$$= 144 \frac{EI}{\gamma} \frac{\int_0^l (x^4 - 4lx^3 + 6l^2x^2 - 4l^3x + l^4) \, dx}{\int_0^l (x^8 - 8lx^7 + 28l^2x^6 - 48l^3x^5 + 36l^4x^4) \, dx}$$

$$= \frac{162}{13} \frac{EI}{l^4\gamma}$$

and

$$\omega_R = \frac{3.530}{l} \sqrt{\frac{EI}{\gamma}}$$

A comparison of this with Eq. 10-96, obtained by the exact solution, shows that the Rayleigh frequency here is in error by less than $\frac{1}{2}\%$.

EXAMPLE 10-9   Solve Example 10-8 based on the assumption that the beam configuration is of cosine form.

SOLUTION   To fit the shape of the first mode, the curve may be expressed as

a.
$$y = y_l \left(1 - \cos \frac{\pi}{2l} x \right)$$

This fits the boundary conditions, as follows:

at $x = 0$,

$$y = 0, \qquad \frac{dy}{dx} = 0, \qquad \frac{d^2y}{dx^2} \neq 0$$

at $x = l$,

$$y = y_l, \qquad \frac{dy}{dx} \neq 0, \qquad \frac{d^2y}{dx^2} = 0$$

Since $M = EI \, d^2y/dx^2$, the last condition listed shows that $M = 0$ at $x = l$. This is important as a restraint here because $M \neq 0$ would represent a

---

† See the subject of deflections in any recognized text on strength of materials.

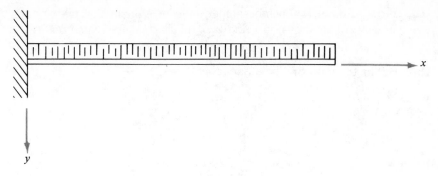

Figure 10-20

condition that would affect the value of the frequency obtained by the Rayleigh method. However, from $V = EI\, d^3y/dx^3$ it is seen that $V \neq 0$ at $x = l$ and this represents a shear restraint that affects the accuracy of the Rayleigh frequency calculated. Proceeding,

b. $\quad T_{max} = \dfrac{\omega^2}{2} \displaystyle\int_0^l \gamma Y^2 \, dx = \dfrac{\gamma\omega^2}{2} \displaystyle\int_0^l \left[y_l \left(1 - \cos \dfrac{\pi}{2l} x \right) \right]^2 dx$

$\qquad = \dfrac{\gamma\omega^2 y_l^2 l}{2} \left(\dfrac{3}{2} - \dfrac{4}{\pi} \right)$

c. $\quad V_{max} = \dfrac{1}{2} \displaystyle\int_0^l EI \left(\dfrac{d^2 Y}{dx^2} \right)^2 dx = \dfrac{EI}{2} \displaystyle\int_0^l \left[y_l \left(\dfrac{\pi}{2l} \right)^2 \cos \dfrac{\pi}{2l} x \right]^2 dx$

$\qquad = \dfrac{EI y_l^2}{l^3} \dfrac{\pi^4}{64}$

Equating the energies gives

d. $\quad \dfrac{\gamma\omega_R^2 y_l^2 l}{2} \left(\dfrac{3}{2} - \dfrac{4}{\pi} \right) = \dfrac{EI y_l^2}{l^3} \dfrac{\pi^4}{64}$

$$\omega_R = \dfrac{3.664}{l^2} \sqrt{\dfrac{EI}{\gamma}}$$

This frequency is 4.24% higher than the correct value.

EXAMPLE 10-10   Solve Example 10-8, assuming that the beam shape is a parabola with vertex at the origin. (This would appear to be a reasonable and proper configuration to assume.)

SOLUTION   The equation of the curve may be written as

$$x^2 = \dfrac{l^2}{y_l} y$$

whence

a.
$$y = \frac{y_l}{l^2} x^2$$

This exhibits the following boundary conditions:

at $x = 0$,

$$y = 0, \qquad \frac{dy}{dx} = 0, \qquad \frac{d^2y}{dx^2} \neq 0$$

at $x = l$,

$$y = y_l, \qquad \frac{dy}{dx} \neq 0, \qquad \frac{d^2y}{dx^2} \neq 0$$

Since $M = EI\, d^2y/dx^2$, this last condition shows that there is a restraining moment at $x = l$, and this will disturb the accuracy of the Rayleigh frequency determined. Proceeding,

b.
$$T_{max} = \frac{\omega^2}{2} \int_0^l \gamma Y^2 \, dx = \frac{\gamma \omega^2}{2} \int_0^l \left(\frac{y_l}{l^2} x^2 \right)^2 dx = \frac{\gamma \omega^2 y_l^2 l}{10}$$

c.
$$V_{max} = \frac{1}{2} \int_0^l EI \left(\frac{d^2Y}{dx^2} \right)^2 dx = \frac{EI}{2} \int_0^l \left(\frac{2y_l}{l^2} \right)^2 dx = \frac{2EI y_l^2}{l^3}$$

Equating the energies results in

d.
$$\frac{\gamma \omega_R^2 y_l^2 l}{10} = \frac{2EI y_l^2}{l^3}$$

$$\omega_R = \frac{4.472}{l^2} \sqrt{\frac{EI}{\gamma}}$$

Compared to the exact value, this frequency is 27.2% high. This large error is due to the constraint which is present in the configuration assumed.

PROBLEMS

10-1. A $\frac{1}{16}$-in. diameter steel wire is stretched between two solid connections located 8 ft apart. The tensile force in the wire is 50 lb, and the material weighs 0.284 lb/in.^{3}. Determine the fundamental frequency (that is, for the first mode) of vibration for the wire.

10-2. The stretched string of a musical instrument is 20 in. long and has a fundamental frequency of 4000 Hz. Determine (a) the frequency of the fourth harmonic (fifth mode), (b) the fundamental frequency if the length is changed to 10 in. by "fingering" the string, and (c) the fundamental frequency if the tension is increased 16%.

10-3. Determine the velocity of wave propagation for the stretched wire described in Prob. 9-1.

10-4. The upper end of a stretched cord is fixed and the lower end is fastened to a pin which can move in a smooth horizontal slot. The cord tension is F and the mass per unit length is γ. Determine the natural frequencies for the principal modes.

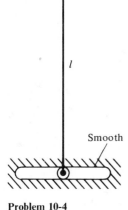

l

Smooth

Problem 10-4

10-5. A cord having mass γ per unit length is stretched with tension F, as shown. The upper end is fixed and the lower end is fastened to mass M which is free to move in a frictionless horizontal slot. Obtain the relation that defines the principal-mode frequencies.

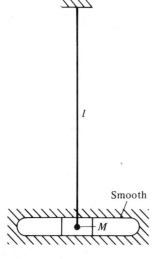

l

Smooth

M

Problem 10-5

10-6. A cord of length l and mass per unit length γ is stretched with tension F. The upper end is fixed and the lower end is fastened to a pin which is free to move in a frictionless horizontal slot. The pin is also connected to springs, as shown. (a) Obtain the equation that defines the principal-mode frequencies. (b) If $F = 1000$ N, $k = 500$ N/m, $l = 2$ m and $\gamma = 0.0035$ kg/m, determine the lowest mode frequency.

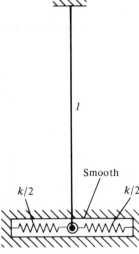

Problem 10-6

10-7. A cord of length l and mass per unit length γ is stretched with tension F. The upper end is fixed and the lower end is fastened to mass M which can move in a frictionless horizontal slot. Mass M is also connected to springs, as shown. Determine the relation that defines the natural frequencies of the principal modes.

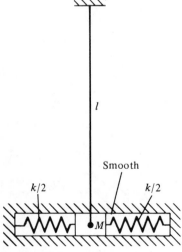

Problem 10-7

10-8. A stretched cord AC is subjected to the initial conditions shown in the accompanying figure, by pulling the midpoint B down a distance y_0 and releasing it. Obtain the particular solution for these conditions.

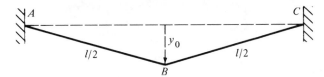

Problem 10-8

10-9. Obtain the solution defining the principal modes of torsional vibration of a circular rod having one end fixed and the other end free. Assume that a principal-mode configuration with no velocity occurs at zero time.

10-10. Determine the fundamental, or first-mode, frequency for the torsional oscillation of a 1-in. diameter steel rod having a length of 5 ft, with both ends free. The material weighs 0.284 lb/in.^{3} and has a modulus of rigidity of 4×10^6 lb/in.^{2}.

10-11. Solve Prob. 10-10 if one end is free and the other is fixed.

10-12. Solve Prob. 10-10 if both ends are fixed.

10-13. Determine the velocity of propagation of the torsional waves along the rod described in Prob. 10-10.

10-14. One end of a circular-section rod is fixed and the other end is free to rotate. The free end is turned through an angle θ_0 and released. Obtain the particular solution for these conditions.

10-15. Obtain the solution defining the principal modes of longitudinal vibration of a slender uniform bar having both ends free. Consider that a principal-mode configuration with no velocity occurs at zero time.

10-16. Solve Prob. 10-15 for a bar having both ends fixed.

10-17. Determine the fundamental frequency for the longitudinal vibration of a uniform, slender steel bar having a length of 3 ft and a cross section 0.5 in. square. The bar is free at both ends. The material has a unit weight of 0.284 lb/in.^{3} and a tension-compression modulus of elasticity of 30×10^6 lb/in.^{2}.

10-18. Determine the velocity of propagation of the longitudinal waves along the bar described in Prob. 10-17.

10-19. Determine the fundamental frequency for the longitudinal vibration of a uniform, slender steel bar having a length of 1 m and a cross section 15 mm square. One end of the bar is fixed and the other end is free. The material has a unit mass of 7861 kg/m^3 and a tension-compression modulus of elasticity of 2×10^{11} Pa $(= \text{N/m}^2)$.

10-20. Determine the velocity of propagation of the longitudinal waves along the bar described in Prob. 10-19.

10-21. The torsional system shown consists of a slender cylindrical rod with disk fastened at the center of the length. (The disk is to be considered as a lumped-inertia part and the shaft as a distributed mass-elastic member of the system.) Write the necessary differential equations and relations and obtain the solution governing the principal modes of torsional vibration for the system.

Consider that both ends of the shaft are fixed and take the maximum angular displacement as θ_0 and the initial velocity as zero.

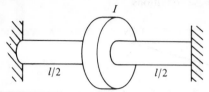

Problem 10-21

10-22. Obtain the solution defining the principal modes of lateral vibration of an elastic bar having both ends fixed. Assume that a principal-mode configuration with no velocity occurs at zero time.

10-23. Obtain the solution defining the principal modes of lateral vibration of an elastic bar having both ends free. Consider that a principal-mode configuration with no velocity occurs at zero time.

10-24. Obtain the solution defining the principal modes of lateral vibration of an elastic bar having simple supports at the ends. Assume that a principal-mode configuration with no velocity occurs at zero time.

10-25. Obtain the solution defining the principal modes of lateral vibration of an elastic bar having one end fixed and the other end simply supported. Assume that a principal-mode configuration with no velocity occurs at zero time.

10-26. Determine the fundamental frequency for the lateral vibration of a 0.5-in. square steel bar having a length of 3 ft, with one end free and the other fixed. The material has a unit weight of 0.284 lb/in.^{3} and a bending modulus of elasticity of 30×10^6 lb/in.^{2}.

10-27. Solve Prob. 10-26 if both ends are free.

10-28. Solve Prob. 10-26 if both ends are fixed.

10-29. Solve Prob. 10-26 if one end is fixed and the other is simply supported.

10-30. Determine the Rayleigh frequency for the first mode of transverse vibration of a uniform beam on simple supports.

10-31. Determine the Rayleigh frequency for the first mode of transverse vibration of a uniform beam with one end fixed and the other simply supported.

chapter
11

Computer Techniques in
Vibration Analysis †

11-1. INTRODUCTION

Modern computer technology has provided a number of powerful tools for the vibration analyst. The tools that now permit not only the rapid and convenient solution to vibration problems but also the analysis of highly complex vibratory systems may be grouped in three broad categories— circuits constructed from electrical analogies, the analog computer (or electronic differential analyzer), and the digital computer.

The analogous behavior of electric circuits and mass-elastic systems has been recognized for many years. The vibratory behavior of complex mechanical systems may be analyzed by series and parallel combinations of resistors, capacitors, and inductors. Systems inputs and responses in the form of voltages or currents can be easily obtained and analyzed. For example, a simple spring-mass-damper system may be represented by a series resistance, inductance, and capacitance $(R\text{-}L\text{-}C)$ circuit, where the force excitation is represented by an input voltage, and the velocity of the mass is observed by monitoring the current. This system of elements—where inductance is analogous to mass, resistance is analogous to viscous damping, and capacitance is analogous to the inverse of stiffness or compliance—is called a force-voltage analogy. It is also possible to utilize a parallel electric circuit; in this case it is termed a force-current analogy. After some experience in

† The material in this chapter was prepared by John G. Bollinger, Professor of Mechanical Engineering, the University of Wisconsin. Modifications for the second edition were made by V. H. Neubert, Professor of Engineering Mechanics, the Pennsylvania State University.

dealing with analogous quantities, it is possible to construct extremely complex electric networks to simulate such mechanical systems as gear trains, automobile suspensions, structures, and almost any system defined by linear differential equations.† An introduction to electromechanical analogs is given in Section 11-2.

There is, however, one fundamental drawback to the utilization of analogies for the solution of vibration problems. The limitation is primarily that an analogy provides a very special computer which will solve only the given physical case at hand. If one wishes to add springs or change the number of masses, expand the system into more degrees of freedom, or make any other modifications in the configuration of the system, it is necessary to construct a new analogous circuit. The study of nonlinear problems, as will be observed later in this chapter, is far less conveniently accomplished by utilizing analogous techniques than by employing the analog computer or the digital computer.

The general-purpose analog computer is a device that is naturally suited for the study of the dynamic behavior of any vibratory system. This computer can be described as a machine consisting of elements that, when properly coupled together, may be used to solve differential equations or sets of differential equations. All variables are represented by voltages; system inputs are represented by voltages, as well as system outputs or responses. The behavior of a system may be observed and the data recorded by using oscilloscopes and electromechanical recorders. The accuracy of analog-computer results is naturally limited by the precision of the components that make up the computer and the ability to measure voltages accurately. Most modern commercial computers are, however, capable of providing results sufficiently accurate for engineering analysis and synthesis.

While the general-purpose analog computer is extremely useful for analyzing most vibratory systems, it has particular value in the study of nonlinear systems. The outstanding flexibility of analog equipment is a result of modern technology and the development of simple-to-use nonlinear function-generation components. Linear systems are defined by linear differential equations, and as has been pointed out throughout this text, many classic solutions are available to the analyst. Furthermore, the principle of superposition for linear analysis provides a degree of organized general solution that is not possible in nonlinear-problem analysis. Thus an analog simulation of a nonlinear problem may be the only practical engineering approach. Because of the great utility of the analog computer in studying vibration problems, the principles of operation and some examples of vibratory systems will be developed in this chapter.

The digital computer is useful in vibration analysis in a number of ways.

---

† For a bibliography on the subject of electromechanical analogies, see Higgins, T. J., "Electroanalogic Methods," *Appl. Mechanics Revs.*, February, 1957, pp. 49–54.

First, in the general case of linear-systems analysis it is often possible to obtain closed solutions to the differential equations defining the behavior of the system. In this case, the digital computer may be used simply as a means for evaluating the response of a system for a wide variety of system parameters. For example, an expression for the transmissibility of a system was represented in Eq. 4-92. This expression may be readily evaluated by using a simple digital program to obtain data for the family of curves shown in Fig. 4-23.

A more sophisticated use of the digital computer is illustrated by the work discussed in Chapter 9. The Holzer method, for example, is an iterative-type procedure, wherein by a series of successive trials the natural frequency is obtained. The Myklestad method for flexural vibration is a similar iterative procedure. In both cases, the logic of the solution and the repetitive operations may be readily programmed on the digital computer for the rapid solution of both torsional and lateral vibration problems. Matrix techniques provide a powerful tool for the analysis of multidegree-of-freedom systems. The generality of the notation and the systematic solution of matrix problems lend themselves extremely well to the application of digital-computer techniques. Complex multidegree-of-freedom problems, such as those encountered in analyzing missile structures, aircraft frames, multibranched torsional systems, and three-dimensional structures, can be readily handled with advanced matrix approaches. This chapter will provide an introduction to the application of the digital computer to vibration problems and will illustrate, with specific example programs, how the logic can be performed.

11-2. ELECTROMECHANICAL ANALOGS AND ELECTRICAL IMPEDANCE

Electrical Circuit Elements

For the electromechanical analog, the elementary passive electric-circuit elements are resistors (R), inductors (L), and capacitors (C). The symbols for these are shown in Fig. 11-1, where e is voltage and i is current. The relationships between voltage and current for resistance, inductance, and capacitance are

$$e_R = Ri_R \tag{11-1}$$

$$e_L = L\frac{di_L}{dt} \tag{11-2}$$

$$e_C = \frac{1}{C}\int_0^t i_C \, dt + e_C(0) \tag{11-3}$$

where t is time and $e_C(0)$ is the initial voltage.

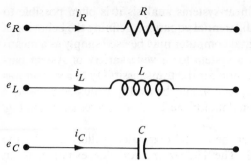

Figure 11-1. Passive electrical circuit elements.

The current-voltage relationships for the elements of Fig. 11-1 are

$$i_R = \frac{1}{R} e_R \tag{11-4}$$

$$i_L = \frac{1}{L} \int_0^t e_L \, dt + i_L(0) \tag{11-5}$$

$$i_C = C \frac{de_C}{dt} \tag{11-6}$$

where $i_L(0)$ is the initial current. The relation between current i_C and the capacitor charge q_C is

$$i_C = \frac{dq_C}{dt} \tag{11-7}$$

Since the capacitor may have an initial charge, the voltage at $t = 0$ is $e_C(0) = q_C(0)/C$. Equation 11-3 may then be written in terms of q_C as follows:

$$e_C = \frac{q_C}{C} + \frac{q_C(0)}{C} \tag{11-8}$$

The differential equation for the one-loop circuit of Fig. 11-2 is written by taking the sum of the voltage drops around the loop, noting that the current is the same in each element. It is assumed that the capacitor has no initial charge. Thus

$$L \frac{di}{dt} + Ri + \frac{1}{C} \int i \, dt = E \tag{11-9}$$

or

$$L \frac{d^2q}{dt^2} + R \frac{dq}{dt} + \frac{1}{C} q = E \tag{11-10}$$

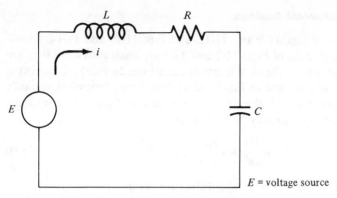

Figure 11-2. Electrical circuit with series connected elements.

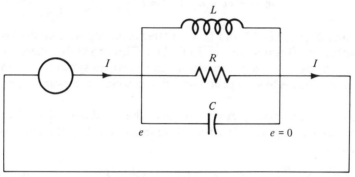

Figure 11-3. Parallel connected electrical elements.

For the parallel circuit shown in Fig. 11-3, a second-order differential equation is obtained by using the relationship between currents at a junction point. Accordingly,

$$i_C + i_R + i_L = I \qquad (11\text{-}11)$$

Then from Eqs. 11-4 through 11-6,

$$C\frac{de}{dt} + \frac{1}{R}e + \frac{1}{L}\int e\,dt = I \qquad (11\text{-}12)$$

The flux ϕ_L in the inductor is related to voltage change by

$$e_L = \frac{d\phi_L}{dt} \qquad (11\text{-}13)$$

Substituting Eq. 11-13 into Eq. 11-12 gives the second-order differential equation

$$C\frac{d^2\phi}{dt^2} + \frac{1}{R}\frac{d\phi}{dt} + \frac{1}{L}\phi = I \qquad (11\text{-}14)$$

Electromechanical Analogs

Comparison of Eqs. 11-9 and 11-12, and Eqs. 11-10 and 11-14, reveals that the electric circuits of Figs. 11-2 and 11-3 are analogous since they are governed by the same form of differential equations. In Eq. 11-9 current is the dependent variable, and in Eq. 11-12 voltage is the dependent variable.

Referring to Chapter 4, the differential equation of motion for a single-degree-of-freedom system may be written as

$$m \frac{d^2x}{dt^2} + c \frac{dx}{dt} + kx = F \tag{11-15}$$

Writing velocity $v = dx/dt$ and substituting results in

$$m \frac{dv}{dt} + cv + k \int v \, dt = F \tag{11-16}$$

By comparing Eqs. 11-9, 11-12, and 11-16, a table of analogous quantities is readily constructed. This is shown in Table 11-1. This shows that there are two different analogies between mechanical and electrical systems—one in which force is analogous to voltage and the other in which force is analogous to current.

As a somewhat more complex case, consider the two-degree-of-freedom mechanical system of Fig. 11-4. In terms of velocities v_1 and v_2, the differential equations are

$$m_1 \frac{dv_1}{dt} + c_1 v_1 + k_1 \int (v_1 - v_2) \, dt = F_1 \tag{11-17}$$

$$m_2 \frac{dv_2}{dt} + c_2 v_2 + k_1 \int (v_2 - v_1) \, dt = 0 \tag{11-18}$$

A third equation may be obtained by adding Eqs. 11-17 and 11-18, which yields

$$m_1 \frac{dv_1}{dt} + c_1 v_1 + m_2 \frac{dv_2}{dt} + c_2 v_2 = F_1 \tag{11-19}$$

This equation may be obtained directly from a free-body diagram in which a cut is made through the two dashpot elements. It is not an independent equation, since it is derived from Eqs. 11-17 and 11-18.

To arrive at one electrical analog for this system, the force-current and mechanical columns of Table 11-1 are consulted and direct substitution of analogous quantities is made in Eqs. 11-17 and 11-18, resulting in

$$C_1 \frac{de_1}{dt} + \frac{1}{R_1} e_1 + \frac{1}{L_1} \int (e_1 - e_2) \, dt = I_1 \tag{11-20}$$

$$C_2 \frac{de_2}{dt} + \frac{1}{R_2} e_2 + \frac{1}{L_1} \int (e_2 - e_1) \, dt = 0 \tag{11-21}$$

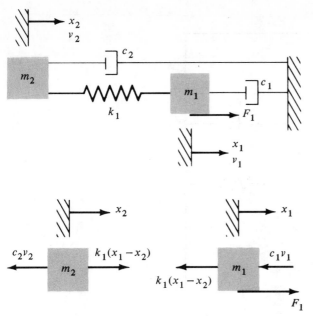

Figure 11-4. Mechanical system.

Table 11-1

Mechanical system	Electrical system	
	Force-voltage	Force-current
m	L	C
c	R	$1/R$
k	$1/C$	$1/L$
v	i	e
x	q	ϕ
F	e	i

The circuit corresponding to the force-current analog may now be deduced, and is shown in Fig. 11-5.

Note that there are three junction points in the circuit in Fig. 11-5, having voltage levels of e_2, e_1, and 0. The first differential equation is obtained by considering currents flowing into the junction point at e_1, the second at e_2. If the analogous equation to Eq. 11-19 were developed, it would correspond to currents flowing into the $e = 0$ junction point.

If the circuit were derived based on the force-voltage analog, it would have two inner loops, and the differential equations would correspond to the sum of voltage drops around these two inner loops. This is left as an exercise for the reader.

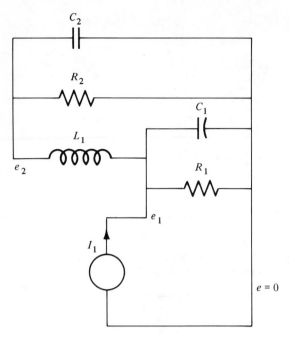

Figure 11-5. Electrical (force-current) analog of mechanical system of Fig. 11-4.

Adding Eqs. 11-20 and 11-21 gives the electrical equation which is analogous to the mechanical equation, Eq. 11-19, as follows:

$$C_1 \frac{de_1}{dt} + \frac{1}{R_1} e_1 + C_2 \frac{de_2}{dt} + \frac{1}{R_2} e_2 = I_1 \qquad (11\text{-}22)$$

By analysis of Fig. 11-5, Eq. 11-22 is obtained directly from the currents flowing into the $e = 0$ junction point.

Electrical Impedance

Electrical impedance Z_{rs} is defined as the ratio of voltage at point r to current at point s; that is,

$$Z_{rs} = \frac{e_r}{i_s} \qquad (11\text{-}23)$$

The voltage and current may be those existing at a particular point, so that $r = s$, or they may be for two different points in a circuit, so that $r \neq s$.

The impedance of the basic passive elements is of interest. For example, the impedance of a resistor is expressed as follows:

$$e = Ri$$

Then

$$Z = \frac{e}{i} = R \qquad (11\text{-}23a)$$

where e is the voltage drop across the resistor. The concept of electrical impedance is applied in the next section to the development of the elements of the analog computer.

11-3. PROGRAMMING PRINCIPLES FOR ANALOG COMPUTATION

The mathematical operations of modern analog computers include integration, algebraic summation, multiplication, and function generation as represented by the symbols of Fig. 11-6. The elements which perform these operations function in such a way that the analog computer is ideal for obtaining the continuous solution of a differential equation, either linear or nonlinear.

The underlying principle of solving a differential equation on an analog computer is simple, but perhaps somewhat deceiving at first. The concept is best grasped by following a simple example.

EXAMPLE 11-1 Set up the computer block diagram required to solve the differential equation

$$\frac{d^2x}{dt^2} + 5\frac{dx}{dt} + 8x = 10t$$

where the initial conditions of the problem are specified as

$$x(0) = 6 \qquad \frac{dx}{dt}(0) = 1$$

SOLUTION The equation may be rearranged to give

$$\frac{d^2x}{dt^2} = -5\frac{dx}{dt} - 8x + 10t$$

which expresses the equation in terms of the highest derivative.

The first step is to assume that the highest derivative is available. It is then possible, with the use of the integrating box of Fig. 11-6(a), to obtain the other terms of the equation. By cascading two integrators, as shown in Fig. 11-7(a), the terms dx/dt and x are obtained. The addition of constant multipliers provides the terms $-5\,dx/dt$ and $-8x$ (Fig. 11-7(b)), and the term $10t$ is obtained by integrating the constant 10 with respect to the independent variable t. Note that the terms at the input of the summation

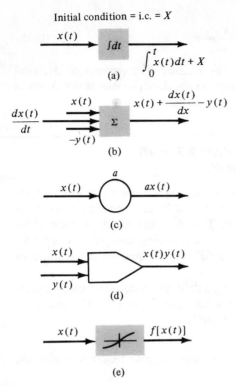

Figure 11-6. Mathematical operations performed by an analog computer. (a) Integration; (b) summation; (c) multiplication by a constant; (d) multiplication of functions; (e) function generation.

block provided are those required to give the second derivative which was initially assumed to be available.

Wherever an integrating box appears, there must also be a symbol (i.c.) indicating the initial value of the output of the integrator at time equals zero. For example, when dx/dt is integrated to get x, then the integrating box must indicate the initial value of x. The complete diagram is shown in Fig. 11-7(b).

In the general case, consider the nth-order differential equation expressed in terms of the highest derivative

$$\frac{d^n x}{dt^n} = -\frac{a_{n-1}}{a_n}\frac{d^{n-1}x}{dt^{n-1}} - \cdots - \frac{a_1}{a_n}\frac{dx}{dt} - \frac{a_0 x}{a_n} + \frac{f(t)}{a_n}$$

The forcing function is $f(t)$, and the initial conditions, for example, are specified as

$$x(0) = X_0, \quad \frac{dx}{dt}(0) = X_0^1, \ldots, \frac{d^{n-1}x}{dt^{n-1}}(0) = X_0^{n-1}$$

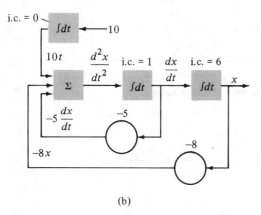

(a)

(b)

Figure 11-7. Computer block diagram for Example 11-1.

The complete generation of this solution is given by the block diagram of Fig. 11-8. Note that each output of the integrating boxes represents the succeeding lower derivative. All the terms required are summed in the summing box to give the highest derivative that was originally assumed available. The procedure may be further illustrated by another simple example.

EXAMPLE 11-2   Consider the problem of solving for the transient response of the spring-dashpot system illustrated in Fig. 11-9(a). The input x_i is a step function X, and the response is x_0. The system is initially in equilibrium.

SOLUTION   First, the differential equation of motion is obtained for point B, as follows:

$$\sum F = 0 = k(x_i - x_o) - c\frac{dx_o}{dt}$$

or

$$c\frac{dx_o}{dt} + kx_o = kX$$

Rearranging yields

$$\frac{dx_o}{dt} = \frac{-k}{c}(x_o - X)$$

The block diagram for the computer solution is given in Fig. 11-9(b).

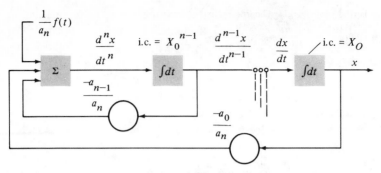

Figure 11-8. Block diagram for the general nth-order equation.

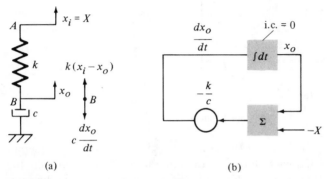

(a)   (b)

Figure 11-9. A spring-dashpot system and the corresponding solution block diagram for Example 11-2.

The solution of differential equations and the generation of functions are thus obtained by the use of operational building blocks. Although operational blocks have been devised which are mechanical, hydraulic, or pneumatic, this chapter will be devoted to the principles of operation and programming of an electronic analog computer. The principles of operation of most commercial analog computers are the same, although there is some deviation from one manufacturer to another in the symbolism used in programming. The symbols adopted here for making computer diagrams are widely used, particularly with the smaller computing machines. Commercial analog computers are available mainly in either 10-volt or 100-volt systems, there being certain advantages to each. In the applications and examples here, the 100-volt system will be assumed.

Analog-Computer Elements

The heart of the analog computer is the d-c operational amplifier. This element, with the proper input and feedback elements,† forms the basic

---

† The feedback elements are connected between the output and the input of the amplifier.

electronic building block for summation and integration. To use an analog computer properly, it is necessary to become familiar with the operation of the amplifier. However, one need not be concerned with the detailed circuitry or design of the amplifier itself.

The Operational Amplifier

The high-gain d-c operational amplifier is represented by the diagram of Fig. 11-10(a). The amplifier is specified as high-gain,† because it is desirable, as will be seen, to have the ratio of the output to the input signal as great as possible. In most commercial computers the gain A ranges from 10^5 to 10^8. For computing purposes, it is desirable to have a linear relationship between the output and input over the entire range of output voltage. It is also desirable to have a flat response (the output of the amplifier is not affected by the input frequency) over a wide range of frequency, commonly from d-c to several thousand cycles per second. Further requirements of a good operational amplifier include zero output voltage for zero input voltage, extremely high internal input impedance, reversal of polarity between output and input voltages, and low noise generated within the amplifier.

Suppose that the amplifier of Fig. 11-10(a) is modified by the addition of an input impedance Z_i, in series with the amplifier, and a feedback impedance Z_f from the output to the amplifier input, as shown in Fig. 11-10(b). The relationship between the output and input voltages may be developed in the following manner: Since the internal input impedance of the amplifier is high, there will be essentially no input current. Therefore

$$i_g \approx 0$$

and by Kirchhoff's law,

$$i_1 = i_2 \tag{11-24}$$

Noting that the current through an electrical impedance is equal to the voltage drop divided by the impedance gives (see Eq. 11-23)

$$i_1 = \frac{e_i - e_g}{Z_i} \quad \text{and} \quad i_2 = \frac{e_g - e_o}{Z_f} \tag{11-25}$$

Substituting Eq. 11-25 into Eq. 11-24,

$$\frac{e_i}{Z_i} = \frac{e_g}{Z_i} + \frac{e_g}{Z_f} - \frac{e_o}{Z_f} \tag{11-26}$$

Noting that the ratio of the output voltage to the grid voltage of the amplifier (Fig. 11-10(a)) is

$$e_o = -Ae_g \tag{11-27}$$

---

† The gain of an amplifier, as used here, is the ratio of the output to the input voltage.

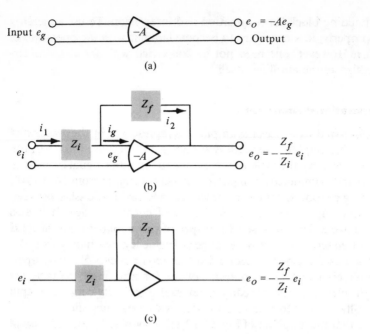

Figure 11-10. The operational amplifier and impedance configurations.

Equation 11-26 becomes, after substitution of Eq. 11-27 to eliminate e_g,

$$\frac{e_i}{Z_i} = -\frac{e_o}{Z_f} - \frac{e_o}{A}\left(\frac{1}{Z_f} + \frac{1}{Z_i}\right) \tag{11-28}$$

In general, $|A| = 10^5$ to 10^8, and for many computers $|e_o| \leq 100$ volts. Also, Z_i and Z_f will be large impedances. Therefore one may neglect the second term on the right-hand side of Eq. 11-28, thus obtaining the simple relationship

$$e_o = -\frac{Z_f}{Z_i} e_i \tag{11-29}$$

Equation 11-29 is the basic transfer relationship of the computer amplifier. The basic computer element is then represented as shown in Fig. 11-10(c). On the computer the ground is common to all amplifiers, thus eliminating the necessity for showing two connections between amplifiers. This simplifies the drawing of computer programs, and the common ground of Fig. 11-10(c) is understood.

   Summation with the Operational Amplifier. The algebraic summation of several variables is accomplished by using resistors as the input and feedback impedances of an operational amplifier. In this process each of the summed variables is multiplied by a constant equal to the ratio of the feedback to the input impedance, in accordance with Eq. 11-29.

EXAMPLE 11-3 Consider the case in which there is a single variable and the input and feedback impedances are the resistors R_i and R_f, respectively, as shown in Fig. 11-11(a). Obtain the equation relating the output to the input voltage.

SOLUTION If $R_f = 1$ megohm, or 10^6 ohms, and $R_i = 0.5$ megohm, or 500 000 ohms, then Fig. 11-11(a) provides the operation

$$e_o = -2e_i$$

The general case of the summation of n input variables is obtained by placing several inputs into the amplifier, each through its own input impedance, as shown in Fig. 11-11(b). The general relationship for the algebraic summation is then

$$e_o = -\left(\frac{R_f}{R_1}e_1 + \frac{R_f}{R_2}e_2 + \cdots + \frac{R_f}{R_n}e_n\right) \tag{11-30}$$

In diagramming the analog computer, the symbols in Fig. 11-11(c) and (d) are generally used to represent the elements of Fig. 11-11(a) and (b), respectively, and will be employed here. Also, some modern computers have the resistor gains built in, and the programmer must choose a particular value, say 0.1, 1, or 10. In either case the diagrams shown in Fig. 11-11(c) and (d) are extremely useful.

EXAMPLE 11-4 Consider the case in which the following resistance values are chosen for the input and feedback resistors, and obtain the equation for the output voltage.

$$R_1 = 0.2 \text{ megohm} \qquad R_3 = 0.1 \text{ megohm}$$
$$R_2 = 1.0 \text{ megohm} \qquad R_f = 0.5 \text{ megohm}$$

SOLUTION

$$e_o = -(2.5e_1 + 0.5e_2 + 5e_3)$$

Integration with the Operational Amplifier. To build an integration device using the operational amplifier, it is necessary to choose a capacitor as the feedback impedance and a resistor for the input impedance, as illustrated in Fig. 11-12(a). An alternate form is shown in Fig. 11-12(c). The basic transfer relationship for this case may be obtained by a derivation similar to Eqs. 11-24 through 11-28. As before, $i_1 = i_2$. Also,

$$i_1 = \frac{e_i - e_g}{R_i} \tag{11-31}$$

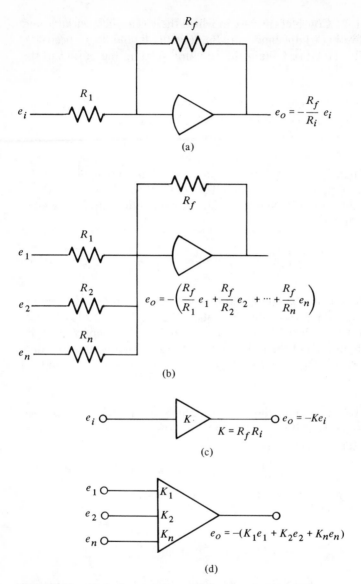

Figure 11-11. Algebraic summation with the operational amplifier.

and

$$i_2 = C_f \frac{d(e_g - e_o)}{dt} \tag{11-32}$$

Therefore

$$\frac{e_i - e_g}{R_i} = C_f \frac{d(e_g - e_o)}{dt} \tag{11-33}$$

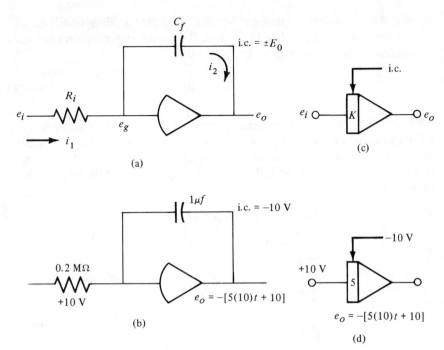

Figure 11-12. Integration with the operational amplifier.

Remembering that $e_o = -Ae_g$

$$\frac{e_i}{R_i} + \frac{e_o/A}{R_i} = C_f \frac{d(-e_o/A - e_o)}{dt} \qquad (11\text{-}34)$$

Neglecting the terms e_o/A, which will be exceedingly small, results in

$$\frac{e_i}{R_i} = -C_f \frac{de_o}{dt} \qquad (11\text{-}35)$$

Integrating both sides with respect to t and rearranging yields

$$e_o = -\frac{1}{R_i C_f} \int_0^t e_i \, dt \qquad (11\text{-}36)$$

which is the expression for the integrating element of the computer. In the course of using this element it may be necessary to place an initial condition on the output variable (e_o has some initial value E_o at time zero). The use of initial conditions on the output of an integrator was illustrated in Examples 11-1 and 11-2. A convenient method of providing this initial condition is to place an initial charge on the feedback capacitor. Thus the *output* of the integrator is equal to the initial condition E_o when the zero time reference is taken.† The complete expression for the integrator is thus obtained by adding the constant E_o to Eq. 11-36, with the proper sign necessary to take into account the sign change in the amplifier.

$$e_o = -\frac{1}{R_i C_f} \int_0^t e_i \, dt \pm E_o \tag{11-37}$$

EXAMPLE 11-5   Consider the integrator in Fig. 11-12(b), or the alternate form shown in Fig. 11-12(d), with a constant input of 10 volts. The input resistance is 0.2 megohm, and the feedback capacitor is 1.0 μF (μF = microfarads or 10^{-6} farad). What is the output voltage in 1 sec if the initial value of e_o is -10 volts?

SOLUTION   The output voltage, using Eq. 11-37, is

$$e_o = -\frac{1}{0.2} \int_0^t 10 \, dt - 10 = -50 - 10 = -60 \text{ volts}$$

In the general case, where there is more than a single input into the integrator, the expression for the integration process is

$$e_o = -\left(\frac{1}{R_1 C_f} \int_0^t e_1 \, dt + \cdots + \frac{1}{R_n C_f} \int_0^t e_n \, dt \right) \pm E_o \tag{11-38}$$

Multiplication by a Constant

The analog computer has a simple method of multiplying by a quantity less than or equal to unity by the use of a voltage divider or potentiometer, as illustrated in Fig. 11-13(a). The expression relating the output voltage to the input voltage is simply

$$e_o = \frac{R_p}{R_t} e_i$$

or where $R_p/R_t = a$, this expression becomes

$$e_o = a e_i \tag{11-39}$$

where $0 \le a \le 1$. The expression is commonly denoted by Fig. 11-13(b).‡

---

† In the case of an integration element it is necessary to have a special switching device to disconnect the input of the integrator and place the initial condition on the capacitor. There are many mechanical and electronic devices for this switching process; thus, for simplicity in programming, one need only specify an initial condition at the output of each integrator. The actual nature of the necessary switching at time zero may be left for the programmer to study in the operation manual of the specific equipment employed.

‡ It should be noted that Eq. 11-39 assumes that there is no loading effect of the next computing element at the output e_o, that is, no additional element with finite impedance from the output to the ground. Unfortunately, this is not always true. However, most commercial computers provide a circuit for the accurate setting of potentiometers after a problem has been completely programmed. Some more elaborate computing installations provide automatic setting of the many potentiometers throughout the computer.

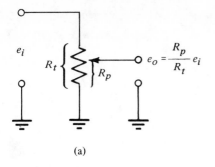

$$e_o = \frac{R_p}{R_t} e_i$$

(a)

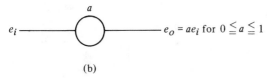

$e_o = ae_i \text{ for } 0 \leq a \leq 1$

(b)

Figure 11-13. The potentiometer.

Multiplication by a constant greater than unity may be accomplished by using a combination of a potentiometer and an impedance ratio at the amplifier. For example, the constant 2.312 may be obtained by a potentiometer setting of 0.2312 and an impedance ratio of 10; that is, $2.132 = 10(0.2312)$.

Multiplication of Two Variables

The multiplication of two variables is a simple matter for the computer programmer. Common commercial function multipliers operate on a variety of basic techniques, each of which requires a special understanding of the principles involved. For a full treatment of the many multiplication methods, the reader is referred to one of the references at the back of this text specifically concerned with analog computing.

If a particular commercial computer is available, the instruction manual will generally illuminate the operating procedure and the multiplier's limitations. Here, it is sufficient for the reader to recognize that the product of two functions may be represented by the symbol of Fig. 11-6(d).

Generation of a Function of a Variable

In the solution of differential equations and in the direct simulation of physical devices (including vibration problems), it is often necessary to generate special functions. With the use of the operational amplifier and special function-generation equipment, it is possible to produce on the computer a wide variety of mathematical functions. The function-generation process is represented by the block of Fig. 11-6(e).

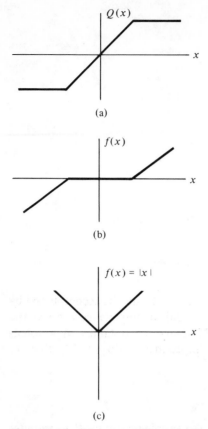

(a)

(b)

(c)

Figure 11-14. Common nonlinear functions.

A common nonlinear function which may be generated is the limited function. A good example is a hydraulic valve that has some limiting flow, regardless of the spool position. The valve flow rate $Q(x)$ versus the spool position x is represented by Fig. 11-14(a). The flow is seen to increase with spool position until a maximum is reached. The flow is then limited, and any further movement will result only in the same flow.

Another frequently used function is that of a dead zone, as shown in Fig. 11-14(b). A computer analysis for a pair of gears with backlash would require the generation of such a function.

A third easily generated function is the absolute value of a function, as illustrated in Fig. 11-14(c).

All of the above are simple functions which may be generated with analog equipment. More elaborate functions, such as powers and roots and trigonometric, exponential, and logarithmic functions, may also be obtained. The reader is again referred to the texts on analog computing listed in the suggested references at the back of this text for a thorough treatment of the various methods and techniques.

Computer-Solution Diagrams

The concept of the solution block diagrams of this section, using the operational building blocks, may now be expanded to build the wiring diagrams or computer-solution diagrams with symbols for the actual computing elements being discussed. The symbols of Fig. 11-6 are used, except that the specific symbols for amplifiers, summers, and integrators used with operational amplifiers are adopted, as shown in Fig. 11-15.

This development is best carried out by a series of simple examples.

EXAMPLE 11-6   The problem is to generate the function $f(t) = t^2$ for the interval of time 0 to 10 sec. (In this text an operational amplifier with a linear output range of ± 100 volts will be assumed. All resistors are given in megohms and all capacitors in microfarads.)

SOLUTION   The block diagram of the solution is given in Fig. 11-16(a). Figure 11-16(c) is the analog-computer-solution diagram. The input is 20 volts d-c, and the output of the first integrator (identified as amplifier 1 by the number within the amplifier figure) is given, by use of Eq. 11-37, as

$$e_o = -\tfrac{1}{2} \int_0^t 20 \, dt + 0 = -10t$$

In the same manner the output of amplifier 2 is given by

$$e_o = -\tfrac{1}{5} \int_0^t -10t \, dt + 0 = t^2$$

One should not fail to recognize that a sign change is associated with each operational amplifier. Fortunately, the output of amplifier 2 will just reach, and not exceed, 100 volts in the time interval desired. If, however, one desires to generate t^2 for 20 sec, the problem must be scaled so that the ± 100-volt limit will not be exceeded. The process of scaling will be developed further in the following section.

EXAMPLE 11-7   Consider the problem of generation of an exponential function, specifically, $f(t) = 50e^{-2t}$.

SOLUTION   The analog-computer diagram is obtained by differentiating $f(t)$ to yield

$$\frac{df(t)}{dt} = -100e^{-2t} = -2f(t)$$

or

$$-\frac{1}{2}\frac{df(t)}{dt} = f(t)$$

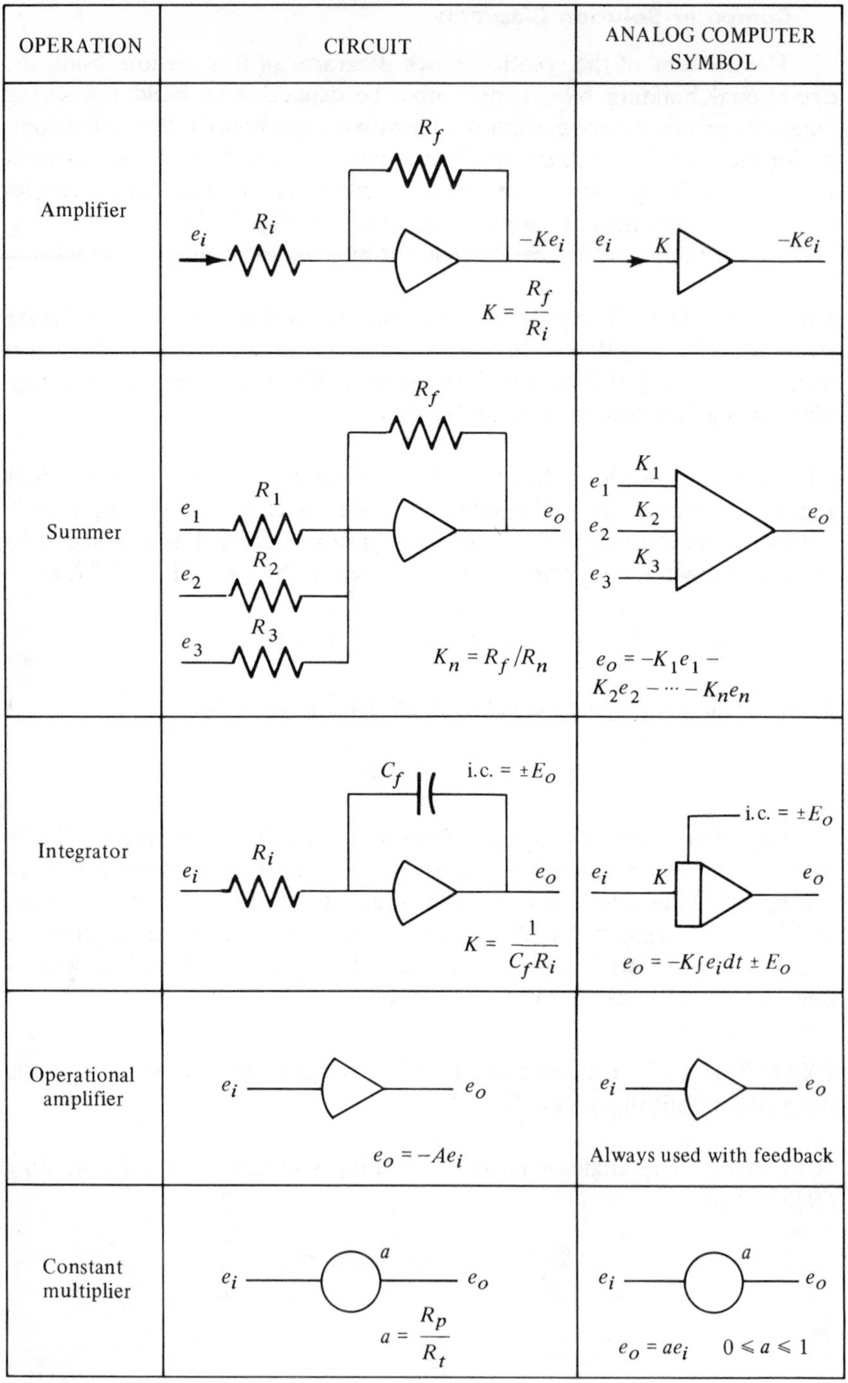

OPERATION	CIRCUIT	ANALOG COMPUTER SYMBOL
Amplifier	e_i R_i R_f $-Ke_i$ $K = \dfrac{R_f}{R_i}$	e_i K $-Ke_i$
Summer	e_1 R_1 R_f e_o e_2 R_2 e_3 R_3 $K_n = R_f/R_n$	e_1 K_1 e_o e_2 K_2 e_3 K_3 $e_o = -K_1 e_1 - K_2 e_2 - \cdots - K_n e_n$
Integrator	C_f i.c. $= \pm E_o$ e_i R_i e_o $K = \dfrac{1}{C_f R_i}$	i.c. $= \pm E_o$ e_i K e_o $e_o = -K \int e_i dt \pm E_o$
Operational amplifier	e_i e_o $e_o = -A e_i$	e_i e_o Always used with feedback
Constant multiplier	e_i a e_o $a = \dfrac{R_p}{R_t}$	e_i a e_o $e_o = a e_i$ $0 \leqslant a \leqslant 1$

Figure 11-15. Computer symbols for amplifier, summer, integrator and constant multiplier.

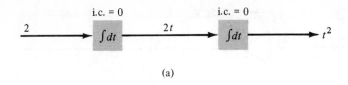

(a)

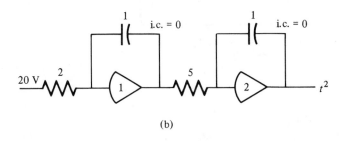

(b)

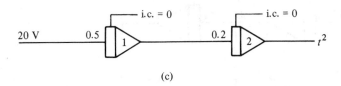

(c)

Figure 11-16. Computer diagram for Example 11-6. (a) Block diagram; (b) circuit diagram; (c) computer diagram.

The analog-computer diagram is given in Fig. 11-17. The only response is due to the initial value of $f(t)$, which is $f(0) = 50$.

EXAMPLE 11-8 Draw the analog-computer solution of the equation of Example 11-1. The equation is

$$\frac{d^2x}{dt^2} = -\left(5\frac{dx}{dt} + 8x - 10t\right)$$

where the initial conditions are

$$x(0) = 6 \quad \text{and} \quad \frac{dx}{dt}(0) = 1$$

SOLUTION A computer diagram for this equation is illustrated in Fig. 11-18. Notice that the negative sign in front of the parentheses on the right-hand side is accounted for in the first summer (amplifier 1). Note also that the output of amplifier 3 is given as $5x$ rather than just x. If x is the quantity to be plotted versus time, then it is necessary only to account for the factor of 5 on the x scale of the data recorded. It may be observed, by actually solving this problem on a computer, that even with the output of amplifiers 2 and 3

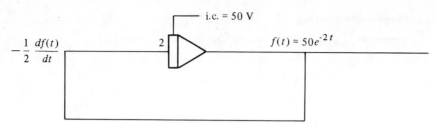

Figure 11-17. Computer diagram for Example 11-7.

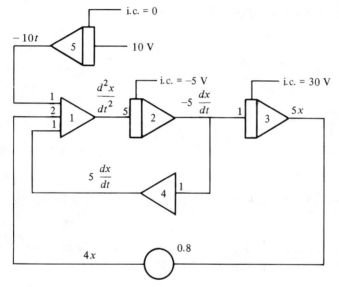

Figure 11-18. Computer diagram for Example 11-8.

multiplied by a factor of 5, none of the amplifiers exceeds ± 100 volts in the course of the solution.

This problem introduces a simple form of magnitude scaling which is discussed further in this section.

Formal Computer Scaling

Besides the formulation of correct computer diagrams, proper scaling is the most important aspect of successful analog computing. In particular, it is necessary to relate the variables of the problem to voltages and times in the computer, so that the solutions are slow or fast enough to be accurately recorded and so that the voltages at any amplifier do not exceed the maximum linear range at any time. In addition, the voltages should not be too small, since the solution will be more susceptible to noise and error.

Magnitude Scaling

The first step of the two fundamental aspects of scaling is magnitude scaling. It is simple to understand that if a problem calls for a displacement of 50 000 ft for the initial condition, 1 volt cannot equal 1 ft. On the other hand, if it could be ascertained that 50 000 ft is the greatest height that will be encountered, then a natural choice would be to equate 50 000 ft to the maximum linear range of the amplifier, say, 100 volts. Then, one must consider the scaling factor of 500 ft/volt relating the computer variable to the actual variable.

Example 11-8 presented a form of magnitude scaling wherein the variables were multiplied by some convenient factor to increase the voltage range of each amplifier. It could have been assumed that 1 volt was equivalent to 1 unit of x. However, working in a lower voltage range on the computer is less accurate, and the solution is not as good.

In computer scaling, as in many other types of engineering problems, it is extremely useful to keep careful account of units. The analog computer operates in the units of volts, regardless of the problem. It is the duty of the programmer to translate the results to the proper physical units by using the proper scale factors. The technique of magnitude scaling is best understood with the aid of an illustrative example.

EXAMPLE 11-9   Consider the problem of generating a sine function, specifically, $x = 40 \sin 2t$. The derivatives are then

$$\dot{x} = 80 \cos 2t$$

$$\ddot{x} = -160 \sin 2t = -4x$$

SOLUTION   From the study of free vibrations without damping, it is known that the solution to the differential equation

$$\ddot{x} = -a^2 x$$

is

$$x(t) = x(0) \cos at + \frac{\dot{x}(0)}{a} \sin at$$

(For this case, $a = 2$). From this the initial conditions needed to generate the sine functions are $x(0) = 0$, $\dot{x}(0)/a = 80/2 = 40$. The resulting analog-computer diagram is shown in Fig. 11-19(a). The difficulty arises in that if one sets 1 volt equal to 1 unit of x, then $\ddot{x} = -160 \sin 2t$ (volts) exceeds the computer limit of ± 100 volts. (Specifically, the maximum value of this term is ± 160 volts.)

There are several alternatives. To avoid having $\ddot{x}$ generated, either of the following differential equations, obtained by rearrangement, may be solved:

$$\frac{\ddot{x}}{a} = -ax$$

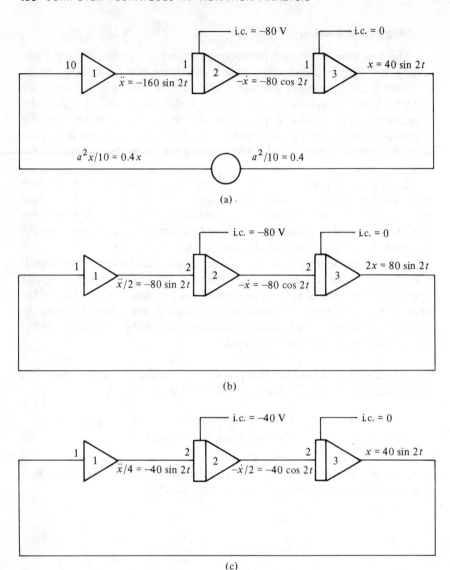

Figure 11-19. Analog computer diagrams for Example 11-9. (a) An unacceptable solution to $\ddot{x} = -a^2 x$, $a = 2$. (b) An acceptable solution, using $\ddot{x}/a = -ax$, $a = 2$. (c) An acceptable solution, using $\ddot{x}/a^2 = -x$, $a = 2$.

where $a = 2$, or

$$\frac{\ddot{x}}{a^2} = -x$$

where $a = 2$. Possible circuits for these equations are shown in Figs. 11-19(b) and 11-19(c), respectively. In the circuit of Fig. 11-19(b) the maximum value of voltage is ± 80, and that for Fig. 11-19(c) is ± 40 volts.

Computer Variables

The variables used in Fig. 11-19(b) and (c) are sometimes called computer variables since they are available at points in the computer circuit. To generalize the concept of scaling, computer variables are introduced for each variable, using a subscript c for computer variable and S_i to represent scale factor. Thus

$$x_c = S_0 x \text{ volts}$$

$$\dot{x}_c = S_1 x \text{ volts}$$

$$\ddot{x}_c = S_2 x \text{ volts}$$

$$\vdots \tag{11-40}$$

The usual derivative relationships exist between derivatives of x. That is

$$\dot{x} = \frac{dx}{dt}, \qquad \ddot{x} = \frac{d\dot{x}}{dt}, \dots$$

However, this is not the case for computer variables. This is shown by

$$\dot{x}_c = S_1 \dot{x} = S_1 \frac{dx}{dt} = S_1 \frac{d}{dt}\left(\frac{x_c}{S_0}\right) = \frac{S_1}{S_0}\frac{dx_c}{dt} \tag{11-41}$$

Similarly,

$$\ddot{x}_c = \frac{S_2}{S_1}\frac{d\dot{x}_c}{dt}$$

EXAMPLE 11-10   Consider Example 11-9, using computer variables and scale factors to determine the analog-computer circuit.

SOLUTION (a)   Consider that the scale factors S_0, S_1, and S_2 are chosen to be equal in magnitude. A convenient value here is $\frac{1}{4}$. Then the scaled computer terms are

$$x_c = 10 \sin 2t$$

$$\dot{x}_c = 20 \cos 2t$$

$$\ddot{x}_c = -40 \sin 2t$$

The initial conditions also scale, so that $\dot{x}_c(0) = \frac{1}{4}\dot{x}(0) = 20$ volts, and $\dot{x}_c(0) = 0$ volts. Because the scale factors are equal,

$$\int \ddot{x}_c \, dt = \dot{x}_c \qquad \text{and} \qquad \int \dot{x}_c \, dt = x_c$$

The differential equation to be solved in terms of computer variables is

$$\ddot{x} = -a^2 x$$

$$\frac{\ddot{x}_c}{S_2} = -a^2 \frac{x_c}{S_0}$$

and since $S_0 = \frac{1}{4} = S_2$, then

$$\ddot{x}_c = -a^2 x_c$$

The complete computer diagram for this case is given in Fig. 11-20(a).

This solution would function, but an undesirable feature is that none of the amplifiers is operating close to the ± 100-volt capacity, since amplifier 1 oscillates between ± 40 volts, amplifier 2 between ± 20 volts, and amplifier 3 between ± 10 volts. The solution is more accurate when all amplifiers are using a large portion of the operating range. To this end, consider choosing unequal scale factors. For example,

SOLUTION (b)   Select

$$S_0 = 2, \qquad S_1 = 1, \qquad S_2 = \tfrac{1}{2}$$

(There is no restriction that the scale factors must be chosen as equal.) Substituting these scale factors into the expressions for the computer variables yields

$$x_c = 80 \sin 2t = 2x$$

$$\dot{x}_c = 80 \cos 2t = \dot{x}$$

$$\ddot{x}_c = -80 \sin 2t = \tfrac{1}{2}\ddot{x}$$

The computer variables have maximum values of ± 80 volts. The initial conditions are $x_c(0) = 0$ and $\dot{x}_c(0) = 80$ volts. Also,

$$\int \ddot{x}_c \, dt = \frac{S_2}{S_1} \dot{x}_c = \tfrac{1}{2}\dot{x}_c$$

$$\int \dot{x}_c \, dt = \frac{S_1}{S_0} x_c = \tfrac{1}{2}x_c$$

The differential equation in terms of computer variables is obtained as follows:

$$\ddot{x} = a^2 x$$

$$2\ddot{x}_c = -a^2 \tfrac{1}{2}x_c$$

$$= -2^2 \cdot \tfrac{1}{2}x_c$$

or

$$\ddot{x}_c = -x_c$$

The computer circuit is shown in Fig. 11-20(b). Note that this is the same circuit as that of Fig. 11-19(b) which was derived by different logic.

At first glance the equations may appear contrary to the derivatives of $\sin 2t$, but it must be realized that the equations with the computer subscript

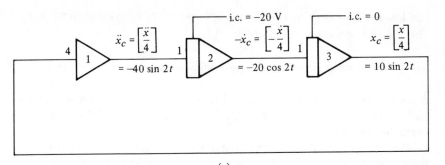

(a)

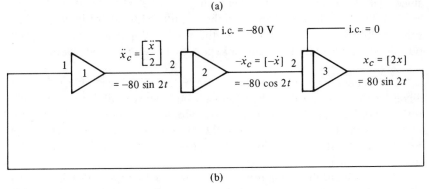

(b)

Figure 11-20. Analog computer diagrams for Example 11-10. (a) Solution (a) of Example 11-10. (b) Solution (b) of Example 11-10.

c are "scaled" equations. On the computer solution, one must satisfy the relation $\ddot{x} = -4x$, and the missing factor of 2 in the computer equations must be accounted for at each integrator on the computer diagram. The final computer diagram is given in Fig. 11-20(b). The mathematical relationship that must be satisfied at amplifier 2 in Fig. 11-20(b) is†

$$\dot{x}_c = \frac{1}{R_i C_f} \int_0^t \ddot{x}_c \, dt \pm \dot{x}_{co}$$

where $\dot{x}_{co}$ is the initial condition. Substituting to obtain the original problem variables yields

$$S_1 \dot{x} = \frac{1}{R_i C_f} \int_0^t S_2 \ddot{x} \, dt + S_1 \dot{x}_o$$

or

$$\dot{x} = \frac{1}{R_i C_f} \frac{S_2}{S_1} \int_0^t \ddot{x} \, dt + \dot{x}_o$$

† Note that the sign change introduced by an integrator, as given in Eq. 11-36, is accounted for on the computer diagram mathematically, and there is no sign reversal.

For this expression to be valid, the coefficient of the integral must be unity. Then, with the chosen scale factors,

$$\frac{1}{R_i C_f} = 2$$

A good choice of computer elements is then a 0.5-megohm resistor and a 1-μF capacitor. In a parallel fashion one may verify the choice of the elements for amplifier 3.

The solution of Fig. 11-20(b) is far better than the previous solution, since all amplifiers are operating between ± 80 volts. Note that if one follows around the loop, the total gain of all the amplifiers [in either Fig. 11-20(a) or Fig. 11-20(b)] is -4. This satisfies the relationship $\ddot{x} = -4x$, and each output is proportional to the actual value by the chosen scale factor.

One general principle that may be derived from this example is that it is advantageous to distribute the total gain between several of the amplifiers rather than have all the gain on a single amplifier. In this case the total gain of 4 is on the first amplifier in Fig. 11-20(a). In the latter solution of the problem the total gain is distributed as a gain of 2 on amplifier 2 and a gain of 2 on amplifier 3.

Another method of magnitude scaling is based on the approximation of the characteristic (or undamped oscillating) frequency of the equation. This method is extremely effective and, after some practice, allows one to scale rapidly and accurately.

EXAMPLE 11-11   Develop the solution to a differential equation. The equation to be solved is that of a simple spring-mass-damper system

$$\frac{d^2 x}{dt^2} + 3.725 \frac{dx}{dt} + 7.312 x = 0$$

with the initial conditions

$$x(0) = 25 \text{ in.} \quad \text{and} \quad \dot{x}(0) = 0 \text{ in./sec}$$

SOLUTION   The undamped resonant frequency of the system is

$$\omega = (7.312)^{1/2}$$

and for the purpose of scaling it may be assumed that $\omega = 3$, approximately. Assuming a sinusoidal solution (it is desired only to make a rough guess of the magnitudes of the variables) gives

$$x = 25 \cos 3t \text{ in.}$$

$$\dot{x} = -75 \sin 3t \text{ in./sec}$$

$$\ddot{x} = -225 \cos 3t \text{ in./sec}^2$$

The method of Example 11-10 may now be applied to choose scale factors for x and its derivatives. In an effort to choose convenient numbers, so that none of the computing variables will exceed ± 100 volts, consider the following scale factors:

$$S_0 = 2 \text{ volts/in.}$$

$$S_1 = 1 \text{ volt/in./sec}$$

$$S_2 = \tfrac{1}{5} \text{ volt/in./sec}^2$$

Rearranging the original equation in terms of the highest derivative and substituting the computer variables with the subscript c yields

$$\frac{1}{S_2}\frac{d^2x_c}{dt^2} = -\left[\frac{10(0.3725)}{S_1}\frac{dx_c}{dt} + \frac{10(0.7312)}{S_0}x_c\right]$$

Substituting the chosen values of the scale factors gives the final equation to be solved as

$$\frac{d^2x_c}{dt^2} = -\left[2(0.3725)\frac{dx_c}{dt} + (0.7312)x_c\right]$$

The computer diagram for this solution is given in Fig. 11-21.

The conditions of the latter equation are satisfied at the summing element, which is amplifier 1. With the introduction of the scale factors, however, the $1/RC$ gains of the integrators must be adjusted to give the proper integrals. The mathematical relationship for the integrator using amplifier 3 is

$$x = \int_0^t \dot{x}\, dt + x(0)$$

Substituting the scaled computer variables yields

$$\frac{x_c}{2} = \int_0^t \frac{\dot{x}_c}{1}\, dt + \frac{50}{2}$$

or

$$x_c = 2\int_0^t \dot{x}_c\, dt + 50$$

The latter equation is expressed in terms of the computer variable x_c. The coefficient of the integral is actually the $1/RC$ gain of the integrator. Thus $1/R_5C_5 = 2$, and the initial condition on amplifier 3 is 50 volts. A good choice for R_5 is then 0.5 megohm, with C_5 being a 1-μF capacitor. (Note that in Example 11-10, the choice of $1/RC$ was based on looking at the integral expressed in terms of the actual variable, whereas in this example the choice was made by looking at the integral in terms of the computer variable.

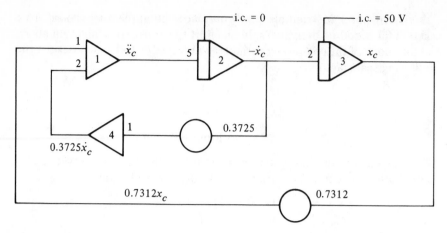

Figure 11-21. Computer diagram for Example 11-11.

Either approach leads to the correct conclusion and may be used with equal success.)

The gain on integrator 2 may be obtained by considering the integral

$$\dot{x} = \int_0^t \ddot{x} \, dt + \dot{x}(0)$$

or scaling

$$\dot{x}_c = 5 \int_0^t \ddot{x}_c \, dt + 0$$

This indicates that $1/R_4 C_4 = 5$, and a good choice for these elements is $C_4 = 1\text{-}\mu\text{F}$ and $R_4 = 0.2$ megohm. The necessary scaling of the problem is complete.

Time Scaling

Time scaling is the second half of the total problem of computer scaling. The purpose here is to relate the independent variable of the problem to the independent variable of the machine when other than a unity relationship is necessary. It was pointed out previously that the speed at which the machine may operate is often limited by the response of the electronic equipment or electromechanical recording equipment. What is, in effect, a change of the independent variable will be adopted here.

The simplest form of time scaling is to speed or slow all the time-dependent components of the computer by the same factor. The basic integrator relationship, expressed by Eq. 11-37, may be modified by recognizing that $1/R_iC_f$ is a constant and may be moved within the integral sign. Thus Eq. 11-37 becomes

$$e_o(t) = -\int_0^t e_i(t)\left(\frac{dt}{R_iC_f}\right) \pm E_o \tag{11-42}$$

Furthermore, one may justify that

$$\frac{dt}{R_iC_f} = d\left(\frac{t}{R_iC_f}\right)$$

The resulting conclusion is that $1/R_iC_f$ may be used as a device to change the independent variable t. For example, let

$$t = Tt_c \tag{11-43}$$

where t is the real time, in seconds; t_c is the computer time, in seconds; and T is a constant, specifically,

$$T = R_iC_f$$

Substituting the change of variable of Eq. 11-43 into Eq. 11-42 yields

$$e_o(t_c) = -\int_0^{Tt_c} e_i(t_c)\,dt_c \pm E_o \tag{11-44}$$

An example illustrating the change of the independent variable will now be examined.

EXAMPLE 11-12 Determine the generation of $f(t) = t^2$, as illustrated in Example 11-6. Suppose that both capacitors of Fig. 11-16(b) are changed to 0.1-μF.

SOLUTION In this case one may consider the input resistors of each integrator as gain factors in front of the integral sign, and take the factor of change of the capacitors $(T = 0.1)$ within the integral. Then

$$t = 0.1t_c$$

This expression states that 1 sec of real time is equivalent to 10 computer seconds. Thus one must modify the length of time for the solution stated in Example 11-6. Rather than taking the solution from 0 to 10 sec in real time, the solution must be taken over 1 sec of real time. Clearly this is equivalent to 10 sec on the computer. The solution has thus been speeded by a factor of 10.

Consider, further, the possibility of replacing all the capacitors of Fig. 11-16 by 10-μF capacitors. All the time-controlling integrators are then related to real time by the relation

$$t = 10t_c$$

Now 1 sec of real time is equivalent to only 1/10 of a computer second. Thus events on the computer are occurring in 10 times the length of real-time seconds, or the solution has been slowed by a factor of 10.

The concepts of time and magnitude scaling may be combined in a method of formal scaling that may be used to scale a wide variety of problems.

Consider the general third-order differential equation with constant coefficients

$$a_0 \frac{d^3x}{dt^3} + a_1 \frac{d^2x}{dt^2} + a_2 \frac{dx}{dt} + a_3 x = f(t) \tag{11-45}$$

One may choose for the change of the independent variable

$$t = Tt_c \tag{11-46}$$

and for the dependent variable

$$x = S_0 x_c \tag{11-47}$$

Equations 11-46 and 11-47 may be substituted in Eq. 11-45 to give a new equation expressed in volts (the units of the dependent variable) and computer seconds (the units of the independent variable).

$$\frac{d^3(S_0 x_c)}{d(Tt_c)^3} + \frac{a_1 d^2(S_0 x_c)}{a_0 d(Tt_c)^2} + \frac{a_2 d(S_0 x_c)}{a_0 d(Tt_c)} + \frac{a_3}{a_0}(S_0 x_c) = \frac{1}{a_0} f(Tt_c)$$

Since both T and S_0 are constant scale factors, they may be brought out in front of each term to give

$$\frac{S_0}{T^3} \frac{d^3 x_c}{dt_c^3} + \frac{a_1}{a_0} \frac{S_0}{T^2} \frac{d^2 x_c}{dt_c^2} + \frac{a_2}{a_0} \frac{S_0}{T} \frac{dx_c}{dt_c} + \frac{a_3}{a_0} S_0 x_c = \frac{1}{a_0} f(Tt_c)$$

Dividing by S_0 and multiplying by T^3 yields the scaled equation

$$\frac{d^3 x_c}{dt_c^3} + \frac{a_1}{a_0} T \frac{d^2 x_c}{dt_c^2} + \frac{a_2}{a_0} T^2 \frac{dx_c}{dt_c} + \frac{a_3}{a_0} T^3 x_c = \frac{T^3}{a_0 S_0} f(Tt_c) \tag{11-48}$$

If one then chooses a value for T such that the coefficients of all the derivatives are close to unity, or at least less than some maximum gain K, then the solution will generally function properly on the computer. In addition, it is necessary to choose a value of S_0 such that a convenient magnitude of the forcing function is obtained—one that will drive all the amplifiers over a large portion of their linear range, yet not cause any amplifier to be overloaded. In general, it is desirable to have K less than 10 and as close to unity as possible. An aid in choosing the value of T that is sometimes helpful involves a value of T near to, and smaller than, the smallest of the following terms:

$$K \frac{a_0}{a_1}, \quad \left(K \frac{a_0}{a_2}\right)^{1/2}, \quad \left(K \frac{a_0}{a_3}\right)^{1/3}$$

where K is near unity. Then

$$\frac{a_1}{a_0} T \le K, \qquad \frac{a_2}{a_0} T^2 \le K, \qquad \frac{a_3}{a_0} T^3 \le K$$

The remaining step is to choose a value of S_0 such that the term $T^3 f(Tt_c)/a_0 S_0$ provides a forcing function of convenient magnitude.

EXAMPLE 11-13   The problem is to scale and draw a computer diagram for the solution of the differential equation

$$1000 \frac{d^3x}{dt^3} + 100 \frac{d^2x}{dt^2} + 20 \frac{dx}{dt} + x = 500 \sin 0.1t$$

with the initial conditions

$$x(0) = 100 \qquad \frac{dx}{dt}(0) = 5 \qquad \frac{d^2x}{dt^2}(0) = 0$$

SOLUTION

$$\frac{d^3x_c}{dt_c^3} + 0.1T \frac{d^2x_c}{dt_c^2} + 0.02T^2 \frac{dx_c}{dt_c} + 0.001T^3 x_c = \frac{0.5T^3 \sin(0.1Tt_c)}{S_0}$$

If K is chosen as 2, then

$$K \frac{a_0}{a_1} = 20, \qquad \left(K \frac{a_0}{a_2}\right)^{1/2} = 10, \qquad \left(K \frac{a_0}{a_3}\right)^{1/3} > 10$$

Therefore let $T = 10$ units/volt. The final scaled equation for computer solution is

$$\frac{d^3x_c}{dt_c^3} + \frac{d^2x_c}{dt_c^2} + 2 \frac{dx_c}{dt_c} + x_c = \frac{500}{S_0} \sin t_c$$

Thus let $S_0 = 10$ units/volt. The final scaled equation for computer solution is

$$\frac{d^3x_c}{dt_c^3} = -\left(\frac{d^2x_c}{dt_c^2} + 2 \frac{dx_c}{dt_c} + x_c - 50 \sin t_c\right)$$

The final computer diagram is given in Fig. 11-22. The initial conditions on the computer are determined by applying the scale factors. Thus

$$x = S_0 x_c = 100 \qquad \text{at } t = 0$$

$$x_c = 10 \text{ volts} \qquad \text{at } t = 0$$

The first derivative is

$$\frac{dx}{dt}(0) = \frac{S_0}{T} \frac{dx_c}{dt_c}(0) = 5$$

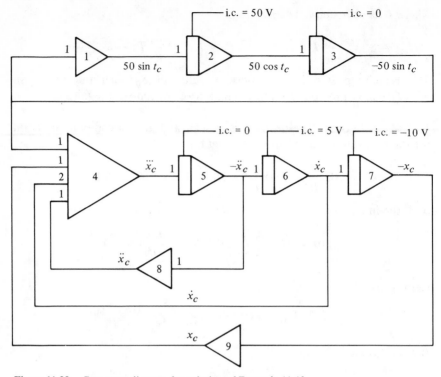

Figure 11-22. Computer diagram for solution of Example 11-13.

or

$$\frac{dx_c}{dt_c}(0) = 5 \text{ volts}$$

and the initial condition on the second derivative is zero.

11-4. EXAMPLES OF VIBRATION ANALYSIS USING THE ANALOG COMPUTER

The application of the analog computer to the analysis of the dynamic behavior of mass-elastic systems is best illustrated by inspection of some examples.

Linear Multidegree-of-Freedom Systems

Figure 11-23 illustrates a two-mass system with damping. An external force $f(t)$ is to be applied to the second mass m_2. It is desired to study the dynamic behavior of the system when different values of damping are incorporated.

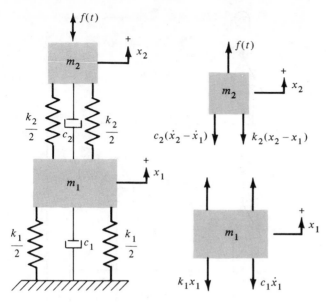

Figure 11-23. Two-mass system.

The equations of motion for the system are

$$m_2 \ddot{x}_2 = +f(t) - k_2(x_2 - x_1) - c_2(\dot{x}_2 - \dot{x}_1)$$
$$m_1 \ddot{x}_1 = +k_2(x_2 - x_1) + c_2(\dot{x}_2 - \dot{x}_1) - k_1 x_1 - c_1 \dot{x}_1$$

Dividing each equation by the respective mass, and rearranging the signs on the right-hand sides, yields the following two equations, which can readily be programmed in the form of a computer diagram.

$$\ddot{x}_2 = -\left[-\frac{f(t)}{m_2} + \frac{k_2}{m_2}(x_2 - x_1) + \frac{c_2}{m_2}(\dot{x}_2 - \dot{x}_1) \right]$$

$$-\ddot{x}_1 = -\left[+\frac{k_2}{m_1}(x_2 - x_1) + \frac{c_2}{m_1}(\dot{x}_2 - \dot{x}_1) - \frac{k_1}{m_1}x_1 - \frac{c_1}{m_1}\dot{x}_1 \right]$$

The analog-computer diagram for the solution of the problem is shown in Fig. 11-24. To simplify the wiring on the diagram, some connections are represented by matching numbers in the small circles. Although no attempt has been made to scale the problem at this point, the formal change of variables could have been introduced into the equations or, with some experience, the scaling may be done directly on the computer diagram.

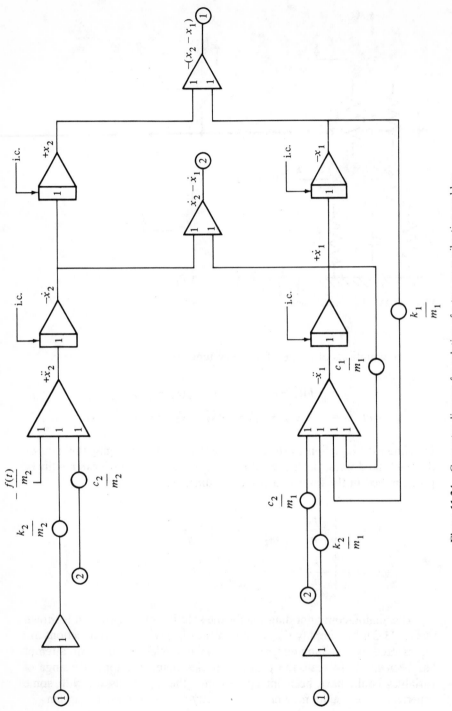

Figure 11-24. Computer diagram for solution of a two-mass vibration problem.

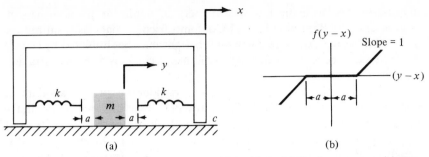

Figure 11-25. A spring-mass-damper system with backlash. (a) Nonlinear system. (b) Nonlinear spring-force function.

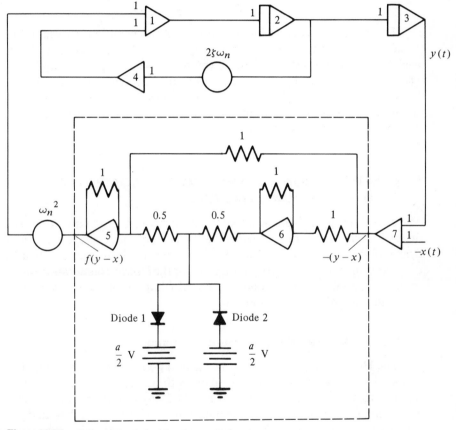

Figure 11-26. Computer solution for the nonlinear mechanical system.

Nonlinear Systems

The general-purpose analog computer is a powerful tool for the analysis of nonlinear vibratory systems. Of particular importance in this area of work is the relative ease of construction of nonlinear functions with electronic elements. Combinations of amplifiers, diodes, resistors, and voltage sources may be utilized to achieve a wide variety of nonlinear phenomena-simulation circuits. Reproductions of Coulomb friction, backlash, saturation, and hysteresis are examples. General functions may be simulated by using a diode function generator which approximates a curve by a series of straight-line segments.

As an example of a nonlinear analysis, consider the spring-mass-damper system shown in Fig. 11-25(a). The nature of this spring force acting on the mass is described by the function $f(y - x)$ illustrated by $f_s = kf(y - x)$. Applying Newton's law yields $m\ddot{y} = -c\dot{y} - kf(y - x)$. Thus the differential equation of motion for the system is $m\ddot{y} + c\dot{y} + kf(y - x) = 0$. The best form of this equation for computer solution is

$$\ddot{y} = -[+2\zeta\omega\dot{y} + \omega^2 f(y - x)]$$

The computer diagram for the solution of this equation is given in Fig. 11-26. The portion of the circuit that generates the dead zone is shown enclosed in the dashed rectangle of the figure. The output of amplifier 5, $f(y - x)$, is obtained by the superposition of a straight line and a limiter formed by diodes 1 and 2. The problem of scaling has not been discussed in the presentation of the solution. Generally, this is not a difficult task, but care must be taken in scaling the nonlinear element.

11-5. APPLICATION OF THE DIGITAL COMPUTER TO VIBRATION ANALYSIS

The digital computer is an ideal tool for any iterative-type computation. The determination of natural frequencies and vibratory-mode shapes for multi-degree-of-freedom systems is that kind of problem. Two examples will be developed in this section: first, a program for the Holzer computation for an in-line torsional system; second, the lateral vibration of a cantilever structure, using the Myklestad approach.

Holzer Method, Digital Computer Program

The Holzer computation† for the natural frequencies of an in-line torsional system may be applied to systems with free or fixed ends. A typical free-free torsional system is shown in Fig. 11-27(a). For this system, the disk at one end is given a prescribed angular rotation with sinusoidal motion at a selected frequency, and the remainder torque at the other end of the system is calculated. By successive selected frequency values and calculations, the remainder torque is brought to a small value. The corresponding frequency is then an acceptable value of a natural frequency for a principal mode. The

---

† Previously discussed in Sections 9-3 and 9-4.

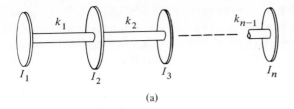

(a)

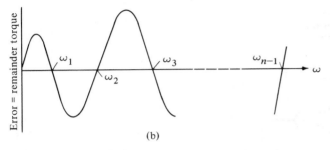

(b)

Figure 11-27. A typical torsional system and remainder-torque curve. (a) Torsional system. (b) Remainder-torque curve.

internal torques are readily determined from an analysis of the system (refer to Section 9-4) and using θ_i for amplitude A_i, as

$$M_{t_1} = I_1 \omega^2 \theta_1$$

$$M_{t_2} = I_1 \omega^2 \theta_1 + I_2 \omega^2 \theta_2$$

$$\cdots$$

$$M_{t_n} = \sum_{i=1}^{n} I_i \omega^2 \theta_i \tag{11-49}$$

where station n is the section just beyond the nth disk. The amplitudes are determined from the relations (see Section 9-4)

$$\theta_2 = \theta_1 - \frac{I_1 \omega^2 \theta_1}{k_1}$$

$$\cdots$$

$$\theta_j = \theta_{j-1} - \frac{1}{k_{j-1}} \sum_{i=1}^{j-1} I_i \omega^2 \theta_i \tag{11-50}$$

The nature of the remainder torque curve is illustrated in Fig. 11-27(b). The programming logic for the computation may be based on the nature of this curve. First, however, it is necessary to develop a flow program for computing the actual Holzer table.

Figure 11-28 is a flow diagram for calculating the essential elements of

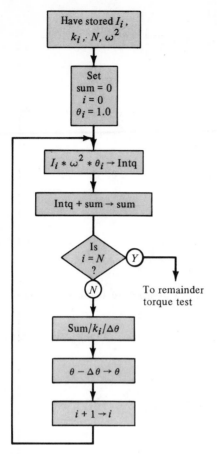

Figure 11-28. Flow diagram for calculating Holzer table. Remainder torque given: N, I_i's, k_i's, and ω^2 in memory.

the classic Holzer table. Given the inertias, stiffnesses, the number of inertias, and a specific value of frequency, this diagram represents the calculation of the error or the remainder torque, for this case.

Figure 11-29 is a flow chart illustrating one possible logic for stepping the program through the frequencies desired. Special attention should be directed to the test on the remainder-torque curve and to the criterion that establishes that a sufficiently close value of ω has been reached. First, the remainder torque is tested for changes from plus to minus; then, after finding the first frequency, the test is reversed to test on changes of sign from minus to plus. The nature of the mode shape at any natural frequency serves as a clue as to which normal mode actually has been obtained. If modes have been skipped, the program may be rerun with a modification of the frequency increment or a rescaling of the data. It can be readily derived that

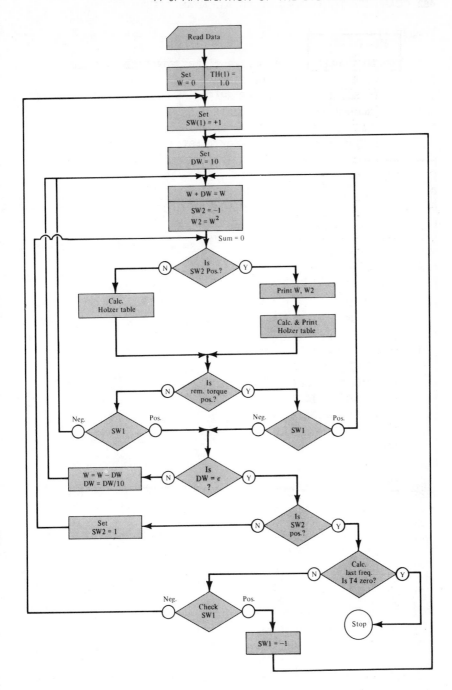

Figure 11-29. Flow chart for Holzer-calculation program.

Program 11-1

```
C       PROGRAM COMPUTES THE HOLZER TABLES FOR FREE—FREE SHAFT
C       FI=MOMENT OF INERTIA OF DISCS
C       FK=TORSIONAL STIFFNESS OF SHAFT
C       N+1=NUMBER OF DISCS (MAX=32)
C       M=NUMBER OF FREQUENCIES TO BE CALCULATED
C       TH=THETA            W=FREQUENCY (RADIANS PER SECOND)
C       DW=INCREMENT IN W
C       STARTING DW :           DW=  10.0    10.0<W<200.0
C                               DW= 100.0   200.0<W.<2000.0
C                               DW=1000.0   2000.0<W
C       SMALLEST DW IS 0.001 FOR W> 10.0
        DIMENSION FI(32),FK(31),TH(32)
        READ(5,100)N,M
  100   FORMAT(2(7X,I3))
        K=N+1
        READ(5,101) (FI(I),I=1,K)
        READ(5,101) (FK(I),I=1,N)
        W=0.0
  101   FORMAT(5E14.7)
        WRITE(6,102) N,(FI(I),I=1,K)
        WRITE(6,103) (FK(I),I=1,N)
  102   FORMAT(3H N=,I2/3H I=,5E14.7)
  103   FORMAT(3H K=,5E14.7)
   25   T1=N
        T4=M
        I=1
        TH(I)=1.
   22   SW1=1.
   24   IF (W-2000.) 203,202,202
  203   IF(W-200.) 204,201,201
  201   DW=100.
        GO TO 13
  202   DW=1000.
        GO TO 13
  204   DW=10.
        GO TO 13
   13   W=W+DW
        SW2=-1.
        W2=W*W
   19   C=0.
        T1=N
        I=1
        IF(SW2)3,3,4
```

Program 11-1 (*Continued*)

```
   4   WRITE(6,104)W,W2
 104   FORMAT(//3H W=,E14.7,2X,4HW*2=,E14.7//1X,3HPOS,
       16X,5HTHETA,12X,7HI*TH*W2,9X,6HTORQUE,9X,8HTORQUE/K)
   3   A=FI(I)*W2
       B=A*TH(I)
       C=C+B
       D=C/FK(I)
       J=I+1
       TH(J)=TH(I)-D
       IF(SW2)5,5,6
   6   WRITE(6,105)I,TH(I),B,C,D
 105   FORMAT(2X,I2,2X,E14.7,2X,E14.7,2X,E14.7,2X,E14.7)
   5   I=I+1
       T1=T1-1.
       IF(T1)7,7,3
   7   A=FI(I)*W2
       B=A*TH(I)
       C=C+B
       IF(SW2)8,8,9
   9   WRITE(6,106) I,TH(I),B,C
 106   FORMAT(2X,I2,2X,E14.7,2X,E14.7,2X,E14.7)
   8   IF(C)10,15,12
  10   IF(SW1)13,13,14
  12   IF(SW1)14,14,13
  14   T3=DW-.01
       IF(T3)15,15,16
  16   W=W-DW
       DW=DW/10.
       GO TO 13
  15   IF(SW2)17,17,18
  17   SW2=1.
       GO TO 19
  18   T4=T4-1.
       IF(T4)20,20,21
  21   IF(SW1)22,22,23
  23   SW1=0.
       GO TO 24
  20   STOP
       END
//DATA.INPUT DD *
        +03         +03
+1.1650000E+00+1.1650000E-02+1.1650000E-02+1.1650000E+00
+2.6200000E+03+2.6200000E+05+2.6200000E+03
```

Table 11-2

N= 3

I= 0.1165000E 01 0.1165000E-01 0.1165000E-01 0.1165000E 01

K= 0.2620000E 04 0.2620000E 06 0.2620000E 04

W= 0.4730995E 02 W*2= 0.2238231E 04

POS	THETA	I*TH*W2	TORQUE	TORQUE/K
1	0.1000000E 01	0.2607539E 04	0.2607539E 04	0.9952440E 00
2	0.4756033E-02	0.1240154E 00	0.2607663E 04	0.9952910E-02
3	-0.5196877E-02	-0.1355106E 00	0.2607528E 04	0.9952395E 00
4	-0.1000436E 01	-0.2608676E 04	-0.1147949E 01	

W= 0.4765972E 03 W*2= 0.2271448E 06

POS	THETA	I*TH*W2	TORQUE	TORQUE/K
1	0.1000000E 01	0.2646237E 06	0.2646237E 06	0.1010014E 03
2	-0.1000014E 03	-0.2646274E 06	-0.3687500E 01	-0.1407443E-04
3	-0.1000014E 03	-0.2646273E 06	-0.2646310E 06	-0.1010042E 03
4	0.1002808E 01	0.2653666E 06	0.7356250E 03	

W= 0.6723359E 04 W*2= 0.4520355E 08

POS	THETA	I*TH*W2	TORQUE	TORQUE/K
1	0.1000000E 01	0.5266213E 08	0.5266213E 08	0.2010005E 05
2	-0.2009905E 05	-0.1058458E 11	-0.1053192E 11	-0.4019817E 05
3	0.2009913E 05	0.1058463E 11	0.5270323E 08	0.2011573E 05
4	-0.1660938E 02	-0.8746849E 09	-0.8219817E 09	

---

the natural frequencies of a multimass system are proportional to $\sqrt{k/I}$, and that if all inertias and stiffnesses are modified by scale factors, the new frequencies are proportional to the square root of the ratio of the factors. A study of the programs presented in this section will reveal that computer time may be conserved by scaling the data in this fashion.

The remainder-torque test, illustrated in the flow chart of Fig. 11-29, is of special interest. It would be a difficult and relatively meaningless approach to attempt to find a frequency where the remainder torque is actually zero. The accuracy to which ω must be carried for this, particularly at higher frequencies where the slope of the remainder-torque curve is steep, would not be of engineering value. The test utilized therefore iterates around a crossing of the zero-torque axis until the frequency increment is carried to a prescribed increment ε. Generally, $\varepsilon = 0.01$ or 0.001 would provide sufficient accuracy.

A computer program for the Holzer computation is given in Program 11-1. The logic follows, in a general way, the flow chart of Fig. 11-29. A sample problem, with results from the program, is given in Table 11-2. Only the most important portion of the Holzer table is reproduced.

Myklestad Method, Digital Computer Program

The Myklestad method for determining the natural frequencies of a massless-elastic beam carrying concentrated masses was outlined in Section 9-11. Such a beam is shown in Fig. 11-30(a), with clamped-free boundary conditions.

For the computation it is convenient to take a beam segment with a mass at the right end as one unit, as shown in Fig. 11-30(b). The superscripts R and L, used to indicate the right and left of the mass, are dropped but it must be remembered here that positions are to the right of the mass. The transfer relation Eq. 9-83 is repeated as follows:

$$\left\{\begin{array}{c} y \\ \theta \\ M \\ V \end{array}\right\}_j = \begin{bmatrix} 1 & l & \dfrac{l^2}{2B} & \dfrac{l^3}{6B} \\ 0 & 1 & \dfrac{l}{B} & \dfrac{l^2}{2B} \\ 0 & 0 & 1 & l \\ m\omega^2 & m\omega^2 l & \dfrac{m\omega^2 l^2}{2B} & \left(1 + \dfrac{m\omega^2 l^3}{6B}\right) \end{bmatrix}_j \left\{\begin{array}{c} y \\ \theta \\ M \\ V \end{array}\right\}_{j-1} \tag{11-51}$$

Referring to Fig. 9-11, in the Myklestad solution for a clamped-free beam, $M_0 = 1$ and the shear V_0 is unknown; also, $y_0 = 0 = \theta_0$. The preceding equation may be written for the first segment as

$$\left\{\begin{array}{c} y \\ \theta \\ M \\ V \end{array}\right\}_1 = \begin{bmatrix} Q \end{bmatrix}_1 \left\{\begin{array}{c} 0 \\ 0 \\ M_0 \\ V_0 \end{array}\right\} \tag{11-52}$$

In order to keep the coefficients of M_0 and V_0 separate in the computer, the right side of Eq. 11-52 is written as two terms, with $M_0 = 1$.

$$\left\{\begin{array}{c} y \\ \theta \\ M \\ V \end{array}\right\}_1 = \begin{bmatrix} Q \end{bmatrix}_1 \left\{\begin{array}{c} 0 \\ 0 \\ 0 \\ 1 \end{array}\right\} V_0 + \begin{bmatrix} Q \end{bmatrix}_1 \left\{\begin{array}{c} 0 \\ 0 \\ 1 \\ 0 \end{array}\right\} \tag{11-53}$$

The symbols used in the computer are given in the following list:

$$l = \text{BL} \qquad \theta = \text{A}$$

$$l/B = \text{SU} \qquad M = \text{BMOM}$$

$$l^2/2B = \text{SL} \qquad V = \text{V}$$

$$l^3/6B = \text{UL} \qquad V_0 = \text{VO}$$

$$y = \text{Y} \qquad \omega = \text{W}$$

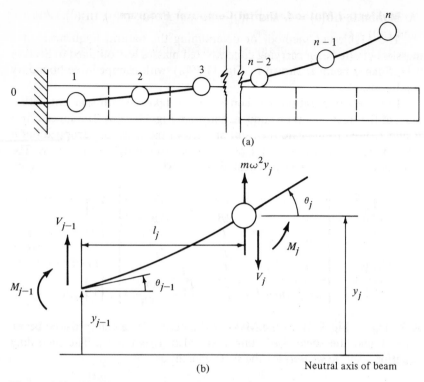

Figure 11-30. Modeling a cantilever beam for a digital computer study. (a) Model for computer study. (b) An element of the beam for computer study.

Then, in computer terms

$$[Q]_j = \begin{bmatrix} 1 & BL(J) & SL(J) & UL(J) \\ 0 & 1 & SU(J) & SL(J) \\ 0 & 0 & 1 & BL(J) \end{bmatrix} \qquad (11\text{-}54)$$

Instead of the last line in the $[Q]$ matrix of Eq. 9-83 to calculate V_j, it is more efficient to use $V_j = m_j \omega^2 y_j + V_{j-1}$, or in computer terms,

$$V(J) = BM(J)*(W**2)*Y(J)+V(J-1) \qquad (11\text{-}55)$$

The state matrices at any station must have computer symbols to identify each element as the multiplication program progresses from station to station.

In computer symbols Eq. 11-53 now becomes

$$\begin{Bmatrix} Y \\ A \\ BMOM \\ V \end{Bmatrix}_j = VO \begin{Bmatrix} DD \\ TT \\ HH \\ GG \end{Bmatrix}_j + \begin{Bmatrix} D \\ T \\ H \\ G \end{Bmatrix}_j \qquad (11\text{-}56)$$

At station 0, GG(0) = 1. and H(0) = 1., and DD(0), TT(0), HH(0), D(0), T(0), and G(0) = 0.

To the right of the nth mass, shear and moment must be zero. Using

$$M_n = 0 = \text{BMOM(N)}$$

then

$$0 = \text{VO*HH(N)+H(N)}$$

or

$$\text{VO} = \text{-H(N)/HH(N)} \tag{11-57}$$

This yields the value of vo, which is then used in Eq. 11-55 to calculate v(N). At a natural frequency, v(N) = 0; at other frequencies there is a remainder shear [that is, v(N) $\neq$ 0] which represents the force which is exciting the system.

If this procedure is repeated for various W, information will be obtained from which a remainder shear curve may be plotted. When the remainder shear goes to zero, a normal mode of vibration has been obtained, and the frequency and mode shape can be punched out.

A flow chart for computation of the lateral-vibration natural frequencies and mode shapes is given in Fig. 11-31. The general form of the logic is clearly similar to the Holzer program. The primary difference is the use of the equations for bending and the manipulation of these equations to solve readily in terms of the boundary conditions of the beam. In this case a remainder slope is used as the test for the selection of the correct frequency. Program 11-2 is an actual computer program for this computation.

In the example for which the data are shown, the number of segments equals 4 and the number of frequencies calculated is 2. Each beam segment is 0.25 long for a beam length of 1. The masses add to 1 if 0.125 mass is also lumped at the clamped end. The value $EI = 1$ for each segment. This is a reasonably good lumped representation of a uniform beam at low frequencies. The complete printout is shown in Table 11-3. The frequency values calculated are $\omega_1 = 3.4200$ rad/sec and $\omega_2 = 20.0999$ rad/sec, compared to values of 3.515 and 22.034 for a clamped-free beam with uniformly distributed mass (see Section 10-9).

Numerical Solution of Forced-Vibration Problem, Digital Computer Program

The numerical solution for the forced response of a mass-elastic system was discussed in Section 5-10. The applicable relations were Eqs. 5-45 through 5-62. Two cases were discussed with regard to the initial value of the forcing function $P(t_1)$, namely, $P(t_1) = 0$ and $P(t_1) \neq 0$. The computer pro-

Program 11-2

```
C      MYKLESTAD METHOD
C      N=NUMBER OF SEGMENTS        M=NUMBER OF FREQUENCIES CALCULATED
C      BL=SEGMENT LENGTH
C      BM=LUMPED MASS AT RIGHT END OF SEGMENT
C      EI=SEGMENT BENDING STIFFNESS
C      W=FREQUENCY (RADIANS PER SECOND)
C      WINC=INCREMENT IN W
C      STARTING W=10.0     WINC=10.0
       DIMENSION BL(40),BM(40),EI(40),A(40),Y(40),V(40),BMOM(40),G(40),
      1GG(40),H(40),HH(40),T(40),TT(40),D(40),DD(40),SU(40),SL(40),UL(40)
       READ(5,90)N,M
   90  FORMAT(2(7X,I3))
       READ(5,101) (BL(I),I=1,N)
       READ(5,102) (BM(I),I=1,N)
       READ(5,102) (EI(I),I=1,N)
  101  FORMAT(5F10.4)
  102  FORMAT(5E14.7)
       DO 14 J=1,N
       SU(J)=BL(J)/EI(J)
       SL(J)=BL(J)*SU(J)/2.
       UL(J)=BL(J)*SL(J)/3.
   14  CONTINUE
       GG(1)=1.
       HH(1)=0.
       TT(1)=0.
       DD(1)=0.
       G(1)=0.
       H(1)=1.
       T(1)=0.
       D(1)=0.
       PI=3.1415927
       MODE=0
       W=0.
       K=N+1
    1  SW=2.
    2  MODE=MODE+1
       WINC=10.
    3  W=W+WINC
       DO 4 J=1,N
       D(J+1)=D(J)+BL(J)*T(J)+SL(J)*H(J)+UL(J)*G(J)
       DD(J+1)=DD(J)+BL(J)*TT(J)+SL(J)*HH(J)+UL(J)*GG(J)
       T(J+1)=T(J)+SU(J)*H(J)+SL(J)*G(J)
       TT(J+1)=TT(J)+SU(J)*HH(J)+SL(J)*GG(J)
       H(J+1)=H(J)+BL(J)*G(J)
       HH(J+1)=HH(J)+BL(J)*GG(J)
       G(J+1)=G(J)+BM(J)*D(J+1)*W**2
```

460

Program 11-2 (*Continued*)

```
        GG(J+1)=GG(J)+BM(J)*DD(J+1)*W**2
    4   CONTINUE
        VO=-H(K)/HH(K)
        DO 5 J=1,K
        Y(J)=D(J)+DD(J)*VO
        A(J)=T(J)+TT(J)*VO
        BMOM(J)=H(J)+HH(J)*VO
        V(J)=G(J)+GG(J)*VO
    5   CONTINUE
        IF(-V(K)) 7,6,6
    6   IF(SW-1.)8,8,3
    7   IF(SW-1.)3,3,8
    8   IF(WINC-.01) 10,10,9
    9   W=W-WINC
        WINC=WINC/10.
        GO TO 3
   10   F=W/(2.*PI)
        WRITE(6,98) N
   98   FORMAT(//5X,21H NUMBER OF STATIONS =,I3)
        WRITE(6,93) MODE,W,F
   93   FORMAT(//5X,6HMODE =,I3,20H NATURAL FREQUENCY =,
       1F9.4,13H RAD PER SEC ,F9.4,4H CPS)
        WRITE(6,94) (A(J),J=1,K)
        WRITE (6,95) (Y(J),J=1,K)
        WRITE (6,96) (V(J),J=1,K)
        WRITE (6,97) (BMOM(J),J=1,K)
   94   FORMAT(//10X,39HSLOPE AT EACH STATION (INCHES PER INCH)
       1//(16X,E14.6))
   95   FORMAT(//10X,35HDEFLECTION AT EACH STATION (INCHES)
       1//(18X,E14.6))
   96   FORMAT(//10X,36HSHEAR FORCE AT EACH STATION (POUNDS)
       1//(16X,E14.6))
   97   FORMAT(//10X,44HBENDING MOMENT AT EACH STATION (INCH POUNDS)
       1//(16X,E14.6))
        IF(MODE-M) 11,13,13
   11   IF (SW-1.) 1,1,12
   12   SW=0.
        GO TO 2
   13   CONTINUE
        STOP
        END
//DATA.INPUT DD *
        +04        +02
+0000.2500+0000.2500+0000.2500+0000.2500
+0.2500000E+00+0.2500000E+00+0.2500000E+00+0.1250000E+00
+0.1000000E+01+0.1000000E+01+0.1000000E+01+0.1000000E+01
```

Table 11-3

NUMBER OF STATIONS = 4
MODE = 1 NATURAL FREQUENCY = 3.4200 RAD PER SEC 0.5443 CPS

    SLOPE AT EACH STATION (INCHES PER INCH)
        0.000000E 00
        0.208078E 00
        0.334849E 00
        0.391736E 00
        0.404931E 00

    DEFLECTION AT EACH STATION (INCHES)
        0.000000E 00
        0.277565E-01
        0.972634E-01
        0.189357E 00
        0.289491E 00

    SHEAR FORCE AT EACH STATION (POUNDS)
       -0.134150E 01
       -0.126033E 01
       -0.975928E 00
       -0.422229E 00
       0.102043E-02

    BENDING MOMENT AT EACH STATION (INCH POUNDS)
        0.100000E 01
        0.664625E 00
        0.349541E 00
        0.105559E 00
        0.953674E-06

---

gram may be written for either circumstance. The initial conditions are

$$x_1 = 0$$

$$\dot{x}_1 = 0$$

$$\ddot{x}_1 = \frac{P(t_1)}{m}$$

If $P(t_1) = 0$, the value of x_2 obtained by solving Eqs. 5-60, 5-61, and 5-62 simultaneously is

$$x_2 = \frac{P(t_2)/m}{\{[6/(\Delta t)^2] + (c/m)[3/(\Delta t)] + (k/m)\}}$$

Table 11-3 (*Continued*)

```
NUMBER OF STATIONS =  4
MODE = 2 NATURAL FREQUENCY = 20.0999 RAD PER SEC    3.1990 CPS

    SLOPE AT EACH STATION (INCHES PER INCH)
            0.000000E 00
            0.107872E 00
           -0.726074E-02
           -0.172565E 00
           -0.239862E 00

    DEFLECTION AT EACH STATION (INCHES)
            0.000000E 00
            0.194060E-01
            0.353523E-01
            0.115949E-01
           -0.427628E-01

    SHEAR FORCE AT EACH STATION (POUNDS)
           -0.454809E 01
           -0.258806E 01
            0.982562E 00
            0.215364E 01
           -0.593567E-02

    BENDING MOMENT AT EACH STATION (INCH POUNDS)
            0.100000E 01
           -0.137023E 00
           -0.784039E 00
           -0.538399E 00
            0.152588E-04
```

Also,

$$\dot{x}_2 = \frac{3x_2}{\Delta t} \quad \text{and} \quad \ddot{x}_2 = \frac{6x_2}{(\Delta t)^2}$$

If $P(t_1) \neq 0$, then

$$x_2 = \frac{\ddot{x}_1 (\Delta t)^2}{2}$$

$$\dot{x}_2 = \ddot{x}_1 (\Delta t)$$

$$\ddot{x}_2 = \frac{P(t_2)}{m} - \frac{c}{m} \dot{x}_2 - \frac{k}{m} x_2$$

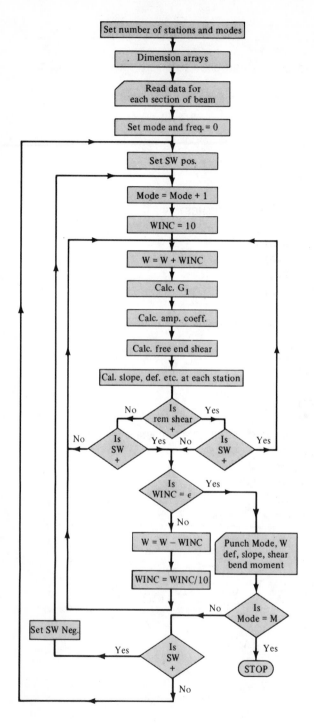

Figure 11-31. Flow chart for Myklestad calculation.

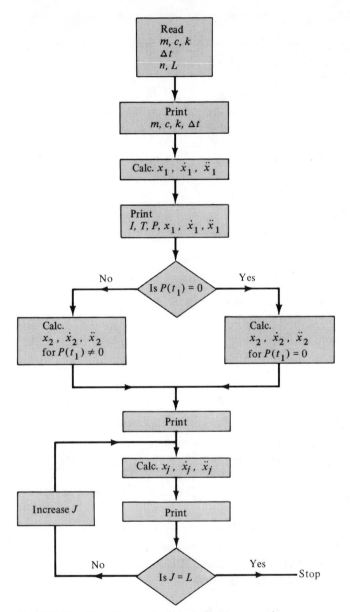

Figure 11-32. Flow diagram for forced vibration example.

For the third and subsequent point ($i \geq 3$), Eqs. 5-47, 5-48, and 5-49 are used to calculate x_i, $\dot{x}_i$, and $\ddot{x}_i$.

The flow diagram is shown in Fig. 11-32. The computer program, designated as Program 11-3 follows. The input data used for an example here are the same as those for Example 5-8 up to $t = 0.05$ sec. The output is given in Table 11-4.

Program 11-3

```
C      NUMERICAL SOLUTION, FORCED TRANSIENT, SINGLE-DEGREE-OF-FREEDOM
C      BM=MASS     C=DAMPING CONSTANT
C      BK=SPRING STIFFNESS
C      DEL=TIME INCREMENT     T=TIME
C      N=NUMBER OF CONSECUTIVE VALUES OF P(T) READ IN
C      L+1=TOTAL NUMBER OF POINTS WHICH WILL BE CALCULATED INCLUDING T=0
       DIMENSION P(500),X(500),V(500),A(500)
       READ (5,101) BM,C,BK
       READ (5,102) DEL
       READ (5,103) N,L
       READ (5,104) (P(J),J=1,N)
 101   FORMAT (3F10.4)
 102   FORMAT (F10.6)
 103   FORMAT (2I5)
 104   FORMAT (8F10.2)
       WRITE (6,105) BM,C,BK,DEL
 105   FORMAT (3H M=,F10.4,2X,2HC=,F10.4,2X,2HK=,F10.4,2X,
      18HDELTA T=,F10.6)
       D=DEL
       D2=D*D
       X(1)=0.
       V(1)=0.
       A(1)=P(1)/BM
       T=0.*D
       J=1
       WRITE (6,106)
 106   FORMAT (4X,1HI,8X,1HT,10X,1HP,14X,1HX,14X,1HV,14X,1HA)
       WRITE (6,107) J,T,P(1),X(1),V(1),A(1)
 107   FORMAT (2X,I4,2X,F10.4,2X,F10.2,2X,E14.6,2X,E14.6,2X,E14.6)
       J=J+1
```

11-6. DIGITAL COMPUTER PROGRAM CSMP FOR ANALOG BLOCK DIAGRAM

Many systems that are simulated on the analog computer may also be studied on the digital computer. The operations available under FORTRAN have been extended to include operations performed on the analog computer, one of the most important being numerical integration. An example is the IBM program designated as CSMP, which is an acronym for Continuous System Modeling Program. The CSMP allows direct digital-computer programming from differential equations or from a block diagram

Program 11-3 (*Continued*)

```
        IF (P(1)) 2,1,2
    1   DEN=(6./D2)+(3.*C/(BM*D))+(BK/BM)
        X(2)=P(2)/(BM*DEN)
        V(2)=3.*X(2)/D
        A(2)=6.*X(2)/D2
        GO TO 3
    2   X(2)=A(1)*D2/2.
        V(2)=A(1)*D
        A(2)=(P(2)/BM)-(C*V(2)/BM)-(BK*X(2)/BM)
        GO TO 3
    3   T=1.*D
        WRITE (6,107) J,T,P(2),X(2),V(2),A(2)
        DO 4 J=2,L
        K=J+1
        IF (N-K) 6,5,5
    6   P(J+1)=0.
    5   X(J+1)=A(J)*D2+2.*X(J)-X(J-1)
        V(J+1)=(3.*X(J+1)-4.*X(J)+X(J-1))/(2.*D)
        A(J+1)=(P(J+1)/BM)-(C*V(J+1)/BM)-(BK*X(J+1)/BM)
        TT=J
        T=TT*D
    4   WRITE (6,107) K,T,P(J+1),X(J+1),V(J+1),A(J+1)
        CONTINUE
        STOP
        END
//DATA.INPUT DD *
+0004.0000+0000.0000+2500.0000
+00.010000
+0006+0020
+000000.00+000400.00+000660.00+000840.00+000960.00+001000.00
```

for analog simulation. Application-oriented input statements simplify programming. A translator converts these statements into a FORTRAN routine. A few examples of the many operations and functions available are given in Table 11-5.

The routines generated are executed with a selected integration routine from a list that includes: (a) fifth-order Milne predictor-corrector, (b) fourth-order Runge-Kutta, (c) Simpson's, (d) second-order Adams, (e) trapezoidal, and (f) rectangular methods. A graphic display feature is also available with CSMP.

As an example of the programming, the differential equation of Example

Table 11-4

M= 4.0000 C= 0.0000 K= 2500.0000 DELTA T= 0.010000

I	T	P	X	V	A
1	0.0000	0.00	0.000000E 00	0.000000E 00	0.000000E 00
2	0.0100	400.00	0.164948E-02	0.494845E 00	0.989691E 02
3	0.0200	660.00	0.131959E-01	0.164948E 01	0.156753E 03
4	0.0300	840.00	0.404175E-01	0.350592E 01	0.184739E 03
5	0.0400	960.00	0.861130E-01	0.549324E 01	0.186179E 03
6	0.0500	1000.00	0.150426E 00	0.736223E 01	0.155984E 03
7	0.0600	0.00	0.230338E 00	0.877109E 01	-0.143961E 03
8	0.0700	0.00	0.295854E 00	0.583175E 01	-0.184909E 03
9	0.0800	0.00	0.342878E 00	0.377793E 01	-0.214299E 03
10	0.0900	0.00	0.368473E 00	0.148797E 01	-0.230296E 03
11	0.1000	0.00	0.371038E 00	-0.894975E 00	-0.231899E 03
12	0.1100	0.00	0.350414E 00	-0.322193E 01	-0.219009E 03
13	0.1200	0.00	0.307888E 00	-0.534758E 01	-0.192430E 03
14	0.1300	0.00	0.246119E 00	-0.713900E 01	-0.153825E 03
15	0.1400	0.00	0.168968E 00	-0.848423E 01	-0.105605E 03
16	0.1500	0.00	0.812566E-01	-0.929920E 01	-0.507854E 02
17	0.1600	0.00	-0.115336E-01	-0.953295E 01	0.720851E 01
18	0.1700	0.00	-0.103603E 00	-0.917089E 01	0.647518E 02
19	0.1800	0.00	-0.189197E 00	-0.823566E 01	0.118248E 03
20	0.1900	0.00	-0.262967E 00	-0.678569E 01	0.164354E 03
21	0.2000	0.00	-0.320300E 00	-0.491164E 01	0.200188E 03

Table 11-5

General Form	Function	
$Y = \text{INTGRL } (\text{IC}, X)$	$Y = \int_0^t X\, dt + IC$	
	$Y(0) = IC$	
$Y = \text{DERIV } (\text{IC}, X)$	$Y = \dfrac{dX}{dt} + IC$	
	$\dfrac{dX}{dt}(0) = IC$	$\cdots$
$Y = \text{LIMIT } (P_1,\ P_2,\ X)$	$Y = P_1 \text{ for } X < P_1$	
	$Y = P_2 \text{ for } X > P_2$	
	$Y = X \text{ for } P_1 \leq X \leq P_2$	
		$\cdots$
$Y = \text{STEP } (P)$	$Y = 0 \text{ for } t < P$	
	$Y = 1 \text{ for } t \geq P$	
		$\cdots$
$Y = \text{RAMP } (P)$	$Y = 0 \text{ for } t < P$	
	$Y = t - P \text{ for } t \geq P$	
		$\cdots$

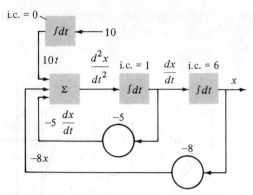

Figure 11-33. Computer block diagram for Example 11-1.

Program 11-4

```
DYNAMIC     X2DOT = TERM1 + TERM2 + TERM3
            XDOT  = INTGRL (1.0, X2DOT)
            X     = INTGRL (6.0, XDOT)
            TERM1 = -C * XDOT
            TERM2 = -K * X
            Y     = RAMP (0.0)
            TERM3 = 10.0 * Y
PARAMETER C       = (1.0, 2.0, 5.0)
PARAMETER K       = 8.0
TIMER     FINTIM = 20.0, OUTDEL=0.2, DELT = 0.05
METHOD    SIMP
```

11-1 is used, along with the block diagram of Fig. 11-7, which is repeated here as Fig. 11-33.

$$\frac{d^2x}{dt^2} = -5\frac{dx}{dt} - 8x + 10t$$

The main portion of the program is given in Program 11-4.

The three values of damping parameter C will result in three separate computer runs, one for each value listed. The FINTIM specifies that the computation will stop when the independent variable, t, reaches a value of 20.0. The t increment for computation is 0.05 and the increment at which output is displayed is 0.2. If no method is specified, the Runge-Kutta numerical integration scheme is used. Here Simpson's rule is specified.

The computation may also be terminated by a FINISH statement, for example:

$$\text{FINISH} \quad X = 0.01, \; XDOT = 0.05$$

The run is then ended when one of the dependent variables reaches or first crosses the specified values, unless FINTIM occurs first.

11-7. SUMMARY

This chapter has outlined some of the possibilities for the application of analog and digital computers in vibration analysis. It was shown that the simulation of vibratory systems, particularly complex machine elements, and nonlinear problems can be accomplished by utilizing general-purpose analog-computer elements and special nonlinear function-generation equipment. If the analyst is primarily interested in transient phenomena, or the general behavior of a system for a variety of inputs, including a random type of input, the analog computer is of particular value. For this reason, considerable emphasis has been placed on analog-computer programming concepts and techniques. A sound foundation in the fundamentals of computer scaling, both magnitude and time scaling, is of particular importance in achieving dependable computer simulation. With sufficient practice, supported by a thorough understanding of the fundamentals, it is possible to bypass the more formal scaling techniques presented in Section 11-3 and scale problems directly on the computer diagram. Special attention, however, should be paid to the scaling of nonlinear elements.

The digital computer is a useful tool for analyzing the normal modes of vibration of machine elements and structures. Examples of the application of digital-computer techniques have been presented: first, an iterative procedure for evaluating in-line torsional systems and, second, a program for analyzing the lateral-vibration characteristics of a cantilever beam. In both cases it is possible to extend the programming technique to consider more complex systems of a multitude of configurations. For example, the torsional program may be extended to consider a branched multimass system that could be used to analyze the torsional vibration of vehicle drive systems or machine-tool drive systems. The lateral-vibration program can readily be extended to consider various beam end-conditions by altering the boundary conditions outlined in the mathematical development. An example of the digital computer program for the solution of a forced-vibration problem is also included. Additionally, the CSMP digital program for systems simulated on the analog computer or block diagram is presented.

The examples presented in this chapter serve to illustrate the current and potential application of computer techniques in the field of vibration analysis. Naturally, many more applications are possible, and the future holds an even greater potential with the widespread application of more sophisticated mathematical techniques.

PROBLEMS

11-1. For a single-degree-of-freedom, forced, damped system, consider the analog-computer method of solution and construct (a) the block diagram and (b) the analog-computer diagram.

11-2. For the two-degree-of-freedom, forced, and intramass-damped system of Fig. 7-20, consider the analog-computer method of solution and construct (a) the block diagram and (b) the analog-computer diagram.

11-3. For any assigned problem of Chapter 9, consider the digital-computer solution method and (a) construct the appropriate flow chart(s) and diagram(s) and (b) write the computer program.

11-4. For any assigned problem in the group Prob. 5-34 through 5-37, consider the digital-computer solution method and (a) construct the appropriate flow chart(s) and diagram(s) and (b) write the computer program.

11-5. Set up the digital computer solution for the problem of a uniform beam with both ends fixed.

Appendix A

Matrices

A-1. INTRODUCTION

A matrix is a rectangular array of elements, in rows and columns, that obey certain rules of multiplication and addition. The elements may be numbers, coefficients, terms, or variables. A matrix alone has no meaning, unless its elements are defined or are associated with other matrices.

A-2. TYPES OF MATRICES, THEIR DESIGNATION AND NOTATION

Rectangular Matrices

The arrangement

$$\begin{bmatrix} a_{11} & a_{12} & a_{13} & \cdots & a_{1n} \\ a_{21} & a_{22} & a_{23} & \cdots & a_{2n} \\ a_{31} & a_{32} & \cdot & \cdots & a_{3n} \\ \cdot & \cdot & \cdot & \cdots & \cdot \\ a_{m1} & a_{m2} & \cdot & \cdots & a_{mn} \end{bmatrix} \tag{A-1}$$

represents the general form of a rectangular matrix of m rows and n columns. The first subscript specifies the *row* and the second subscript denotes the *column* in which the element appears. The matrix may also be designated as $[A]$, or by $[a_{ij}]$ which, in reality, designates the general element of the matrix.

473

Thus

$$[A] = \begin{bmatrix} a_{11} & a_{12} & a_{13} & \cdots \\ a_{21} & a_{22} & \cdot & \cdots \\ \cdot & \cdot & \cdot & \cdots \\ \cdot & \cdot & \cdot & \cdots \end{bmatrix} = [a_{ij}]$$ (A-2)

The elements may be numbers. For example,

$$\begin{bmatrix} 4 & -5 & 3 & 2 \\ 1 & 7 & 0 & 6 \\ 8 & 0 & -9 & 10 \end{bmatrix}$$

is a 3 × 4 or 3 by 4 matrix—that is, 3 rows and 4 columns.

Column and Row Matrices

The arrangement

$$\{x\} = \begin{Bmatrix} x_1 \\ x_2 \\ x_3 \\ \cdot \\ \cdot \\ x_m \end{Bmatrix}$$ (A-3)

represents a column matrix of m elements. In effect, it is a rectangular matrix of m rows and 1 column, or $m \times 1$.

The array

$$\lfloor y \rfloor = \lfloor y_1 \quad y_2 \quad y_3 \quad \cdots \quad y_n \rfloor$$ (A-4)

is a row matrix of n elements. It can be considered as a rectangular matrix of 1 row and n columns, or $1 \times n$.

Square, Diagonal, and Unit Matrices

A rectangular matrix in which $m = n$ represents a square matrix. Thus

$$[B] = \begin{bmatrix} b_{11} & b_{12} & b_{13} & \cdots & b_{1n} \\ b_{21} & b_{22} & \cdot & \cdots & b_{2n} \\ b_{31} & \cdot & \cdot & \cdots & \cdot \\ \cdot & \cdot & \cdot & \cdots & \cdot \\ b_{n1} & \cdot & \cdot & \cdots & b_{nn} \end{bmatrix}$$ (A-5)

is a square matrix. This type frequently occurs in matrix representation.

If all of the elements in a square matrix are zero except the diagonal terms (that is, $c_{ij} = 0$ for $i \neq j$), the array is called a diagonal matrix. Thus

$$[C] = \begin{bmatrix} c_{11} & 0 & 0 & \cdots & 0 \\ 0 & c_{22} & 0 & \cdots & 0 \\ 0 & & \cdot & \cdots & \cdot \\ \cdot & & & & \\ 0 & 0 & \cdot & \cdots & c_{nn} \end{bmatrix} \tag{A-6}$$

is a diagonal matrix.

A diagonal matrix in which the diagonal elements are equal to unity $(c_{ii} = 1)$ is termed a unit matrix or identity matrix $[I]$. Thus

$$[I] = \begin{bmatrix} 1 & 0 & 0 & \cdots & 0 \\ 0 & 1 & 0 & \cdots & 0 \\ 0 & 0 & 1 & \cdots & \cdot \\ \cdot & & \cdot & \cdots & \cdot \\ 0 & 0 & \cdot & \cdots & 1 \end{bmatrix} \tag{A-7}$$

Transpose of a Matrix

The transpose of matrix $[A]$ may be designated as $[A]^T$ and is obtained by interchanging columns and rows in order. Thus

$$[A]^T = \begin{bmatrix} a_{11} & a_{12} & \cdots & a_{1n} \\ a_{21} & a_{22} & \cdots & a_{2n} \\ \cdot & \cdot & \cdots & \cdot \\ \cdot & \cdot & \cdots & \cdot \\ a_{m1} & \cdot & \cdots & a_{mn} \end{bmatrix}^T = \begin{bmatrix} a_{11} & a_{21} & \cdots & a_{m1} \\ a_{12} & a_{22} & \cdots & a_{m2} \\ \cdot & \cdot & \cdots & \cdot \\ \cdot & \cdot & \cdots & \cdot \\ a_{1n} & \cdot & \cdots & a_{mn} \end{bmatrix} \tag{A-8}$$

The transpose of a column matrix is a row matrix and vice versa. That is,

$$\{x\}^T = \lfloor x \rfloor \qquad \text{and} \qquad \lfloor y \rfloor^T = \{y\} \tag{A-9}$$

Symmetric and Skew-Symmetric Matrices

If a square matrix is symmetric about its diagonal—that is, if $a_{ij} = a_{ji}$—the matrix is said to be symmetric. For example,

$$[A] = \begin{bmatrix} 4 & 1 & 3 \\ 1 & 5 & 2 \\ 3 & 2 & 0 \end{bmatrix}$$

is a symmetric matrix. For a symmetric matrix $[A_s]$, it is noted that $[A_s]^T = [A_s]$.

A skew-symmetric matrix is a square matrix in which $a_{ij} = -a_{ji}$ and $a_{ii} = 0$. Thus

$$[B] = \begin{bmatrix} 0 & -1 & 3 \\ 1 & 0 & -2 \\ -3 & 2 & 0 \end{bmatrix}$$

is a skew-symmetric matrix.

Singular Matrix

A matrix is said to be singular if its determinant is zero.

Null Matrix

A null matrix is one in which all of the elements are zero. Thus

$$[0] = \begin{bmatrix} 0 & 0 & \cdots & 0 \\ 0 & \cdot & \cdots & 0 \\ \cdot & \cdot & \cdots & \cdot \\ 0 & \cdot & \cdots & 0 \end{bmatrix} \tag{A-10}$$

is a null rectangular matrix. Similarly, $\{0\}$ and $\lfloor 0 \rfloor$ are null column and null row matrices, respectively.

A-3. SUMMATION OF MATRICES

Addition and Subtraction of Matrices

Addition and subtraction of two matrices may be done if they are of the same order. Thus

$$[A] \pm [B] = [C] \tag{A-11}$$

is accomplished by adding or subtracting corresponding elements as represented by

$$a_{ij} \pm b_{ij} = c_{ij} \tag{A-12}$$

This may be illustrated by

$$\begin{bmatrix} 2 & 1 \\ 4 & 3 \end{bmatrix} + \begin{bmatrix} 1 & 3 \\ 5 & -7 \end{bmatrix} = \begin{bmatrix} 3 & 4 \\ 9 & -4 \end{bmatrix}$$

and

$$\begin{bmatrix} 2 & 1 \\ 4 & 3 \end{bmatrix} - \begin{bmatrix} 1 & 3 \\ 5 & -7 \end{bmatrix} = \begin{bmatrix} 1 & -2 \\ -1 & 10 \end{bmatrix}$$

Equality of Matrices

Two matrices are equal if, and only if, they are of the same order and the corresponding elements are equal. Thus

$$[A] = [B] \tag{A-13}$$

providing

$$a_{ij} = b_{ij} \tag{A-13a}$$

for each pair of subscripts i and j. This is made evident by subtracting matrix $[B]$ from both sides of Eq. A-13, resulting in

$$[A] - [B] = [0]$$

which is correct only if the requirement A-13a is fulfilled for the equal-order matrices.

Commutative and Associative Laws of Addition

Matrix addition or subtraction is a commutative process. Thus

$$[A] + [B] = [B] + [A] \tag{A-14}$$

since

$$a_{ij} + b_{ij} = b_{ij} + a_{ij}$$

The addition of matrices is also associative.

$$[A] + ([B] + [C]) = ([A] + [B]) + [C] \tag{A-15}$$

as

$$a_{ij} + (b_{ij} + c_{ij}) = (a_{ij} + b_{ij}) + c_{ij}$$

A-4. MULTIPLICATION OF MATRICES

Matrices may be multiplied providing they meet certain requirements, which will be considered later. Thus

$$[A][B] = [C] \tag{A-16}$$

indicates that matrix $[C]$ is the product of matrices $[A]$ and $[B]$. In the multiplying procedure, the c_{ij} element of matrix $[C]$ is obtained as the sum of the products of the elements of *row* i of matrix $[A]$ and the corresponding elements of *column* j of matrix $[B]$. To illustrate this process, consider the matrices

$$[A] = \begin{bmatrix} a_{11} & a_{12} & a_{13} \\ a_{21} & a_{22} & a_{23} \end{bmatrix} \qquad [B] = \begin{bmatrix} b_{11} & b_{12} & b_{13} & b_{14} \\ b_{21} & b_{22} & b_{23} & b_{24} \\ b_{31} & b_{32} & b_{33} & b_{34} \end{bmatrix}$$

Then

$$[A][B] = \begin{bmatrix} a_{11} & a_{12} & a_{13} \\ a_{21} & a_{22} & a_{23} \end{bmatrix} \begin{bmatrix} b_{11} & b_{12} & b_{13} & b_{14} \\ b_{21} & b_{22} & b_{23} & b_{24} \\ b_{31} & b_{32} & b_{33} & b_{34} \end{bmatrix}$$

$$= \begin{bmatrix} (a_{11}b_{11} + a_{12}b_{21} + a_{13}b_{31}) & (a_{11}b_{12} + a_{12}b_{22} + a_{13}b_{32}) \\ (a_{21}b_{11} + a_{22}b_{21} + a_{23}b_{31}) & (a_{21}b_{12} + a_{22}b_{22} + a_{23}b_{32}) \end{bmatrix}$$

$$\qquad \begin{matrix} (a_{11}b_{13} + a_{12}b_{23} + a_{13}b_{33}) & (a_{11}b_{14} + a_{12}b_{24} + a_{13}b_{34}) \\ (a_{21}b_{13} + a_{22}b_{23} + a_{23}b_{33}) & (a_{21}b_{14} + a_{22}b_{24} + a_{23}b_{34}) \end{matrix}$$

$$= [C] \tag{A-17}$$

From the foregoing, the general rule for multiplication of matrices may be stated as

$$c_{ij} = \sum_{k=1}^{n} a_{ik} b_{kj} \tag{A-18}$$

where k is a dummy index.

A-5. REQUIREMENTS AND CONDITIONS PERTAINING TO MATRIX MULTIPLICATION

Two matrices may be multiplied only if they are conformable—that is, if the number of columns of the first matrix is equal to the number of rows of the second matrix. Thus

$$[A] \quad [B] \quad = \quad [C]$$
$$m \times n \; n \times p \quad m \times p \tag{A-19}$$

That is, if $[A]$ is an $m \times n$ matrix and $[B]$ is an $n \times p$ matrix, then the multiplication of $[A]$ by $[B]$ results in the $m \times p$ matrix $[C]$.

In general, matrix multiplication is not commutative, so that

$$[A][B] \neq [B][A] \tag{A-20}$$

It then becomes necessary to specify the order in which the multiplication is carried out. This is done in two ways: Eq. A-19 is described as matrix $[B]$ premultiplied by $[A]$, or matrix $[A]$ postmultiplied by $[B]$.

If a square matrix $[A]$ is multiplied by a unit matrix $[I]$ of the same order, the result is commutative and also equal to matrix $[A]$. That is,

$$[A][I] = [I][A] = [A] \tag{A-21}$$

as can be readily demonstrated.

Multiplication of matrices is an associative process. Thus if

$$[A][B] = [D]$$

and

$$[B][C] = [E]$$

then

$$[A][B][C] = [D][C] = [A][E] = [F] \qquad \text{(A-22)}$$

The matrix product

$$[A][B] = [0] \qquad \text{(A-23)}$$

does not necessarily require or imply that either $[A]$ or $[B]$ or both, are null matrices, although this could be the case. Consider the example

$$\begin{bmatrix} 3 & 3 \\ 3 & 3 \end{bmatrix} \begin{bmatrix} -2 & 2 \\ 2 & -2 \end{bmatrix} = \begin{bmatrix} 0 & 0 \\ 0 & 0 \end{bmatrix}$$

For a scalar multiplier of a matrix, each element of the matrix is multiplied by the scalar. Thus

$$c \begin{bmatrix} a_{11} & a_{12} & a_{13} & \cdots & a_{1n} \\ a_{21} & a_{22} & a_{23} & \cdots & a_{2n} \\ \cdot & \cdot & \cdot & \cdots & \cdot \\ a_{n1} & \cdot & \cdot & \cdots & a_{nn} \end{bmatrix} = \begin{bmatrix} ca_{11} & ca_{12} & ca_{13} & \cdots & ca_{1n} \\ ca_{21} & ca_{22} & ca_{23} & \cdots & ca_{2n} \\ \cdot & \cdot & \cdot & \cdots & \cdot \\ ca_{n1} & \cdot & \cdot & \cdots & ca_{nn} \end{bmatrix} \qquad \text{(A-24)}$$

If $[B]$ is a symmetric matrix, then

$$\{A\}^T[B]\{C\} = \{C\}^T[B]\{A\} \qquad \text{(A-25)}$$

To illustrate this, let

$$\{A\} = \begin{Bmatrix} a_1 \\ a_2 \\ a_3 \end{Bmatrix}, \quad [B] = \begin{bmatrix} b_{11} & b_{12} & b_{13} \\ b_{21} & b_{22} & b_{23} \\ b_{31} & b_{32} & b_{33} \end{bmatrix}, \quad \{C\} = \begin{Bmatrix} c_1 \\ c_2 \\ c_3 \end{Bmatrix}$$

Then

$$\{A\}^T[B]\{C\} = \lfloor a_1 \quad a_2 \quad a_3 \rfloor \begin{bmatrix} b_{11} & b_{12} & b_{13} \\ b_{21} & b_{22} & b_{23} \\ b_{31} & b_{32} & b_{33} \end{bmatrix} \begin{Bmatrix} c_1 \\ c_2 \\ c_3 \end{Bmatrix}$$

$$= \lfloor a_1 \quad a_2 \quad a_3 \rfloor \begin{Bmatrix} (b_{11}c_1 + b_{12}c_2 + b_{13}c_3) \\ (b_{21}c_1 + b_{22}c_2 + b_{23}c_3) \\ (b_{31}c_1 + b_{32}c_2 + b_{33}c_3) \end{Bmatrix}$$

$$= \lfloor a_1(b_{11}c_1 + b_{12}c_2 + b_{13}c_3) + a_2(b_{21}c_1 + b_{22}c_2 + b_{23}c_3)$$
$$+ a_3(b_{31}c_1 + b_{32}c_2 + b_{33}c_3) \rfloor \qquad \text{(A-26)}$$

Similarly,

$$\{C\}^T[B]\{A\} = \lfloor c_1 \quad c_2 \quad c_3 \rfloor \begin{bmatrix} b_{11} & b_{12} & b_{13} \\ b_{21} & b_{22} & b_{23} \\ b_{31} & b_{32} & b_{33} \end{bmatrix} \begin{Bmatrix} a_1 \\ a_2 \\ a_3 \end{Bmatrix}$$

$$= \lfloor c_1(b_{11}a_1 + b_{12}a_2 + b_{13}a_3) + c_2(b_{21}a_1 + b_{22}a_2 + b_{23}a_3)$$

$$+ c_3(b_{31}a_1 + b_{32}a_2 + b_{33}a_3) \rfloor \tag{A-27}$$

If $[B]$ is symmetric, then

$$b_{12} = b_{21}, \qquad b_{13} = b_{31}, \qquad \text{and} \qquad b_{23} = b_{32}$$

and Eqs. A-26 and A-27 are the same, which verifies the general relation, Eq. A-25.

A-6. DETERMINANT OF A SQUARE MATRIX

Consider the square matrix

$$[A] = \begin{bmatrix} a_{11} & a_{12} & a_{13} & \cdots & a_{1n} \\ a_{21} & a_{22} & a_{23} & \cdots & a_{2n} \\ \cdot & \cdot & \cdot & \cdots & \cdot \\ a_{n1} & a_{n2} & a_{n3} & \cdots & a_{nn} \end{bmatrix} = [a_{ij}] \tag{A-28}$$

The determinant of $[A]$ is designated in the following manner:

$$|A| = \begin{vmatrix} a_{11} & a_{12} & a_{13} & \cdots & a_{1n} \\ a_{21} & a_{22} & a_{23} & \cdots & a_{2n} \\ \cdot & \cdot & \cdot & \cdots & \cdot \\ a_{n1} & a_{n2} & a_{n3} & \cdots & a_{nn} \end{vmatrix} \tag{A-29}$$

By the determinant is meant the sum of all possible products, with prescribed signs, formed by choosing one and only one element from each row and from each column in the determinant. The manner of expanding a determinant will be subsequently explained.

The minor M_{ij} of any element a_{ij} of determinant $|A|$ of order n is the $(n-1)$-order subdeterminant that remains when the row and column containing a_{ij} are deleted from the determinant. Thus

$$M_{11} = \begin{vmatrix} a_{22} & a_{23} & \cdots & a_{2n} \\ a_{32} & a_{33} & \cdots & a_{3n} \\ \cdot & \cdot & \cdots & \cdot \\ a_{n2} & a_{n3} & \cdots & a_{nn} \end{vmatrix} \qquad M_{12} = \begin{vmatrix} a_{21} & a_{23} & \cdots & a_{2n} \\ a_{31} & a_{33} & \cdots & a_{3n} \\ \cdot & \cdot & \cdots & \cdot \\ a_{n1} & a_{n3} & \cdots & a_{nn} \end{vmatrix}$$

$$M_{21} = \begin{vmatrix} a_{12} & a_{13} & \cdots & a_{1n} \\ a_{32} & a_{33} & \cdots & a_{3n} \\ \cdot & \cdot & \cdots & \cdot \\ a_{n2} & a_{n3} & \cdots & a_{nn} \end{vmatrix} \qquad M_{22} = \begin{vmatrix} a_{11} & a_{13} & \cdots & a_{1n} \\ a_{31} & a_{33} & \cdots & a_{3n} \\ \cdot & \cdot & \cdots & \cdot \\ a_{n1} & a_{n3} & \cdots & a_{nn} \end{vmatrix} \tag{A-30}$$

The cofactor C_{ij} of any element a_{ij} of determinant $|A|$ is $(-1)^{i+j}$ times the minor of a_{ij}. That is, the cofactor C_{ij} is the signed minor of a_{ij}. Thus

$$C_{ij} = (-1)^{i+j}M_{ij} \tag{A-31}$$

The expansion of a determinant is the sum of the products of the elements of *any row or column* multiplied by the corresponding cofactors. Thus if the determinant is expanded using the first row,

$$|A| = a_{11}C_{11} + a_{12}C_{12} + a_{13}C_{13} + \cdots + a_{1n}C_{1n}$$

$$= a_{11}M_{11} - a_{12}M_{12} + a_{13}M_{13} - \cdots + (-1)^{(1+n)}M_{1n} \tag{A-32}$$

As an illustration, obtain the determinant of the matrix

$$[A] = \begin{bmatrix} 1 & 4 & 2 \\ 3 & 0 & -1 \\ -2 & 3 & -4 \end{bmatrix} \tag{A-33}$$

by expanding on the elements of the first row and their cofactors. Thus

$$|A| = \begin{vmatrix} 1 & 4 & 2 \\ 3 & 0 & -1 \\ -2 & 3 & -4 \end{vmatrix}$$

$$= 1 \times \begin{vmatrix} 0 & -1 \\ 3 & -4 \end{vmatrix} - 4 \times \begin{vmatrix} 3 & -1 \\ -2 & -4 \end{vmatrix} + 2 \times \begin{vmatrix} 3 & 0 \\ -2 & 3 \end{vmatrix}$$

$$= 1 \times 3 - 4 \times (-14) + 2 \times 9 = 77 \tag{A-34}$$

A-7. MATRIX REPRESENTATION OF A SYSTEM OF EQUATIONS

Consider the set of n nonhomogeneous linear equations in n unknowns

$$a_{11}x_1 + a_{12}x_2 + a_{13}x_3 + \cdots + a_{1n}x_n = b_1$$

$$a_{21}x_1 + a_{22}x_2 + a_{23}x_3 + \cdots + a_{2n}x_n = b_2$$

$$\vdots$$

$$a_{n1}x_1 + a_{n2}x_2 + a_{n3}x_3 + \cdots + a_{nn}x_n = b_n \tag{A-35}$$

Using the principles and rules of matrix multiplication outlined above, this system of equations can be written in matrix form as

$$\begin{bmatrix} a_{11} & a_{12} & a_{13} & \cdots & a_{1n} \\ a_{21} & a_{22} & a_{23} & \cdots & a_{2n} \\ \cdot & \cdot & \cdot & \cdots & \cdot \\ a_{n1} & a_{n2} & \cdot & \cdots & a_{nn} \end{bmatrix} \begin{Bmatrix} x_1 \\ x_2 \\ \cdot \\ x_n \end{Bmatrix} = \begin{Bmatrix} b_1 \\ b_2 \\ \cdot \\ b_n \end{Bmatrix} \tag{A-36}$$

or more succinctly as

$$[a_{ij}]\{x_i\} = \{b_i\} \tag{A-37}$$

or

$$[A]\{x\} = \{B\} \tag{A-38}$$

A-8. SOLUTION OF NONHOMOGENEOUS EQUATIONS

The set of nonhomogeneous equations listed as Eq. A-36 has a unique solution which may be determined by Cramer's rule providing the determinant of the coefficient matrix

$$|A| = \begin{vmatrix} a_{11} & a_{12} & a_{13} & \cdots & a_{1n} \\ a_{21} & a_{22} & & \cdots & \cdot \\ \cdot & \cdot & \cdot & \cdots & \cdot \\ a_{n1} & \cdot & \cdot & \cdots & a_{nn} \end{vmatrix} \tag{A-39}$$

is different from zero—that is, the matrix $[A]$ is nonsingular. The solution is then given by

$$x_1 = \frac{|A_1|}{|A|}, \quad x_2 = \frac{|A_2|}{|A|}, \quad \ldots, \quad x_n = \frac{|A_n|}{|A|} \tag{A-40}$$

wherein $|A_j|$ is the determinant obtained by replacing the elements of the jth column of the a's by the column of the b's.

A-9. ADJOINT OF A SQUARE MATRIX

If $[A] = [a_{ij}]$ is a square matrix and if $C_{ij} = (-1)^{i+j}M_{ij}$ is its cofactor in the determinant of $[A]$, then the transpose of matrix C_{ij} is the adjoint matrix of $[A]$. Thus

$$[\text{adj } A] = [C_{ij}]^T \tag{A-41}$$

To illustrate this, determine the adjoint of the following matrix:

$$[A] = \begin{bmatrix} 1 & 4 & 2 \\ 3 & 0 & -1 \\ -2 & 3 & -4 \end{bmatrix} \tag{A-42}$$

(which is the same matrix as Eq. A-33)

$$[\text{adj } A] = \begin{bmatrix} \begin{vmatrix} 0 & -1 \\ 3 & -4 \end{vmatrix} & -\begin{vmatrix} 3 & -1 \\ -2 & -4 \end{vmatrix} & \begin{vmatrix} 3 & 0 \\ -2 & 3 \end{vmatrix} \\[8pt] -\begin{vmatrix} 4 & 2 \\ 3 & -4 \end{vmatrix} & \begin{vmatrix} 1 & 2 \\ -2 & -4 \end{vmatrix} & -\begin{vmatrix} 1 & 4 \\ -2 & 3 \end{vmatrix} \\[8pt] \begin{vmatrix} 4 & -2 \\ 0 & -1 \end{vmatrix} & -\begin{vmatrix} 1 & 2 \\ 3 & -1 \end{vmatrix} & \begin{vmatrix} 1 & 4 \\ 3 & 0 \end{vmatrix} \end{bmatrix}^T$$

$$= \begin{bmatrix} 3 & 14 & 9 \\ 22 & 0 & -11 \\ -4 & 7 & -12 \end{bmatrix}^T = \begin{bmatrix} 3 & 22 & -4 \\ 14 & 0 & 7 \\ 9 & -11 & -12 \end{bmatrix} \qquad \text{(A-43)}$$

A-10. INVERSE, OR RECIPROCAL, OF SQUARE MATRIX

The inverse, or reciprocal, of a square matrix $[A]$ is denoted as $[A]^{-1}$ and is defined by

$$[A][A]^{-1} = [I] = [A]^{-1}[A] \qquad \text{(A-44)}$$

where $[I]$ is the unit matrix of the same order as $[A]$.

Some of the properties of the unit matrix are noted below. From Eq. A-44 it follows that

$$[I][I]^{-1} = [I] \qquad \text{(A-45)}$$

Also, from Eq. A-21 wherein $[A]$ is a square matrix,

$$[A][I] = [A]$$

Postmultiplying this by $[I]^{-1}$ gives

$$[A][I][I]^{-1} = [A][I]^{-1}$$

which, by virtue of Eq. A-45, becomes

$$[A][I] = [A][I]^{-1}$$

from which it is concluded that

$$[I] = [I]^{-1} \qquad \text{(A-46)}$$

Then

$$[I][I] = [I][I]^{-1}$$
$$= [I] \qquad \text{(A-47)}$$

It is also reasonably apparent that the inverse of a matrix inverse is the matrix itself. Thus

$$[[A]^{-1}]^{-1} = [A] \tag{A-48}$$

A-11. DETERMINATION OF INVERSE MATRIX

In order to obtain a relation for determining the inverse matrix, the product of the matrix $[A]$ and its adjoint $[\text{adj } A]$ will first be obtained, as follows

$$[A][\text{adj } A] = \begin{bmatrix} a_{11} & a_{12} & a_{13} & \cdots & a_{1n} \\ a_{21} & a_{22} & a_{23} & \cdots & a_{2n} \\ a_{31} & \cdot & \cdot & \cdots & a_{3n} \\ \cdot & \cdot & \cdot & \cdots & \cdot \\ a_{n1} & \cdot & \cdot & \cdots & a_{nn} \end{bmatrix}$$

$$\times \begin{bmatrix} M_{11} & -M_{21} & M_{31} & \cdots & (-1)^{1+n}M_{n1} \\ -M_{12} & M_{22} & -M_{32} & \cdots & (-1)^{2+n}M_{n2} \\ M_{13} & \cdot & \cdot & \cdots & (-1)^{3+n}M_{n3} \\ \cdot & \cdot & \cdot & \cdots & \cdot \\ (-1)^{1+n}M_{1n} & \cdot & \cdot & \cdots & (-1)^{n+n}M_{nn} \end{bmatrix} \tag{A-49}$$

$$= \begin{bmatrix} |A| & 0 & 0 & \cdot & 0 \\ 0 & |A| & 0 & \cdot & 0 \\ 0 & 0 & |A| & 0 & \cdot \\ \cdot & \cdot & \cdot & \cdot & \cdot \\ 0 & \cdot & \cdot & 0 & |A| \end{bmatrix} \tag{A-50}$$

$$= |A| \begin{bmatrix} 1 & 0 & 0 & \cdots & 0 \\ 0 & 1 & 0 & \cdots & 0 \\ 0 & 0 & 1 & \cdots & 0 \\ \cdot & \cdot & \cdot & \cdots & \cdot \\ 0 & \cdot & \cdot & \cdots & 1 \end{bmatrix}$$

$$= |A|[I] \tag{A-51}$$

The elements of matrix Eq. A-50 resulting from the multiplication of the matrices here are not readily apparent. However, careful consideration reveals that each diagonal term is precisely the determinant $|A|$. For example, grouping the first row and first column gives

$$a_{11}M_{11} - a_{12}M_{12} + a_{13}M_{13} - a_{14}M_{14} + \cdots + (-1)^{1+n}a_{1n}M_{1n} = |A|$$

which is the first diagonal term. Likewise, grouping the second row and second column gives

$$-a_{21}M_{21} + a_{22}M_{22} - a_{23}M_{23} + a_{24}M_{24} - \cdots + (-1)^{2+n}a_{2n}M_{2n} = |A|$$

which is the second diagonal term, and so on.

That the remaining terms of the matrix are zero results from the following theorem: The sum of the products formed by multiplying the elements of one row (or column) of a determinant by the cofactors of the corresponding elements of *another* row (or column) is zero.† This is exactly the procedure involved in the determination of each nondiagonal term of Eq. A-50. For example, grouping the first row and second column of Eq. A-49 gives

$$-a_{11}M_{21} + a_{12}M_{22} - a_{13}M_{23} + a_{14}M_{24} - \cdots + (-1)^{2+n}a_{1n}M_{2n} = 0$$

Similarly, grouping the first row and third column gives

$$a_{11}M_{31} - a_{12}M_{32} + a_{13}M_{33} - a_{14}M_{34} + \cdots + (-1)^{3+n}a_{1n}M_{3n} = 0$$

and so on.

Here, Eq. A-51 may then be written as

$$|A|[I] = [A][\text{adj } A]$$

and dividing by $|A|$, which is scalar, gives

$$[I] = \frac{[A][\text{adj } A]}{|A|}$$

$$[A][A]^{-1} = \frac{[A][\text{adj } A]}{|A|}$$

$$[A]^{-1} = \frac{[\text{adj } A]}{|A|} \tag{A-52}$$

which may be used for determining the inverse of a matrix. To illustrate the foregoing, consider the following example: Determine the product of the matrix

$$[A] = \begin{bmatrix} 1 & 4 & 2 \\ 3 & 0 & -1 \\ -2 & 3 & -4 \end{bmatrix}$$

and its adjoint

$$[\text{adj } A] = \begin{bmatrix} 3 & 22 & -4 \\ 14 & 0 & 7 \\ 9 & -11 & -12 \end{bmatrix}$$

† See C. R. Wylie, Jr., *Advanced Engineering Mathematics*, 3rd. ed. (New York: McGraw-Hill, 1966), *Theorem 11*, p. 409.

(refer to Eqs. A-33, A-34, and A-43)

$$
[A][\text{adj } A] = \begin{bmatrix} 1 & 4 & 2 \\ 3 & 0 & -1 \\ -2 & 3 & -4 \end{bmatrix} \begin{bmatrix} 3 & 22 & -4 \\ 14 & 0 & 7 \\ 9 & -11 & -12 \end{bmatrix}
$$

$$
= \begin{bmatrix} (1 \times 3 + 4 \times 14 + 2 \times 9) \\ (3 \times 3 + 0 \times 14 - 1 \times 9) \\ (-2 \times 3 + 3 \times 14 - 4 \times 9) \end{bmatrix}
$$

$$
(1 \times 22 + 4 \times 0 + 2 \times -11)
$$
$$
(3 \times 22 + 0 \times 0 - 1 \times -11)
$$
$$
(-2 \times 22 + 3 \times 0 - 4 \times -11)
$$

$$
\begin{bmatrix} (1 \times -4 + 4 \times 7 + 2 \times -12) \\ (3 \times -4 + 0 \times 7 - 1 \times -12) \\ (-2 \times -4 + 3 \times 7 - 4 \times -12) \end{bmatrix}
$$

$$
= \begin{bmatrix} 77 & 0 & 0 \\ 0 & 77 & 0 \\ 0 & 0 & 77 \end{bmatrix} = 77 \begin{bmatrix} 1 & 0 & 0 \\ 0 & 1 & 0 \\ 0 & 0 & 1 \end{bmatrix} \qquad \text{(A-53)}
$$

Which agrees with Eq. A-51. (To show that the 77 is the determinant for this case, refer to Eq. A-34.)

Also, determine the inverse of the above matrix $[A]$. The inverse matrix $[A]^{-1}$ will be

$$
[A]^{-1} = \frac{[\text{adj } A]}{|A|}
$$

$$
= \frac{\begin{bmatrix} 3 & 22 & -4 \\ 14 & 0 & 7 \\ 9 & -11 & -12 \end{bmatrix}}{77\dagger}
$$

$$
= \begin{bmatrix} \dfrac{3}{77} & \dfrac{2}{7} & \dfrac{-4}{77} \\ \dfrac{2}{11} & 0 & \dfrac{1}{11} \\ \dfrac{9}{77} & -\dfrac{1}{7} & \dfrac{-12}{77} \end{bmatrix} \qquad \text{(A-54)}
$$

It can be readily shown that $[A][A]^{-1} = [I]$ for this case.

† See Eq. A-34.

Appendix B

Derivation of Lagrange's Equations

Lagrange's equations will be derived for a holonomic† system of μ mass particles having n degrees of freedom. The derivation will be based on the principle that for an incremental change in configuration of such a system, the change in kinetic energy is equal to the work done by the forces acting on the system. Thus

$$dT = dU \tag{B-1}$$

Before employing this principle it is necessary to establish certain fundamental concepts and relations.

A set of *independent* coordinates which properly and completely define the configuration of a system and whose number is equal to the number of degrees of freedom is called the *generalized coordinates*. A single generalized coordinate is thus associated with each degree of freedom, and a change in any one coordinate does not require a change in any other coordinate. The generalized coordinates may be linear, or angular, or of some other nature. The n generalized coordinates will be designated here as $q_1, q_2, \ldots, q_k, \ldots, q_n$.

---

† Systems having equations of constraint containing only coordinates or coordinates and time are called *holonomic systems*. If the equations of constraint contain velocities, the systems are then *nonholonomic*.

The n *generalized forces*, which may be forces, moments, or some other quantity, are defined by

$$Q_1 = \frac{dU_1}{dq_1}, \qquad Q_2 = \frac{dU_2}{dq_2}, \qquad \dots,$$

$$Q_k = \frac{dU_k}{dq_k}, \qquad \dots, \qquad Q_n = \frac{dU_n}{dq_n}$$

(B-2)

where dU_k is the work done by the force system in an incremental displacement dq_k of the generalized coordinate q_k only. Thus the generalized forces are not usually actual or observable forces acting on the system, but some component of a combination of such forces.

It is also necessary to consider a set of x, y, z coordinates. If there are v mass particles, then $n = 3v$, and the configuration of the mass system can be expressed by $x_1, x_2, \dots, x_j, \dots, x_v; y_1, y_2, \dots, y_j, \dots, y_v; z_1, z_2, \dots, z_j, \dots, z_v$. The relations between the x, y, z coordinates and the generalized coordinates are given by

$$x_j = x_j(q_1, q_2, \dots, q_n)$$

$$y_j = y_j(q_1, q_2, \dots, q_n)$$

(B-3)

$$z_j = z_j(q_1, q_2, \dots, q_n)$$

where $j = 1, 2, \dots, v$. The kinetic energy T of the system may then be expressed as

$$T = \frac{1}{2} \sum_{j=1}^{v} m_j(\dot{x}_j^2 + \dot{y}_j^2 + \dot{z}_j^2)$$

(B-4)

In order to express T in terms of the generalized coordinates, the derivatives of Eqs. B-3 with respect to time are needed. To clarify this process the total derivative of x_j is first written. Thus

$$dx_j = \frac{\partial x_j}{\partial q_1} dq_1 + \frac{\partial x_j}{\partial q_2} dq_2 + \cdots + \frac{\partial x_j}{\partial q_n} dq_n$$

(B-5)

Dividing by dt results in

$$\dot{x}_j = \frac{\partial x_j}{\partial q_1} \dot{q}_1 + \frac{\partial x_j}{\partial q_2} \dot{q}_2 + \cdots + \frac{\partial x_j}{\partial q_n} \dot{q}_n = \sum_{k=1}^{n} \frac{\partial x_j}{\partial q_k} \dot{q}_k$$

(B-6)

Similarly,

$$\dot{y}_j = \frac{\partial y_j}{\partial q_1} \dot{q}_1 + \frac{\partial y_j}{\partial q_2} \dot{q}_2 + \cdots + \frac{\partial y_j}{\partial q_n} \dot{q}_n = \sum_{k=1}^{n} \frac{\partial y_j}{\partial q_k} \dot{q}_k$$

$$\dot{z}_j = \frac{\partial z_j}{\partial q_1} \dot{q}_1 + \frac{\partial z_j}{\partial q_2} \dot{q}_2 + \cdots + \frac{\partial z_j}{\partial q_n} \dot{q}_n = \sum_{k=1}^{n} \frac{\partial z_j}{\partial q_k} \dot{q}_k$$

Substituting these into Eq. B-4 results in

$$T = \frac{1}{2} \sum_{j=1}^{v} m_j \left[\left(\sum_{k=1}^{n} \frac{\partial x_j}{\partial q_k} \dot{q}_k \right)^2 + \left(\sum_{k=1}^{n} \frac{\partial y_j}{\partial q_k} \dot{q}_k \right)^2 + \left(\sum_{k=1}^{n} \frac{\partial z_j}{\partial q_k} \dot{q}_k \right)^2 \right] \quad \text{(B-7)}$$

Taking the derivative of this with respect to a particular $\dot{q}_k$, and then multiplying by $\dot{q}_k$ gives

$$\frac{\partial T}{\partial \dot{q}_k} \dot{q}_k = \sum_{j=1}^{v} m_j \left[\left(\frac{\partial x_j}{\partial q_k} \dot{q}_k \right)^* \sum_{k=1}^{n} \frac{\partial x_j}{\partial q_k} \dot{q}_k \right.$$
$$\left. + \left(\frac{\partial y_j}{\partial q_k} \dot{q}_k \right)^* \sum_{k=1}^{n} \frac{\partial y_j}{\partial q_k} \dot{q}_k + \left(\frac{\partial z_j}{\partial q_k} \dot{q}_k \right)^* \sum_{k=1}^{n} \frac{\partial z_j}{\partial q_k} \dot{q}_k \right] \quad \text{(B-8)}$$

In expressing this derivative, note that

$$\frac{\partial}{\partial \dot{q}_k} \left(\sum_{k=1}^{n} \frac{\partial x_j}{\partial q_k} \dot{q}_k \right)^2 = \left[\frac{\partial x_j}{\partial q_k} \frac{\partial \dot{q}_k}{\partial \dot{q}_k} + \dot{q}_k \frac{\partial}{\partial \dot{q}_k} \left(\frac{\partial x_j}{\partial q_k} \right) \right] \cdot 2 \sum_{k=1}^{n} \left(\frac{\partial x_j}{\partial q_k} \dot{q}_k \right)$$
$$= 2 \frac{\partial x_j}{\partial q_k} \sum_{k=1}^{n} \left(\frac{\partial x_j}{\partial q_k} \dot{q}_k \right) \quad \text{(B-9)}$$

This is because x_j is a function of q_k only, so that $\partial x_j / \partial q_k$ is also a function of q_k only, and then its partial derivative with respect to $\dot{q}_k$ is zero. Likewise,

$$\frac{\partial}{\partial \dot{q}_k} \left(\frac{\partial y_j}{\partial q_k} \right) = 0 \quad \text{and} \quad \frac{\partial}{\partial \dot{q}_k} \left(\frac{\partial z_j}{\partial q_k} \right) = 0$$

Also note that in Eq. B-8 the terms designated by ( )* are common terms which are not within the summation relative to k. Next, summing Eq. B-8 gives

$$\sum_{k=1}^{n} \frac{\partial T}{\partial \dot{q}_k} \dot{q}_k = \sum_{j=1}^{v} m_j \left[\left(\sum_{k=1}^{n} \frac{\partial x_j}{\partial q_k} \dot{q}_k \right)^2 + \left(\sum_{k=1}^{n} \frac{\partial y_j}{\partial q_k} \dot{q}_k \right)^2 + \left(\sum_{k=1}^{n} \frac{\partial z_j}{\partial q_k} \dot{q}_k \right)^2 \right]$$

$$\text{(B-10)}$$

Comparison of Eqs. B-7 and B-10 reveals that

$$2T = \sum_{k=1}^{n} \frac{\partial T}{\partial \dot{q}_k} \dot{q}_k \quad \text{(B-11)}$$

Differentiating this with respect to time gives

$$2 \frac{dT}{dt} = \sum_{k=1}^{n} \left[\frac{d}{dt} \left(\frac{\partial T}{\partial \dot{q}_k} \right) \dot{q}_k + \frac{\partial T}{\partial \dot{q}_k} \frac{d\dot{q}_k}{dt} \right] \quad \text{(B-12)}$$

Studying Eqs. B-3 and B-6 reveals that since x_j, y_j, and z_j are functions of the q's, then $\dot{x}_j$, $\dot{y}_j$, and $\dot{z}_j$ are functions of the q's and $\dot{q}$'s. Then from Eq. B-7 it may be stated that T is a function of the q's and $\dot{q}$'s, and the total time derivative of T would be written as

$$\frac{dT}{dt} = \sum_{k=1}^{n} \left(\frac{\partial T}{\partial q_k} \frac{dq_k}{dt} + \frac{\partial T}{\partial \dot{q}_k} \frac{d\dot{q}_k}{dt} \right) \tag{B-13}$$

Subtracting this from Eq. B-12 and multiplying by dt gives

$$dT = \sum_{k=1}^{n} \left[\frac{d}{dt} \left(\frac{\partial T}{\partial \dot{q}_k} \right) dq_k - \frac{\partial T}{\partial q_k} dq_k \right] \tag{B-14}$$

From the definition of generalized force,

$$dU = Q_1 \, dq_1 + Q_2 \, dq_2 + \cdots + Q_n \, dq_n = \sum_{k=1}^{n} Q_k \, dq_k \tag{B-15}$$

Then using Eqs. B-14 and B-15 and equating dT and dU in accordance with Eq. B-1, gives

$$\sum_{k=1}^{n} \left[\frac{d}{dt} \left(\frac{\partial T}{\partial \dot{q}_k} \right) - \frac{\partial T}{\partial q_k} - Q_k \right] dq_k = 0 \tag{B-16}$$

The variations identified by dq_k are arbitrary since $q_1, q_2, \ldots, q_n$ are independent coordinates without constraint. Therefore the term within the brackets must vanish. Hence

$$\frac{d}{dt} \left(\frac{\partial T}{\partial \dot{q}_k} \right) - \frac{\partial T}{\partial q_k} = Q_k \tag{B-17}$$

wherein $k = 1, 2, \ldots, n$.

The generalized force Q_k can be considered to be composed of (a) non-potential force Q_k' which alters the energy of the system, that is, either removes energy or adds energy to the system, and (b) force Q_k'' which has potential and does not alter the energy of the system. Thus

$$Q_k = Q_k' + Q_k'' \tag{B-18}$$

Now for the displacement dq_k the change in potential energy, dV, is $(\partial V / \partial q_k) \cdot dq_k$. This is equal to the negative of the work done by the potential force, which is $Q_k'' \, dq_k$. Thus

$$\frac{\partial V}{\partial q_k} dq_k = -Q_k'' \, dq_k$$

and

$$Q_k'' = -\frac{\partial V}{\partial q_k} \tag{B-19}$$

Substituting Eqs. B-18 and B-19 into Eq. B-17 yields

$$\frac{d}{dt}\left(\frac{\partial T}{\partial \dot{q}_k}\right) - \frac{\partial T}{\partial q_k} + \frac{\partial V}{\partial q_k} = Q'_k \tag{B-20}$$

This is known as *Lagrange's equation* or *equations* (since it represents n relations as k takes on the values from 1 to n). Defining $L = T - V$, where L is commonly termed the *Lagrangian function*, enables Eq. B-20 to be written as

$$\frac{d}{dt}\left(\frac{\partial L}{\partial \dot{q}_k}\right) - \frac{\partial L}{\partial q_k} = Q'_k \tag{B-21}$$

since $\partial V/\partial \dot{q}_k = 0$ because V is a function of q_k only.

For the special case of a conservative system, $Q'_k = 0$ and Lagrange's equation becomes

$$\frac{d}{dt}\left(\frac{\partial T}{\partial \dot{q}_k}\right) - \frac{\partial T}{\partial q_k} + \frac{\partial V}{\partial q_k} = 0 \tag{B-22}$$

or

$$\frac{d}{dt}\left(\frac{\partial L}{\partial \dot{q}_k}\right) - \frac{\partial L}{\partial q_k} = 0 \tag{B-23}$$

Appendix C

Units and Dimensional Values

Some of the guidelines for the use of SI units are given here.†

C-1. UNIT PREFIXES

In general, numerical values should be kept between 0.1 and 1000. For this purpose, the following *prefixes* are used:

Amount	Multiple	Prefix	Symbol
1 000 000	10^6	mega	M
1 000	10^3	kilo	k
100	10^2	hecto	h
10	10	deka	da
0.1	10^{-1}	deci	d
0.01	10^{-2}	centi	c
0.001	10^{-3}	milli	m
0.000 001	10^{-6}	micro	μ

Thus a length of 0.032 74 m is written as 3.274 cm or 32.74 mm; a mass of 4358 g is written as 4.358 kg; and a force of 0.007 32 N is written as 7.32 mN.

---

† See *SI Metric Reference Manual*, 3rd Ed. (New York: IBM, March 1974). Also see J. L. Meriam, *Dynamics*, 2nd Ed., SI-Version (New York: Wiley, 1975), pp. 5–10.

C-2. UNIT DESIGNATIONS

In order to avoid confusion, a dot is used where units are multiplied together. For example, the unit for the moment of a force, which is newton-meter, is written as N · m (the unit mN would have the meaning of millinewton). A compound unit such as that for stress, which is formed by division, would be written as N/m^2 or $N \cdot m^{-2}$ and not as N/m/m which is ambiguous. Also, prefixed units are generally not used in the denominator. That is, $37 \ N/cm^2$ would usually be written as $370 \ kN/m^2$.

C-3. NUMBER GROUPING

A space instead of a comma is used to separate groups of numbers in powers of 10^3. This is shown, for example, by the number 3 478 562.091 82, where the space is used on both sides of the decimal point. The space is generally omitted for four digit numbers only, such as 0.0427 or 2893.

C-4. CONVERSION FACTORS

The following table shows factors for converting non-SI units to the SI system, for dimensions involved in vibrational calculations. Factors marked with a dagger (†) are exact; others are correct to the number of places given.

To convert from	To	Multiply by
Acceleration		
foot/second^2 (ft/s^2)	meter/second^2 (m/s^2)	$3.048\,000 \times 10^{-1}$†
inch/second^2 (in./s^2)	meter/second^2 (m/s^2)	$2.540\,000 \times 10^{-2}$†
Area		
foot^2 (ft^2)	meter^2 (m^2)	$9.290\,304 \times 10^{-2}$†
inch^2 (in.^{2})	meter^2 (m^2)	$6.451\,600 \times 10^{-4}$†
Damping coefficient		
pound second/inch (lbf · sec/in.)	newton sec/m (N · s/m)	$1.751\,268 \times 10^2$
Density		
pound mass/foot^3 (lbm/ft^3)	kilogram/meter^3 (kg/m^3)	$1.601\,846 \times 10$
pound mass/inch^3 (lbm/in.^{3})	kilogram/meter^3 (kg/m^3)	$2.767\,990 \times 10^4$
Energy, work		
foot-pound-force (ft · lbf)	joule (J) = N · m	$1.355\,818$
inch-pound-force (in · lbf)	joule (J) = N · m	$1.129\,848 \times 10^{-1}$
Flexibility coefficient		
inch/pound (in./lbf)	meter/newton (m/N)	$5.710\,148 \times 10^{-3}$

(*Continued overleaf*)

Table (*Continued*)

To convert from	To	Multiply by
Force		
kip (1000 lbf)	newton (N)	$4.448\,222 \times 10^3$
pound force (lbf)	newton (N)	$4.448\,222$
ounce force	newton (N)	$2.780\,139 \times 10^{-1}$
Length		
foot (ft)	meter (m)	$3.048\,000 \times 10^{-1}$†
inch (in.)	meter (m)	$2.540\,000 \times 10^{-2}$†
mile (mi) [U.S. statute]	meter (m)	$1.609\,344 \times 10^3$†
Mass		
pound mass (lbm)	kilogram (kg)	$4.535\,924 \times 10^{-1}$
slug (lbf · s²/ft.)	kilogram (kg)	$1.459\,390 \times 10$
pound force · second²/inch (lbf · s²/in.)	kilogram (kg)	$1.751\,268 \times 10^2$
Modulus of elasticity, Modulus of rigidity		
pound/inch² (lbf/in.² = psi)	pascal = newton/meter² (Pa = N/m²)	$6.894\,757 \times 10^3$
Moment, bending moment, torque		
pound foot (lbf · ft)	newton meter (N · m)	$1.355\,818$
pound inch (lbf · in.)	newton meter (N · m)	$1.129\,848 \times 10^{-1}$
Moment of inertia of area		
inch⁴ (in.⁴)	meter⁴ (m⁴)	$4.162\,314 \times 10^{-7}$
Moment of inertia of mass		
pound inch second² (lbf · in · s²)	kilogram meter² (kg · m²)	$1.129\,848 \times 10^{-1}$
pound foot second² or slug feet² (lbf · ft · s² or slug ft²)	kilogram meter² (kg · m²)	$1.355\,818$
Power		
horsepower (550 ft · lbf/s)	watt (W)	$7.456\,999 \times 10^2$
Stiffness coefficient		
pound/inch (lbf/in.)	newton/meter (N/m)	$1.751\,268 \times 10^2$
Stress, pressure		
pound force/inch² (lbf/in.² = psi)	pascal = newton/meter² (Pa = N/m²)	$6.894\,757 \times 10^3$
pound force/inch² (lbf/in.² = psi)	kilogram force/millimeter² (kgf/mm²)	$7.030\,696 \times 10^{-4}$
pound force/inch² (lbf/in.² = psi)	kilogram force/centimeter² (kgf/cm²)	$7.030\,696 \times 10^{-2}$

Table (*Continued*)

To convert from	To	Multiply by
Torsional stiffness coefficient		
pound-inch/radian (lbf · in./rad)	newton meter/rad (N · m/rad)	$1.129\,848 \times 10^{-1}$
Velocity		
foot/second (ft/s)	meter/sec (m/s)	$3.048\,000 \times 10^{-1}$†
inch/second (in./s)	meter/sec (m/s)	$2.540\,000 \times 10^{-2}$†
mile*/hour (mi/h = mph)	meter/sec (m/s)	$4.470\,400 \times 10^{-1}$†
Volume		
foot³ (ft³)	meter³ (m³)	$2.831\,685 \times 10^{-2}$
inch³ (in.³)	meter³ (m³)	$1.638\,706 \times 10^{-5}$

* [U.S. statute]

Answers to Selected Problems

CHAPTER 2

2-2. 8.27 Hz

2-3. 0.666 sec

2-5. 61.88 lb/in.

2-6. 8.95 Hz

2-8. 24.85 cm

2-10. 6 Hz

2-12. $k = 7.843$ lb/in., $W = 3.067$ lb

2-13. 1.237 in.

2-15. (a) 2.5 Hz, (b) 4.712 in./sec, (c) 74.02 in./sec²

2-18. (a) 0.5129 in., (b) 12.89 in./sec

2-21. $f = \dfrac{1}{2\pi}\sqrt{\dfrac{g(l/2 - a)}{(l^3/3 - la + a^2)}}$

2-23. $f = \dfrac{l}{2\pi}\sqrt{\dfrac{k_1 k_2}{(k_1 l^2 + k_2 a^2)m}}$

2-24. $f = \dfrac{1}{2\pi}\sqrt{\dfrac{ka^2 - Wb}{mb^2}}$

2-27. $f = \dfrac{1}{2\pi}\sqrt{\dfrac{k + \rho A}{m}}$

2-29. $\tau = \pi\sqrt{8l/3g}$

2-31. $\tau = 8\sqrt{\dfrac{2\pi I}{G}\dfrac{l_1 D_2^4 + l_2 D_1^4}{(D_1 D_2)^4}}$

2-33. $f = \dfrac{1}{2\pi}\sqrt{\dfrac{kl^3 + 3EI}{ml^3}}$

2-35. $x = m_2\sqrt{\dfrac{2gh}{(m_1 + m_2)k}}\sin\sqrt{\dfrac{k}{m_1 + m_2}}\,t$

2-37. $f = (1/2\pi)\sqrt{\mu g/a}$
2-38. $f = 0.3703$ Hz
2-41. $I_G = mgb/(2\pi f)^2 - m(r - b)^2$
2-43. $I_c = 0.8820$ kg $\cdot$ m^2

2-45. $f = \dfrac{1}{2\pi}\sqrt{\dfrac{K(n^2 + 1)}{I_B + n^2 I_A}}$

2-46. 0.4538 s
2-47. $f = (1/2\pi)(a/l)\sqrt{3g/b}$
2-49. $f = (1/2\pi)\sqrt{2g/3(R - r)}$
2-51. $m\ddot{x} + (k_2 + Q/b)x + (k_1/2b^2)x^3 = 0$
2-52. (b) $f = (2/\pi)\sqrt{6EIg/Wl^3}$

CHAPTER 3

3-1. 0.390 lb sec/in.
3-2. 2.049 lb sec/in.
3-4. 436.46 N $\cdot$ s/m
3-5. (a) $\zeta = 0.3150$, (b) $f = 2.807$ Hz, $f_d = 2.664$ Hz
3-7. (a) $f_d = 7.943$ Hz, (b) $\tau_d = 0.1259$ sec
3-9. $c = 4\pi m\sqrt{f_1^2 - f_2^2}$
3-10. $f_d = (1/4\pi l^2 m)\sqrt{4b^2 l^2 mk - c^2 a^4}$
3-12. (b) $c_c = \frac{2}{3}(l/a)^2\sqrt{3mk}$
3-17. (a) $c = 1.732$ lb sec/in.; (b) 1.904 in., 0.2795 in., 0.0000 in.
3-18. 7.358 mm
3-22. (a) $\delta = 0.7862$, (b) $x_1/x_2 = 2.195$
3-25. 0.046 16
3-28. 1.195 N $\cdot$ s/m
3-30. $c = 12.11$ N $\cdot$ s/m
3-32. (a) $\dot{x}_0 = 165.5$ ft/sec (b) $t = 0.4481$ sec (c) $x = 0.0390$ in.
3-36. (a) 1.6 cm, (b) 2.000 Hz
3-37. $b = 0.002\,86$, $x = Xe^{-0.0299t}\sin(20.94t + \phi)$

CHAPTER 4

4-2. 4.125 in.
4-3. 10.97 cm
4-5. 26 632 N/m for $m\omega_f^2 < k$; 8898 N/m for $m\omega_f^2 > k$

4-7. 12.22 lb for $k < m\omega_f^2$; 110.0 lb for $k > m\omega_f^2$

4-8. 1.111 in.

4-12. $x = (\dot{x}_0/\omega - Xr) \sin \omega t + x_0 \cos \omega t + X \sin \omega_f t$, where $X = X_0/(1 - r^2)$

4-14. (a) 1.257 cm, (b) 6.283 cm, (c) 12.57 cm

4-15. 4.615 sec

4-20. (g) $P_D = c\omega_f Y \cos (\omega_f t - \lambda)$, where $\tan \lambda = c\omega_f/k$

4-25. $1/\sqrt{1 - \zeta^2}$

4-28. 0.1531

4-30. (a) 9.298 Hz, (b) 445.6 N · m

4-32. (a) 0.0410 in., (b) 39.14 lb, (c) 34.46 deg

4-34. (a) 35.01 N · s/m, (b) 81.54 g · cm

4-37. $c = 0.5865$ lb sec/in., $k = 12.03$ lb/in.

4-39. (a) $k = 24.12$ lb/in., clearance $> 0.011\,73$ in.

4-41. 0.7454 lb sec/in.

4-43. (a) 57.93 lb, (b) 0.3342 lb sec/in.

4-45. 16.985 in.

4-46. $\gamma = 2.128$, $\beta = 0.013\,11$ (in.)$^{-0.13}$

CHAPTER 5

5-2. $P = \dfrac{4B}{\pi}\left(\sin z + \dfrac{\sin 3z}{3} + \cdots\right)$

5-5. $P = \dfrac{-2B}{\pi}\left(\sin z + \dfrac{\sin 2z}{2} + \dfrac{\sin 3z}{3} + \cdots\right)$

5-8. $P = B\left[\dfrac{1}{2} - \dfrac{4}{\pi^2}\left(\cos z + \dfrac{\cos 3z}{3^2} + \cdots\right)\right]$

5-11. $P = B\left[\dfrac{1}{\pi} + \dfrac{\sin z}{2} - \dfrac{2}{\pi}\left(\dfrac{\cos 2z}{3} + \dfrac{\cos 4z}{15} + \dfrac{\cos 6z}{35} + \cdots\right)\right]$

5-14. $x_b = \dfrac{4B}{\pi k}\left[\dfrac{\sin (\omega_f t - \psi_1)}{\sqrt{(1 - r^2)^2 + (2\zeta r)^2}} + \dfrac{\sin (3\omega_f t - \psi_3)}{3\sqrt{(1 - 9r^2)^2 + (6\zeta r)^2}} + \cdots\right]$

where $r = \dfrac{\omega_f}{\omega}$, $\tan \psi_1 = \dfrac{2\zeta r}{1 - r^2}$, $\tan \psi_3 = \dfrac{6\zeta r}{1 - 9r^2}, \cdots$

5-17. $x_b = \dfrac{-2B}{\pi k}\left[\dfrac{\sin (\omega_f t - \psi_1)}{\sqrt{(1 - r^2)^2 + (2\zeta r)^2}} + \dfrac{\sin (2\omega_f t - \psi_2)}{2\sqrt{(1 - 4r^2)^2 + (4\zeta r)^2}} + \cdots\right]$

where $r = \dfrac{\omega_f}{\omega}$, $\tan \psi_1 = \dfrac{2\zeta r}{1 - r^2}$, $\tan \psi_2 = \dfrac{4\zeta r}{1 - 4r^2}, \cdots$

5-20. $x_b = \dfrac{B}{k}\left\{\dfrac{1}{2} - \dfrac{4}{\pi^2}\left[\dfrac{\cos (\omega_f t - \psi_1)}{\sqrt{(1 - r^2)^2 + (2\zeta r)^2}} + \dfrac{\cos (3\omega_f t - \psi_3)}{9\sqrt{(1 - 9r^2)^2 + (6\zeta r)^2}} + \cdots\right]\right\}$

where $r = \dfrac{\omega_f}{\omega}$, $\tan \psi_1 = \dfrac{2\zeta r}{1 - r^2}$, $\tan \psi_3 = \dfrac{6\zeta r}{1 - 9r^2}, \cdots$

5-23. $x_b = \dfrac{B}{k} \left\{ \dfrac{1}{\pi} + \dfrac{\sin (\omega_f t - \psi_1)}{2\sqrt{(1 - r^2)^2 + (2\zeta r)^2}} - \dfrac{2}{\pi} \left[\dfrac{\cos (2\omega_f t - \psi_2)}{3\sqrt{(1 - 4r^2)^2 + (4\zeta r)^2}} \right. \right.$

$\left. \left. + \dfrac{\cos (4\omega_f t - \psi_4)}{15\sqrt{(1 - 16r^2)^2 + (8\zeta r)^2}} \right. \right.$

$\left. \left. + \dfrac{\cos (6\omega_f t - \psi_6)}{35\sqrt{(1 - 36r^2)^2 + (12\zeta r)^2}} + \cdots \right] \right\}$

where $r = \dfrac{\omega_f}{\omega}$, $\tan \psi_1 = \dfrac{2\zeta r}{1 - r^2}$, $\tan \psi_2 = \dfrac{4\zeta r}{1 - 4r^2}$, $\tan \psi_4$

$= \dfrac{8\zeta r}{1 - 16r^2}$, $\tan \psi_6 = \dfrac{12\zeta r}{1 - 36r^2}$, $\cdots$

5-25. $b_3 = 0.4023$

5-27. $a_3 = -0.040\,47$

5-29. (a) $x = \dfrac{B}{k} \left(\dfrac{\omega t}{2\pi} - \dfrac{\sin \omega t}{2\pi} \right)$, (b) $x = \dfrac{B}{k} \left(2 - \dfrac{\omega t}{2\pi} + \dfrac{\sin \omega t}{2\pi} \right)$

5-31. (a) $x = \dfrac{B}{k} \left(\dfrac{\omega t}{2\pi} - \dfrac{\sin \omega t}{2\pi} \right)$, (b) $x = \dfrac{B}{k} \cos \omega t$

5-34. $x_2 = 0.000\,750$ in., $\ldots$, $x_7 = 0.008\,793\,5$ in., $\ldots$

5-36. $x_2 = 0.000\,103\,802$ in., $\ldots$, $x_8 = 0.009\,436\,07$ in., $\ldots$

CHAPTER 6

6-2. (a) 12.5% at $r = 3$, (b) 2.59% at $r = 3$, (c) 0.02% at $r = 7.0711$

6-4. (a) 56.2% at $r = 0.6$, (b) 1.79% at $r = 0.6$, (c) -6.16% at $r = 0.6$

6-6. 0.6942 in.

6-7. 19.76 in./sec^2

6-9. $k = 5.527$ lb/in., $c = 0.1436$ lb sec/in.

6-11. (a) $\zeta = 0.6556$, (b) $0 \leq r \leq 0.5834$

6-13. $k = 13.521$ lb/in., $c = 0.002\,424$ lb sec/in.

6-15. $k = 20\,000$ N/m, $m = 0.0698$ kg

6-17. (a) 0 to 18.90 Hz, (b) 0 to 46.22 Hz

6-19. $z = 0.955 \sin (8\pi t - 173.9°) + 0.457 \sin (16\pi t - 177.1°)$
$+ 0.302 \sin (24\pi t - 178.1°)$

6-21. $\ddot{z} = -\{5.570 \sin (7.5\pi t - 11.28°) + 15.155 \sin (22.5\pi t - 36.26°)\}$

6-22. $t_1 = 0.008\,36$ sec, $t_2 = 0.008\,95$ sec

CHAPTER 7

7-1. $\omega_1 = 0.618\sqrt{k/m}$; $A_1 = 1$, $A_2 = 1.618$
$\omega_2 = 1.621\sqrt{k/m}$; $A_1 = 1$, $A_2 = -0.618$

7-7. $\omega_1 = 0.7962\sqrt{k/I}$; $A_1 = 1$, $A_2 = 1.366$
$\omega_2 = 1.538\sqrt{k/I}$; $A_1 = 1$, $A_2 = -0.366$

7-11. $\omega_1 = 0.7655\sqrt{g/l}$; $A_1 = 1$, $A_2 = \sqrt{2}$
$\omega_2 = 1.8478\sqrt{g/l}$; $A_1 = 1$, $A_2 = -\sqrt{2}$

7-15. (a) 1.627 Hz, (b) 0.9027 Hz

7-17. (a) $\omega_1 = 0$; $A_1 =$ any value, including variable, and $A_2 = 0$ (Upper pendulum is in constant-velocity equilibrium or in static equilibrium, and lower pendulum hangs vertically.)

$$\omega_2 = \sqrt{2g/l}; \ A_1 = 1, \ A_2 = -2$$

(b) $\omega_1 =$ imaginary, and motion of the upper pendulum is initially unstable

$$\omega_2 = \sqrt{\frac{1+\sqrt{2}}{2}\left(\frac{g}{l}\right)}; \ A_1 = 1, \ A_2 = -(3 + 2\sqrt{2})$$

7-19. (b) $\omega_1 = \sqrt{\dfrac{3-\sqrt{5}}{2}\dfrac{g}{l}}$; $A_1 = 1$, $A_2 = \dfrac{\sqrt{5}-1}{2l}$

$$\omega_2 = \sqrt{\frac{3+\sqrt{5}}{2}\frac{g}{l}}; \ A_1 = 1, \ A_2 = -\left(\frac{\sqrt{5}+1}{2l}\right)$$

7-21. $\omega_1 = \sqrt{k/m}$; $A_1 = 1$, $A_2 = 1$
$\omega_2 = \sqrt{2k/m}$; $A_1 = 1$, $A_2 = -1$

7-24. (b) $\omega_1 = \sqrt{\tfrac{2}{3}k/m}$; $A_1 = 1$, $A_2 = 4$
$\omega_2 = \sqrt{3k/m}$; $A_1 = 1$, $A_2 = -3$

7-25. (b) $\omega_1 = 0.2853\sqrt{k/m}$; $\omega_2 = 3.325\sqrt{k/m}$

7-26. $\omega_1^2 = \dfrac{2-\sqrt{2}}{3}\dfrac{k}{m}$; $\theta = 1$, $X = \sqrt{2}a$

$$\omega_2^2 = \frac{2+\sqrt{2}}{3}\frac{k}{m}; \ \theta = 1, \ X = -\sqrt{2}a$$

7-29. $I_G m\omega^4 - [I_G k + m(k_T + kb^2)]\omega^2 + kk_T = 0$

7-31. $\omega_1 = 30.88$ rad/sec; $A_1 = 1$ in., $A_2 = 0.0622$ rad
$\omega_2 = 42.84$ rad/sec; $A_1 = 1$ in., $A_2 = -0.0638$ rad

7-33. $\omega_1 = 3.750\sqrt{EI/ml^3}$; $A_1 = 1$, $A_2 = 1 + \sqrt{2}$
$\omega_2 = 9.052\sqrt{EI/ml^3}$; $A_1 = 1$, $A_2 = 1 - \sqrt{2}$

7-34. $x_1 = (\dot{x}_0/\omega_1)\sin\omega_1 t = x_2$

7-36. $x_1 = (\dot{x}_0/\omega_1)\sin\omega_1 t + x_0\cos\omega_2 t$
$x_2 = (\dot{x}_0/\omega_1)\sin\omega_1 t - x_0\cos\omega_2 t$

7-38. $X_1 = 0.313$ in., $X_2 = 0.625$ in.

7-40. $X_1 = -1.0000$ in., $X_2 = 0.3333$ in.

7-42. $s^4 + 0.6s^3 + 1000s^2 + 220s + 90\,000 = 0$

7-44. $X_2 = 6.40$ in.

7-46. $X_1 = 0.2836$ in., $X_2 = 0.4855$ in.

7-48. System is stable for $c > P_0$

CHAPTER 8

8-1.
$$\begin{bmatrix} (k_1 + k_2) & -k_2 & 0 \\ -k_2 & (k_2 + k_3) & -k_3 \\ 0 & -k_3 & (k_3 + k_4) \end{bmatrix} \begin{Bmatrix} x_1 \\ x_2 \\ x_3 \end{Bmatrix} + \begin{bmatrix} m_1 & 0 & 0 \\ 0 & m_2 & 0 \\ 0 & 0 & m_3 \end{bmatrix} \begin{Bmatrix} \ddot{x}_1 \\ \ddot{x}_2 \\ \ddot{x}_3 \end{Bmatrix} = \begin{Bmatrix} 0 \\ 0 \\ 0 \end{Bmatrix}$$

8-3. $\omega_1 = 0.5284\sqrt{k/m}$; $A_1 = 1$, $A_2 = 1.7208$, $A_3 = 1.4805$
$\omega_2 = \sqrt{k/m}$; $A_1 = 1$, $A_2 = 1$, $A_3 = -1$
$\omega_3 = 1.545\sqrt{k/m}$; $A_1 = 1$, $A_2 = -0.3874$, $A_3 = 0.07505$

8-6. $\omega_1 = (1/l)\sqrt{k/m}$; $A_1 = 1$, $A_2 = 1$, $A_3 = 1$, $A_4 = 1$
$\omega_2 = (1.122/l)\sqrt{k/m}$; $A_1 = 1$, $A_2 = 0.4821$, $A_3 = -0.1607$, $A_4 = -0.7203$
$\omega_3 = (1.414/l)\sqrt{k/m}$; $A_1 = 1$, $A_2 = -1$, $A_3 = -2$, $A_4 = 1$
$\omega_4 = (1.890/l)\sqrt{k/m}$; $A_1 = 1$, $A_2 = -4.148$, $A_3 = 1.381$, $A_4 = -0.199$

8-8. $\omega_1 = 0.3473\sqrt{k/I}$; $A_1 = 1$, $A_2 = 1.8794$, $A_3 = 2.5321$, $A_4 = 2.8793$
$\omega_2 = \sqrt{k/I}$; $A_1 = 1$, $A_2 = 1$, $A_3 = 0$, $A_4 = -1$
$\vdots$

8-9. $m\ddot{x} + k_{11} x + k_{12} y + k_{13} z = 0$

where $k_{11} = k_1 \cos \alpha_1 \cos \alpha_1 + k_2 \cos \alpha_2 \cos \alpha_2 + k_3 \cos \alpha_3 \cos \alpha_3$
$k_{12} = k_1 \cos \alpha_1 \cos \beta_1 + k_2 \cos \alpha_2 \cos \beta_2 + k_3 \cos \alpha_3 \cos \beta_3$
$\vdots$

8-12. (a) $k_{11} = k_1 + k_2$, $k_{12} = -k_2$, ...
$k_{21} = -k_2$, $k_{22} = k_2 + k_3$, ...
$\vdots$

8-16. $\omega_1 = (1.148/l)\sqrt{k/m}$; $A_1 = 1$, $A_2 = 1.366$, $A_3 = 1.299$
$\vdots$
$\omega_3 = (1.874/l)\sqrt{k/m}$; $A_1 = 1$, $A_2 = -3.023$, $A_3 = 0.5471$

8-19. (a)
$$[k] = k \begin{bmatrix} 5 & -2 & 0 \\ -2 & 3 & -1 \\ 0 & -1 & 1 \end{bmatrix} \qquad [m] = m \begin{bmatrix} 1 & 0 & 0 \\ 0 & 1 & 0 \\ 0 & 0 & 1 \end{bmatrix}$$

where $k = 24EI/l^3$

8-24. (a)
$$[a] = (l^3/12EI) \begin{bmatrix} 9 & 11 & 7 \\ 11 & 16 & 11 \\ 7 & 11 & 9 \end{bmatrix}$$

8-27. (a)
$$[k] = \frac{3}{13} \frac{EI}{l^3} \begin{bmatrix} 7 & -16 & 12 \\ -16 & 44 & -46 \\ 12 & -46 & 80 \end{bmatrix}$$

8-31. $T = \frac{1}{2}\{[m_1 a^2 + m_2(b - l)^2]\dot{\theta}_1^2 + m_2 l^2 \dot{\theta}_2^2 + 2m_2(bl - l^2)\dot{\theta}_1 \dot{\theta}_2\}$
$V = \frac{1}{2}\{(W_2 b - W_1 a)\theta_1^2 + W_2 l\theta_2^2\}$
$\vdots$

8-41. (a) Unstable, (b) unstable

CHAPTER 9

9-1. $\omega_R = 0.5033\sqrt{Q/ml}$

9-3. $\omega_R = 2.828\sqrt{EI/ml^3}$

9-6. $\omega = 0.8165\sqrt{k/m}$; $A_1 = 1$, $A_2 = 0.3333$, $A_3 = -1$

9-10. $\omega = 1.879\sqrt{k/I}$; $A_1 = 1$, $A_2 = -1.532$, $A_3 = 1.347$, $A_4 = -0.5321$

9-14. $\omega_1 = 0.4420\sqrt{EI/ml^3}$; $A_1 = 1$, $A_2 = 1.289$, $A_3 = 1.108$

9-22. $\omega_1 = 58.38$ rad/sec; $A_1 = 1$, $A_2 = 0.320\,47$

CHAPTER 10

10-2. $20\,000$ Hz, 8000 Hz, 4308 Hz

10-4. $\omega = (n - \tfrac{1}{2})\pi(c/l)$; $n = 1, 2, 3, \ldots$
$c = \sqrt{F/\gamma}$

10-7. $\tan\left(\dfrac{\omega l}{c}\right) = \dfrac{(F/kl)(\omega l/c)}{(Mc^2/kl^2)(\omega l/c)^2 - 1}$
$c = \sqrt{F/\gamma}$

10-8. $y = \dfrac{8y_0}{\pi^2} \displaystyle\sum_{n=1}^{\infty} \dfrac{(-1)^{n-1}}{(2n-1)^2} \sin\dfrac{(2n-1)\pi}{l} x \cos\dfrac{(2n-1)\pi}{l}\sqrt{\dfrac{F}{\gamma}}\,t$

10-11. $f_1 = 307.2$ Hz

10-13. $73\,733$ in./sec

10-14. $\theta = \dfrac{8\theta_0}{\pi^2} \displaystyle\sum_{n=1}^{\infty} \dfrac{(-1)^{n-1}}{(2n-1)^2} \sin\left(n - \dfrac{1}{2}\right)\dfrac{\pi}{l} x \cos\left(n - \dfrac{1}{2}\right)\dfrac{\pi}{l}\sqrt{\dfrac{G}{\rho}}\,t$

10-17. 2804 Hz

10-18. $201\,923$ in./sec

10-19. 2522 Hz

10-20. 5044 m/s

10-22. $y = C_2\{\cosh \lambda x - \cos \lambda x - \beta(\sinh \lambda x - \sin \lambda x)\} \cos\left[\lambda^2 \sqrt{\dfrac{EI}{\gamma}}\,t\right]$

where

Mode =	1	2	3
λl =	4.730	7.8532	10.9956
$\beta(= -A/B)$ =	0.9825	1.0008	0.999\,97

(C_2 depends on assigned initial amplitude value)

10-24. $y = C \sin (n\pi x/l) \cos [(n\pi/l)^2\sqrt{EI/\gamma}\,t]$

where $n = 1, 2, 3, \ldots$

10-26. 12.58 Hz

10-28. 80.08 Hz

10-30. $\omega_R = (9.8766/l^2)\sqrt{EI/\gamma}$

Suggested References

GENERAL REFERENCE TEXTS ON VIBRATIONS

Anderson, R. A. *Fundamentals of Vibrations*. New York: Macmillan, 1967.

Chen, Y. *Vibrations: Theoretical Methods*. Reading, Mass.: Addison-Wesley, 1966.

Crafton, P. A. *Shock and Vibration in Linear Systems*. New York: Harper & Row, 1961.

Church, A. H. *Mechanical Vibrations*, 2d ed. New York: Wiley, 1963.

DenHartog, J. P. *Mechanical Vibrations*, 4th ed. New York: McGraw-Hill, 1956.

Haberman, C. M. *Vibration Analysis*, Columbus, Ohio: Merril, 1968.

Jacobsen, L. S. and Ayre, R. S. *Engineering Vibrations*. New York: McGraw-Hill, 1958.

Manley, R. G. *Fundamentals of Vibration Study*. New York: Wiley, 1942.

Meirovitch, L. *Elements of Vibration Analysis*, New York: McGraw-Hill, 1975.

Myklestad, N. O. *Fundamentals of Vibration Analysis*. New York: McGraw-Hill, 1956.

Steidel, R. F. Jr. *An Introduction to Mechanical Vibrations*. New York: Wiley, 1971.

Thomson, W. T. *Mechanical Vibrations*, 2d ed. Englewood Cliffs, N.J.: Prentice-Hall, 1953.

Thomson, W. T. *Theory of Vibration with Applications*. Englewood Cliffs, N.J.: Prentice-Hall, 1972.

Timoshenko, S. *Vibration Problems in Engineering*. 2d ed. Princeton, N.J.: VanNostrand, 1937.

Timoshenko, S., Young, D. H., and Weaver, W., Jr. *Vibration Problems in Engineering*. 4th ed. New York: Wiley, 1974.

Tong, K. N. *Theory of Mechanical Vibration*. New York: Wiley, 1960.

Tse, F. S., Morse, I. E., and Hinkle, R. T. *Mechanical Vibrations*. Boston: Allyn and Bacon, 1963.

REFERENCE BOOKS ON SPECIALIZED OR ADVANCED TOPICS IN VIBRATIONS

Andronow, A. A., and Chaikin, C. E. *Theory of Oscillations* (translated by S. Lefschetz). Princeton, N.J.: Princeton University Press, 1949.

Crandall, S. H., et al. *Random Vibration.* Cambridge, Mass.: The Technology Press of M.I.T., and New York: Wiley, 1958.

Crandall, S. H., et al. *Random Vibration*, vol. 2. Cambridge, Mass.: The M.I.T. Press, 1963.

Crandall, S. H., and Mark, W. D. *Random Vibration in Mechanical Systems.* New York: Academic Press, 1963.

Crede, C. E. *Vibration and Shock Isolation.* New York: Wiley, 1951.

Crede, C. E. *Shock and Vibration Concepts in Engineering Design.* Englewood Cliffs, N.J.: Prentice-Hall, 1965.

Cunningham, W. J. *Introduction to Nonlinear Analysis.* New York: McGraw-Hill, 1958.

Fertis, D. G. *Dynamics and Vibration of Structures.* New York: Wiley, 1973.

Haag, J. *Oscillatory Motions*, vols. I and II. Belmont, Calif.: Wadsworth, 1962.

Harris, C. M., and Crede, C. E., eds. *Shock and Vibration Handbook*, 2d ed. New York: McGraw-Hill, 1976.

Manley, R. G. *Waveform Analysis.* New York: Wiley, 1945.

Meirovitch, L. *Analytical Methods in Vibrations.* New York: Macmillan, 1967.

Minorsky, N. *Nonlinear Oscillations.* Princeton, N.J.: VanNostrand, 1962.

Rayleigh, Lord. *The Theory of Sound*, vol. I. New York: Dover, 1945.

Snowdon, J. C. *Vibration and Shock in Damped Mechanical Systems.* New York: Wiley, 1968.

Stoker, J. J. *Nonlinear Vibrations.* New York: Wiley, 1950.

Wilson, W. K. *Practical Solution of Torsional Vibration Problems*, 2d ed. New York: Wiley, 1948.

REFERENCE BOOKS ON MATHEMATICS AND ANALYSIS

Crandall, S. H. *Engineering Analysis.* New York: McGraw-Hill, 1956.

Frazer, R. A., Duncan, W. J., and Collar, A. R. *Elementary Matrices.* England: Cambridge University Press, 1963.

Gaskell, R. E. *Engineering Mathematics.* New York: Dryden Press, 1958.

Gere, J. M., and Weaver, W., Jr. *Matrix Algebra for Engineers.* VanNostrand, 1965.

Hohn, F. E. *Elementary Matrix Algebra*, 2d ed. New York: Macmillan, 1964.

Pestel, E. C., and Leckie, F. A. *Matrix Methods in Elastomechanics.* New York: McGraw-Hill, 1963.

von Karman, T., and Biot, M. A. *Mathematical Methods in Engineering.* New York: McGraw-Hill, 1940.

Wylie, C. R., Jr. *Advanced Engineering Mathematics*, 3rd ed. New York: McGraw-Hill, 1966.

REFERENCE BOOKS ON COMPUTER METHODS

Harrison, H. L., and Bollinger, J. G. *Introduction to Automatic Controls*, 2d ed. New York: IEP, 1969.

MacNeal, R. H. *Electric Circuit Analogies for Elastic Structures*, New York: Wiley, 1962.

Pilkey, Walter, and Pilkey, Barbara. *Shock and Vibration Computer Programs*. Washington, D.C.: Shock and Vibration Information Center, 1975.

James, M. L., Smith, G. M., and Wolford, J. C. *Analog and Digital Computer Methods in Engineering Analysis*. New York: IEP, 1964.

James, M. L., Smith, G. M., and Wolford, J. C. *Applied Numerical Methods for Digital Computation with FORTRAN*, 2d ed. New York: IEP, 1977.

Soroka, W. W. *Analog Methods in Computation and Simulation*. New York: McGraw-Hill, 1954.

Index